STUDENT SOLUTIONS MANUAL

John Olmsted
California State University, Fullerton

David Robichaud
California State University, Fullerton

CHEMISTRY

Fourth Edition

JOHN OLMSTED
California State University, Fullerton

GREG WILLIAMS
University of Oregon

We acknowledge with special thanks the excellent work of Dr. Richard S. Treptow, Professor Emeritus of Chemistry at Chicago State University, who acted as a reviewer and accuracy checker for the solutions in this manual. Dr. Treptow's thorough and thoughtful comments helped us immensely in achieving greater accuracy and improved clarity.

WILEY

JOHN WILEY & SONS, INC.

To order books or for customer service, please call 1-800-CALL-WILEY (225-5945).

ISBN 978-0-471-69838-8

Printed in the United States of America

10 9 8 7 6 5 4

Printed and bound by Courier Kendallville, Inc.

Table of Contents

1.1 Political issues that involve chemical reactions include the following: What are appropriate standards for "clean" air and water? What should be the regulations for testing new drugs before they are approved for human use? How can toxic waste sites best be cleaned up? Should chlorofluorocarbons be banned because of potential damage to the ozone layer? Are the consequences of global warming sufficient to require curbs on the use of fossil fuels?

1.3 A pharmacist deals in chemical substances, most of which are toxic when incorrectly administered. Among the chemistry–related skills that a pharmacist uses are the following: weighing, volume measurements and unit conversions; knowing chemical compatibility of different drugs; identifying similar or chemically equivalent drugs; and protecting drugs from degradation caused by exposure to adverse conditions.

1.5 Your criticism should be based on the experimental nature of chemistry. When experimental results fail to match theoretical expectations, the experiment may be flawed or the theory may be incorrect. The chemist should redo the experiment, adjusting conditions if necessary, to determine whether or not the results are correct. If repeated measurements show that the results consistently differ from what theory predicts, then the chemist should examine how to revise the theory to accommodate the results.

1.7 You must commit to memory the correspondence between names and symbols of various elements, but remembering them is simplified by the fact that most English names and elemental symbols are related: (a) H; (b) He; (c) Hf; (d) N; (e) Ne; and (f) Nb.

1.9 Associating an element's name with its symbol requires memorization of both names and symbols. The examples in this problem all begin with the letter "A:" (a) arsenic; (b) argon; (c) aluminum; (d) americium; (e) silver; (f) gold; (g) astatine; and (h) actinium.

1.11 To convert a molecular picture into a molecular formula, count atoms of each type and consult (or recall) the color scheme used for the elements. See Figure 1–4 of your textbook for the color scheme used in this and many other texts: (a) Br_2; (b) HCl; (c) C_2H_5I; and (d) C_3H_6O.

1.13 In writing a chemical formula, remember to use elemental symbols and subscripts for the number of atoms: (a) CCl_4; (b) H_2O_2; (c) P_4O_{10}; and (d) Fe_2S_3.

1.15   The next element after the end of a row is the first element in the succeeding row, cesium.

1.17   Elements in the same column of the periodic table share similar chemical properties. For sulfur, these are its vertical neighbors, oxygen (O) and selenium (Se). The horizontal neighbors are phosphorus (P) and chlorine (Cl).

1.19   Metals from Group 1 react in a 1:1 ratio with Br from Group 17: Li, Na, K, Rb, and Cs. Other metals that react in a 1:1 ratio with bromine are Cu, Ag, and Au.

1.21   Consult the periodic table: lithium, Li; beryllium, Be; boron, B; carbon, C; oxygen, O; fluorine, F; and neon, Ne.

1.23   Remember that a pure substance contains a single chemical element or compound, and a solution is a homogeneous mixture of substances: (a) pure substance; (b) solution of various substances in water; (c) heterogeneous mixture (indicated by its opacity); (d) solution of nitrogen, oxygen, and other gases; (e) heterogeneous mixture of gases and particulate matter; and (f) heterogeneous (indicated by the presence of grains).

1.25   Remember that gases and liquids both flow, but gases are much less dense than liquids: (a) liquid; (b) solid; (c) solid; and (d) gas.

1.27   In a chemical transformation, one substance is converted into another. In a physical transformation the substance remains the same; only the state (gas, liquid, solid) changes: (a) physical transformation (water vapor to ice); (b) physical transformation (liquid water to vapor); and (c) chemical transformation.

1.29   A mixture contains more than one substance, while a compound is a single substance containing more than one element: (a) mixture of water, suspended solids, and dissolved substances; (b) pure compound, H_2O; (c) mixture of water and one or more solids; (d) single element (He); (e) mixture of isopropanol and water; and (f) mixture of a solvent and various pigments.

1.31   Scientific notation expresses any number as a value between 1 and 10 times a power of ten. Trailing zeros are retained only when they are significant: (a) 1.00000×10^5; (b) 1.0×10^4; (c) 4.00×10^{-4}; (d) 3×10^{-4}; and (e) 2.753×10^2.

1.33   To do unit conversions, multiply by a ratio that cancels the unwanted unit(s). Refer to Table 1–3 for the SI base units:

(a) $432 \text{ kg} = 4.32 \times 10^2 \text{ kg}$

(b) $624 \text{ ps} \left(\dfrac{10^{-12} \text{ s}}{1 \text{ ps}} \right) = 6.24 \times 10^{-10} \text{ s}$

(c) $1024 \text{ ng} \left(\dfrac{10^{-9} \text{ g}}{1 \text{ ng}} \right) \left(\dfrac{10^{-3} \text{ kg}}{1 \text{ g}} \right) = 1.024 \times 10^{-9} \text{ kg}$

(d) $93,000 \text{ km} \left(\dfrac{10^3 \text{ m}}{1 \text{ km}} \right) = 9.300 \times 10^7 \text{ m}$

(e) $1 \text{ day} \left(\dfrac{24 \text{ hr}}{1 \text{ day}} \right) \left(\dfrac{60 \text{ min}}{1 \text{ hr}} \right) \left(\dfrac{60 \text{ sec}}{1 \text{ min}} \right) = \text{exactly } 8.64 \times 10^4 \text{ s}$ (assuming exactly 1 day)

(f) $0.0426 \text{ in} \left(\dfrac{2.54 \text{ cm}}{1 \text{ in}} \right) \left(\dfrac{1 \text{ m}}{100 \text{ cm}} \right) = 1.08 \times 10^{-3} \text{ m}$

1.35   This is a unit–conversion problem involving summation and unusual units. First convert all masses into kg, then put them into the same power of ten and add the masses:

$$m(\text{diamonds}) = 5.0 \times 10^{-1} \text{ carat} \left(\dfrac{3.168 \text{ grains}}{1 \text{ carat}} \right) \left(\dfrac{1 \text{ g}}{15.4 \text{ grains}} \right) \left(\dfrac{10^{-3} \text{ kg}}{1 \text{ g}} \right) = 1.0 \times 10^{-4} \text{ kg}$$

$$m(\text{gold}) = 7.00 \text{ g} \left(\dfrac{10^{-3} \text{ kg}}{1 \text{ g}} \right) = 7.00 \times 10^{-3} \text{ kg}$$

$$m(\text{total}) = (7.00 \times 10^{-3} \text{ kg}) + (0.10 \times 10^{-3} \text{ kg}) = 7.10 \times 10^{-3} \text{ kg}$$

1.37   Because the quart is a volume measure, density must be used to convert from volume to mass. The density of water is 1.00 g/cm^3 (Table 1–4). Assume exactly one quart:

$$1 \text{ quart} \left(\dfrac{1 \text{ L}}{1.057 \text{ quart}} \right) \left(\dfrac{10^3 \text{ mL}}{1 \text{ L}} \right) \left(\dfrac{1 \text{ cm}^3}{1 \text{ mL}} \right) \left(\dfrac{1.00 \text{ g}}{1 \text{ cm}^3} \right) = 9.46 \times 10^2 \text{ g}$$

1.39   Density is mass divided by volume, and the volume of a block is $V = lwh$:

$V = (15.5 \text{ cm})(4.6 \text{ cm})(1.75 \text{ cm}) = 1.25 \times 10^2 \text{ cm}^3$

$\rho = \dfrac{m}{V} = \dfrac{98.456 \text{ g}}{1.25 \times 10^2 \text{ cm}^3} = 0.79 \text{ g/cm}^3$ (This result has two significant figures because one dimension is known to only two significant figures.)

3

1.41   To convert from mass to volume, divide by density. See Table 1–4 for densities:

$$V = \frac{m}{\rho} = 15.4 \text{ g}\left(\frac{1 \text{ cm}^3}{2.70 \text{ g}}\right) = 5.70 \text{ cm}^3$$

1.43   The question asks for the density of water expressed in SI units (kg/m^3). Begin by analyzing the given information. The mass of the container is given before and after the water has been added. Thus, the mass of water can be obtained from the difference between the masses of the filled and empty container:

$m = 270.064 \text{ g} - 93.054 \text{ g} = 177.010 \text{ g } H_2O$

In SI units, $m = 177.010 \text{ g}\left(\dfrac{10^{-3} \text{ kg}}{1 \text{ g}}\right) = 0.177010 \text{ kg}$

Convert the units from inches to meters before computing the volume:

$r = 0.875 \text{ in}\left(\dfrac{2.54 \text{ cm}}{1 \text{ in}}\right)\left(\dfrac{10^{-2} \text{ m}}{1 \text{ cm}}\right) = 0.022225 \text{ m}$

$h = 4.500 \text{ in}\left(\dfrac{2.54 \text{ cm}}{1 \text{ in}}\right)\left(\dfrac{10^{-2} \text{ m}}{1 \text{ cm}}\right) = 0.1143 \text{ m}$

$V = \pi r^2 h = (3.1416)(0.022225 \text{ m})^2(0.1143 \text{ cm}) = 1.7737 \times 10^{-4} \text{ m}^3$

Calculate the density by dividing the mass of water by the volume:

$\rho = \dfrac{m}{V} = \dfrac{0.177010 \text{ kg}}{1.7737 \times 10^{-4} \text{ m}^3} = 9.98 \times 10^2 \text{ kg/m}^3 \text{ } or \text{ } 0.998 \text{ g/cm}^3$ (Round to three

significant figures because the radius of the cylinder is known to only three significant figures.)

1.45   The question asks for a comparison of the masses of two objects of different densities and shapes. For each object, $m = \rho V$:

Au sphere: $\rho = 19.3 \text{ g cm}^{-3}$, $V = \left(\dfrac{4}{3}\right)\pi r^3$, and $r = \text{diameter}/2 = 1.00 \text{ cm}$

$V = \dfrac{4\pi(1.00 \text{ cm})^3}{3} = 4.19 \text{ cm}^3 \quad and \quad m(\text{gold}) = 4.19 \text{ cm}^3\left(\dfrac{19.3 \text{ g}}{1 \text{ cm}^3}\right) = 80.9 \text{ g}$

Ag cube: $\rho = 10.50 \text{ g cm}^{-3}$, $V = lwh$, and $l = w = h = 2.00 \text{ cm}$

$V = (2.00 \text{ cm})^3 = 8.00 \text{ cm}^3 \quad and \quad m(\text{silver}) = 8.00 \text{ cm}^3\left(\dfrac{10.50 \text{ g}}{1 \text{ cm}^3}\right) = 84.0 \text{ g}$

The silver cube has more mass than the gold sphere.

1.47   Volume and mass are related through density:

$$V = \frac{m}{\rho} = \left(\frac{36.5 \text{ g}}{3.10 \text{ g/mL}} \right) = 11.8 \text{ mL}$$

1.49   Consult the periodic table and/or the alphabetical listing at the end of the text to answer questions about elemental symbols. Eight elements have symbols beginning with "T": Ti (titanium), Tc (technetium), Te (tellurium), Ta (tantalum), Tl (thallium), Tb (terbium), Tm (thulium), and Th (thorium).

1.51   Both molecular fluorine and molecular chlorine are diatomic (two atoms per molecule) species. To have a total of 20 molecules in a 3:1 ratio there should be 5 molecules of chlorine (dark) and 15 molecules of fluorine (light).

1.53   "Do not match" means that the letters in the symbol are not related to the English name of the element. See Table 1–1 in your textbook: lead, Pb.

1.55   Although it may not at first appear to be, this is a unit–conversion problem. The speed of the athlete in SI units is:

$$\text{speed} = \left(\frac{100 \text{ yards}}{10.17 \text{ s}} \right) \left(\frac{0.9144 \text{ m}}{1 \text{ yard}} \right) = 8.991 \text{ m/s (assuming exactly 100 yards)}$$

The time of the 100–meter dash can be found using the definition of speed:

$$\text{speed} = \frac{\text{distance}}{\text{time}}$$

$$\text{time} = 100 \text{ m} \left(\frac{1 \text{ s}}{8.991 \text{ m}} \right) = 11.12 \text{ s (assuming exactly 100 meters)}$$(The result has four significant figures, because the time has four significant figures.)

1.57   The periodic table is organized with elements that have similar chemical properties in the same column. Gold shares Group 11 with silver and copper. All three are shiny, non–reactive, soft, and good conductors of heat and electricity.

1.59   In a molecular picture, each atom is
        shown as a sphere and different
        elements are represented by labeling,
        shading, or the use of color coding.
        Here we show with shading:

1.61   The value of an extensive property changes with the amount of the substance, while the
        value of an intensive property is independent of the amount of the substance: (a), (c), and
        (e) are intensive; (b) and (d) are extensive.

1.63   Kelvin–Celsius conversions involve addition or subtraction rather than multiplication or
        division, because the two scales have the same unit size but different zero points:
        $$T(K) = T(^oC) + 273.15 \qquad T = -11.5\ ^oC + 273.15 = 261.7\ K$$

1.65   The number of significant figures in a result of multiplication/division is determined by
        the number that contains the least number of significant figures:

        (a) $\dfrac{\left(6.531 \times 10^{13}\right)\left(6.02 \times 10^{23}\right)}{(435)(2.000)} = 4.52 \times 10^{34}$

        (b) $\dfrac{4.476 + (3.44)(5.6223) + 5.666}{(4.3)\left(7 \times 10^{4}\right)} = 1 \times 10^{-4}$

1.67   The key to this question is to recognize how solids, liquids, and gases are organized at the
        molecular level: A solid has a rigid, constant shape because its molecules are in fixed,
        regular arrangements. Thus, (b) matches (d). A liquid is dense yet able to flow easily,
        because its molecules, though close together, are not in fixed positions. Thus, (c) matches
        (e). A gas has low density and flows easily because its molecules are separated by much
        empty space. Thus, (a) matches (f).

1.69   Determine a molecular formula from a model by counting the atoms of each kind and
        referring to the color code in use for the elements: (a) $CHClF_2$; (b) CH_2O_2; (c) BrF_3; and
        (d) C_4H_6.

1.71   Do unit conversions by multiplying by the appropriate ratios:

(a) $454 \text{ in}^3 \left(\dfrac{2.54 \text{ cm}}{1 \text{ in}}\right)^3 \left(\dfrac{10^{-2} \text{ m}}{1 \text{ cm}}\right)^3 = 7.44 \times 10^{-3} \text{ m}^3$

(b) $\left(\dfrac{35 \text{ miles}}{1 \text{ hr}}\right)\left(\dfrac{1.609 \text{ km}}{1 \text{ mile}}\right)\left(\dfrac{10^3 \text{ m}}{1 \text{ km}}\right)\left(\dfrac{1 \text{ hr}}{60 \text{ min}}\right)\left(\dfrac{1 \text{ min}}{60 \text{ s}}\right) = 16 \text{ m s}^{-1}$

(c) $\left[6 \text{ ft} \left(\dfrac{12 \text{ in}}{1 \text{ ft}}\right) + 9 \text{ in}\right]\left(\dfrac{2.54 \text{ cm}}{1 \text{ in}}\right)\left(\dfrac{10^{-2} \text{ m}}{1 \text{ cm}}\right) = 2.1 \text{ m}$

(d) $227 \text{ lb} \left(\dfrac{1 \text{ kg}}{2.205 \text{ lb}}\right) = 1.03 \times 10^2 \text{ kg}$

1.73   Make use of the periodic table to match these correctly: (a) Alkaline earth metals are in Group 2: Sr; (b) Elements in the same column as Al have similar chemical properties: In; (c) Elements in Columns 16 and 17 react with K: either O or Br is a correct answer; (d) Transition metals lie in the *d* block: Co; (e) Noble gases are in Group 18: Ne; and (f) Actinides are in the *f* block, $n = 5$: Pu.

1.75   To draw molecular models, make use of the color-coded, scaled atoms shown in Figure 1-4 in your textbook. Here, we use shading to indicate different elements:

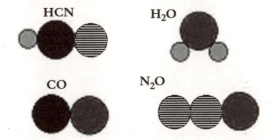

1.77   Time is related to distance through time = distance/speed:

$\text{time} = 2786 \text{ mile}\left(\dfrac{1.609 \text{ km}}{1 \text{ mile}}\right)\left(\dfrac{1 \text{ hr}}{685 \text{ km}}\right)\left(\dfrac{60 \text{ min}}{1 \text{ hr}}\right) = 393 \text{ min}$

1.79   Physical properties do not involve other substances, while chemical properties do: Physical properties: appearance, melting point, softness, and density; chemical properties: reaction with chlorine and reaction with water.

1.81   Consult the periodic table to find examples of various classes of elements: (a) halogens (Group 17) are fluorine (F), chlorine (Cl), bromine (Br), and iodine (I); (b) alkaline earth metals (Group 2) are beryllium (Be), magnesium (Mg), calcium (Ca), strontium (Sr), and barium (Ba); (c) actinides occupy the second long block, for example uranium (U) and berkelium (Bk); and (d) noble gases (Group 18) are helium (He), neon (Ne), argon (Ar), krypton (Kr), xenon (Xe), and radon (Rn).

1.83   Distance (d) is related to speed through the equation, $speed = \dfrac{d}{t}$:

$d = (speed)\, t$

The speed of light (see inside back cover of your textbook) is 2.9979×10^8 m/s

$$d = \left(\frac{2.9979 \times 10^8 \text{ m}}{1 \text{ s}}\right)\left(\frac{10^{-3} \text{ km}}{1 \text{ m}}\right)\left(\frac{60 \text{ s}}{1 \text{ min}}\right)\left(\frac{60 \text{ min}}{1 \text{ hr}}\right)\left(\frac{24 \text{ hr}}{1 \text{ day}}\right)\left(\frac{365.24 \text{ day}}{1 \text{ yr}}\right)(1 \text{ year})$$

$d = 9.4604 \times 10^{12}$ km

1.85   Consult the periodic table to match elemental names and symbols: (a) copper, Cu; sulfur, S; and oxygen, O; (b) $CuSO_4$.

1.87   Each atom contributes a distance equal to its diameter, so the number of atoms is the total distance divided by the diameter of an atom. A unit conversion is required between inches and picometers:

$$\text{\# of atoms} = 1.0 \text{ in} \left(\frac{2.54 \text{ cm}}{1 \text{ in}}\right)\left(\frac{1 \text{ m}}{100 \text{ cm}}\right)\left(\frac{10^{12} \text{ pm}}{1 \text{ m}}\right)\left(\frac{1 \text{ atom}}{200 \text{ pm}}\right) = 1.3 \times 10^8 \text{ atoms}$$

1.89   This is a unit–conversion problem. Multiply by dimensional ratios to convert units:

$$\text{In feet: } 20{,}000 \text{ leagues} \left(\frac{3 \text{ mi}}{1 \text{ league}}\right)\left(\frac{10 \text{ cable}}{1 \text{ mi}}\right)\left(\frac{100 \text{ fathoms}}{1 \text{ cable}}\right)\left(\frac{6 \text{ ft}}{1 \text{ fathom}}\right) = 3.6 \times 10^8 \text{ ft}$$

$$\text{In kilometers: } 3.6 \times 10^8 \text{ ft} \left(\frac{0.3048 \text{ m}}{1 \text{ ft}}\right)\left(\frac{10^{-3} \text{ km}}{1 \text{ m}}\right) = 1.1 \times 10^5 \text{ km}$$

1.91   Your list might include questions about which you have a special interest. Our list includes pollution questions, such as: how to reverse the processes that form the ozone hole; health questions, such as: how to design new drugs to combat bacterial infections and how important for our health are trace elements in our diets; biochemical questions, such as: what molecular processes take place when we smell an odor; and technological questions, such as: how to design materials that are high–temperature superconductors.

1.93   Relative densities can be determined by observing whether materials float or sink or by comparing weights of objects of similar sizes: (a) Oil floats on vinegar, so vinegar has the higher density. (b) Table salt added to water sinks, so table salt has the higher density. (c) Aluminum is much lighter in weight than iron, so iron has the higher density.

1.95   This is a unit–conversion problem. Multiply by dimensional ratios to convert units:

$$100 \text{ yr} \left(\frac{365.24 \text{ day}}{1 \text{ yr}} \right) \left(\frac{24 \text{ hr}}{1 \text{ day}} \right) \left(\frac{60 \text{ min}}{1 \text{ hr}} \right) \left(\frac{60 \text{ s}}{1 \text{ min}} \right) = 3.1557 \times 10^9 \text{ s}$$

The number of days in a year is not exact, so the result has the same number of significant figures as the conversion factor for days/yr.

2.1   Molecular pictures must show the structures of individual particles (atoms, molecules, ions) and the differences between phases. All these particles have monatomic units. A gas is mostly empty space, a liquid is tightly packed but not entirely regular, and a solid has a regular repeating structure:

(a) gaseous helium   (b) solid tungsten   (c) liquid gallium

 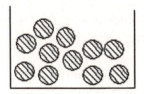

2.3   There are four features of atomic theory. They are:
(1) All matter is composed of tiny particles called atoms. In this reaction, all starting materials and products are made from atoms of carbon and oxygen.
(2) All atoms of a given element have identical chemical properties. All oxygen atoms are in diatomic molecules in the starting materials and CO molecules in the products; likewise, C atoms behave all in the same way.
(3) Atoms form chemical compounds by combining in whole number ratios. C and O combine in a 1:1 ratio to give CO molecules.
(4) Atoms can change the ways they are combined, but they are neither created nor destroyed. There are the same number of oxygen and carbon atoms in the reactants and products. However, they are bonded differently in the products (carbon bonded to oxygen) than in the reactants (oxygen bonded to oxygen, carbon to carbon).

2.5   The law of conservation of mass applies to the entire system, not to any particular portion of the system. The system in this problem is the magnesium strip *and* the surrounding air. When magnesium burns in air, magnesium atoms combine with oxygen from the air to form magnesium oxide (the solid residue). The mass of the residue is the mass of the magnesium strip plus the mass of the oxygen that reacted with the magnesium. The mass of the solid increases, but the mass of solid plus gas (the system) remains constant.

2.7   Molecular pictures must show the structures of individual molecules and the differences between phases. All of these pictures should contain diatomic molecules of bromine. A solid has a regular repeating structure, a liquid is tightly packed but not entirely regular, and a gas is mostly empty space:

solid bromine   liquid bromine   gaseous bromine
(temp < 266K)   (266K < temp < 332K)   (temp > 332K)

2.9　The molecules that give roses their aroma continuously evaporate from the surface of the flower. Once in the gas phase, they move slowly away from the rose until, when they reach a nose, they trigger our olfactory sensors.

2.11　"Dynamic equilibrium" refers to the fact that molecules continue to be transferred from one phase or form to another even though no net change is taking place. In this example, iodine molecules escape from the crystals at the bottom of the flask into the gas phase, where their presence imparts a pale violet color to the gas. Some of these molecules then condense on the surfaces of the flask, forming crystals. At any time, there is a constant number of molecules in the gas phase, but molecules are escaping into the gas from the crystals at the same rate as molecules are condensing onto the crystals.

2.13　Charge per electron and mass per electron can be found in Table 2–1 of your textbook. The charge of the beam is the sum of the charges of the individual electrons, so determine the number of electrons by dividing the total charge by the charge of a single electron:

(a) $\#_{electrons} = \dfrac{\text{total charge}}{\text{charge per } e^-} = \dfrac{-1.00 \times 10^{-6} \text{ C}}{-1.6022 \times 10^{-19} \text{ C/}e^-} = 6.24 \times 10^{12}$ electrons

(b) To determine the combined mass of all of the electrons, multiply the number of electrons by the mass of a single electron:

$$m_{electrons} = (\#)(m_{electron}) = 6.24 \times 10^{12} \text{ electrons} \left(\dfrac{9.1094 \times 10^{-31} \text{ kg}}{1 \text{ electron}} \right) = 5.68 \times 10^{-18} \text{ kg}$$

2.15　In the Millikan experiment, gravitational force was balanced by electrical force for some of the charged particles. Changing the polarity would reverse the direction of the electrical force: negatively–charged particles would be attracted downward, but now positively–charged particles would be attracted upward. Under these conditions, the proper electrical field would cause some of the positively–charged particles to be suspended in space. Because each of these particles had lost one or more electrons, the charge on the electron could be determined from these observations.

2.17　Charge per proton and mass per proton can be found in Table 2–1 of your textbook:

$$\#_{protons} = 1.5 \text{ g} \left(\dfrac{10^{-3} \text{ kg}}{1 \text{ g}} \right) \left(\dfrac{1 \text{ proton}}{1.6726 \times 10^{-27} \text{ kg}} \right) = 9.0 \times 10^{23} \text{ protons}$$

To determine the charge, multiply the number of protons by the charge per proton:
Charge = $(9.0 \times 10^{23} \text{ protons})(1.6022 \times 10^{-19} \text{ C/p}) = 1.4 \times 10^{5}$ C

2.19　The left superscript in an isotopic symbol is the mass number, A, which is the sum of protons and neutrons; the left subscript is the atomic number, Z, which is the number of protons, and a right superscript is the number of protons minus the number of electrons, when this quantity is non–zero:
(a) 8 protons, 8 neutrons, 10 electrons; (b) 5 p, 6 n, 5 e; (c) 25 p, 30 n, 22 e; (d) 17 p, 18 n, 18 e; and (e) 17 p, 20 n, 16 e

2.21   An isotopic symbol has the elemental symbol prefaced by a superscript giving the mass number, A (the sum of protons and neutrons), and a subscript giving the atomic number, Z (the number of protons). Use the periodic table in your text to determine which elemental symbol corresponds to the atomic number:

(a) # protons $= Z = 26$ (iron, Fe), and $A = 26 + 30 = 56$: $^{56}_{26}\text{Fe}$

(b) From the periodic table, the atomic number for uranium is 92: $^{236}_{92}\text{U}$

(c) Argon (Ar) has 18 protons and 20 neutrons, $A = 18 + 20 = 38$: $^{38}_{18}\text{Ar}$

(d) $Z = 9$. $A = 9 + 10 = 19$, the element with 9 protons is fluorine (F): $^{19}_{9}\text{F}$

2.23   A mass spectrum should have the mass number along the x axis and the intensity (or relative abundance) along the y axis. There are two isotopes, so the mass spectrum has two peaks. The mass–11 peak is 4 times as intense as the mass–10 peak:

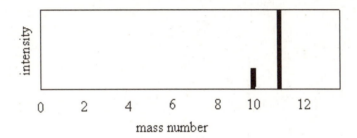

2.25   A pie chart is a circular chart with each piece of the pie proportional in size to the percentage of the isotope that it represents:

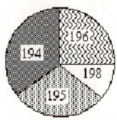

2.27   Nuclei in the "belt of stability" are stable. Instability occurs if a nucleus has too few neutrons, too many neutrons, or $Z > 83$:
(a) unstable, too few neutrons; (b) unstable, $Z > 83$; and (c) stable

2.29   Names, symbols, and atomic numbers of elements all are unique, so there is only one possibility for each entry in the table:

Name	Symbol	Z
Tungsten	W	74
Cobalt	Co	27
Mercury	Hg	80

2.31　The molar mass of a naturally occurring element can be calculated by summing the product of the fractional abundance of each isotope times its isotopic molar mass:

^{36}Ar: 35.96755 g/mol $\left(\dfrac{0.337\%}{100\%}\right) = 0.121$ g/mol

^{38}Ar: 37.96272 g/mol $\left(\dfrac{0.063\%}{100\%}\right) = 0.024$ g/mol

^{40}Ar: 39.96238 g/mol $\left(\dfrac{99.600\%}{100\%}\right) = 39.803$ g/mol

Elemental molar mass $= 0.121 + 0.024 + 39.803 = 39.948$ g/mol

2.33　Mass–mole conversions require the use of masses in grams and molar masses in grams per mole. To determine the number of moles, convert the mass into grams and divide by the molar mass:　　　$n = \dfrac{m}{MM}$

(a) $n = 7.85$ g $\left(\dfrac{1\ \text{mol}}{55.85\ \text{g}}\right) = 0.141$ mol

(b) $n = 65.5\ \mu g \left(\dfrac{10^{-6}\ \text{g}}{1\ \mu g}\right)\left(\dfrac{1\ \text{mol}}{12.01\ \text{g}}\right) = 5.45 \times 10^{-6}$ mol

(c) $n = 4.68$ mg $\left(\dfrac{10^{-3}\ \text{g}}{1\ \text{mg}}\right)\left(\dfrac{1\ \text{mol}}{28.09\ \text{g}}\right) = 1.67 \times 10^{-4}$ mol

(d) $n = 1.46$ ton $\left(\dfrac{10^{3}\ \text{kg}}{1\ \text{ton}}\right)\left(\dfrac{10^{3}\ \text{g}}{1\ \text{kg}}\right)\left(\dfrac{1\ \text{mol}}{26.98\ \text{g}}\right) = 5.41 \times 10^{4}$ mol

2.35　To calculate the number of atoms in a mass, convert the mass to grams, divide by molar mass to obtain moles and multiply by Avogadro's number to obtain number of atoms:

(a) # $= 5.86$ mg $\left(\dfrac{10^{-3}\ \text{g}}{1\ \text{mg}}\right)\left(\dfrac{1\ \text{mol}}{9.012\ \text{g}}\right)\left(\dfrac{6.022 \times 10^{23}\ \text{atoms}}{1\ \text{mol}}\right) = 3.92 \times 10^{20}$ atoms

(b) # $= 5.86$ mg $\left(\dfrac{10^{-3}\ \text{g}}{1\ \text{mg}}\right)\left(\dfrac{1\ \text{mol}}{30.97\ \text{g}}\right)\left(\dfrac{6.022 \times 10^{23}\ \text{atoms}}{1\ \text{mol}}\right) = 1.14 \times 10^{20}$ atoms

(c) # $= 5.86$ mg $\left(\dfrac{10^{-3}\ \text{g}}{1\ \text{mg}}\right)\left(\dfrac{1\ \text{mol}}{91.22\ \text{g}}\right)\left(\dfrac{6.022 \times 10^{23}\ \text{atoms}}{1\ \text{mol}}\right) = 3.87 \times 10^{19}$ atoms

(d) # $= 5.86$ mg $\left(\dfrac{10^{-3}\ \text{g}}{1\ \text{mg}}\right)\left(\dfrac{1\ \text{mol}}{238.0\ \text{g}}\right)\left(\dfrac{6.022 \times 10^{23}\ \text{atoms}}{1\ \text{mol}}\right) = 1.48 \times 10^{19}$ atoms

2.37 Cations are positively charged, anions are negatively charged, and neutral species carry no charge. Charge is designated by the right superscript (no superscript indicates a neutral species) and a reaction to form an ion requires addition or removal of electrons: cation: Cr^{3+}; $Cr \rightarrow Cr^{3+} + 3 e^-$; anion: Cl^-, $Cl + e^- \rightarrow Cl^-$; and neutral: C.

2.39 The chemical formula for a charged species has a right superscript charge:
(a) Cl^-; (b) Na^+; and (c) O^{2-}

2.41 Elements in Group 1 easily lose one electron, and elements in Group 2 easily lose two electrons. Those in Group 16 easily gain two electrons, and those in Group 17 easily gain one electron: (a) Rb^+; (b) F^-; and (c) Ba^{2+}

2.43 Cations and anions combine to form neutral compounds, in which the total amount of positive charge must match the total amount of negative charge. The compounds are RbF and BaF_2.

2.45 Any ionic compound must be electrically neutral. Remember that Group 16 and 17 elements form –2 and –1 ions, respectively. The charges on the ions are Al = +3, O = –2, and F = –1. Two Al ions balance the charge of three O ions and three F ions balance the charge of one Al ion: Aluminum oxide is Al_2O_3, Aluminum fluoride is AlF_3.

2.47 Before we can determine the number of atoms in a sample, we first know the number of moles. This, in turn, requires that we know the mass. We know the density of gold, and we can calculate the volume of the ingot from its dimensions:
$V = l\,w\,h = (15 \text{ cm})(7.5 \text{ cm})(2.0 \text{ cm}) = 225 \text{ cm}^3$

$$\rho = \frac{m}{V} \qquad so \qquad m = V\rho = (225 \text{ cm}^3)(19.32 \text{ g/cm}^3) = 4347 \text{ g}$$

$$n = \frac{m}{MM} = \frac{4347 \text{ g}}{196.97 \text{ g/mol}} = 22.07 \text{ mol}$$

$\# = n\,N_A = (22.07 \text{ mol})(6.022 \times 10^{23} \text{ atoms/mol}) = 1.3 \times 10^{25} \text{ atoms}$
We round the final result to two significant figures to match the dimensions of the ingot.

2.49 The elemental symbol identifies the value of Z, and the left superscript is $A = Z + N$:

Part:	(a)	(b)	(c)	(d)	(e)	(f)
Z	3	20	92	52	10	82
A	6	43	238	130	20	205
N	3	23	146	78	10	123

2.51 Atomic number symbols have the elemental symbol accompanied by a left superscript denoting A and a left subscript denoting Z:
(a) ^4_2He ; (b) $^{184}_{74}\text{W}$; (c) $^{60}_{28}\text{Ni}$; and (d) $^{26}_{12}\text{Mg}$

2.53   Use Avogadro's number to convert from moles to number of atoms. Multiply the total number of atoms by the atomic diameter to find the total length:

$$l = 1.000 \text{ mol} \left(\frac{6.022 \times 10^{23} \text{ atoms}}{1 \text{ mol}} \right) \left(\frac{127.8 \text{ pm}}{1 \text{ atom}} \right) \left(\frac{10^{-12} \text{ m}}{1 \text{ pm}} \right) = 7.696 \times 10^{13} \text{ m}$$

2.55   Because the ratio of nitrogen to oxygen is 4:1, your picture should show eight nitrogen molecules and two oxygen molecules, randomly distributed in the space available:

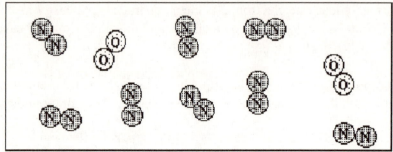

2.57   A mass spectrum contains a peak at each isotopic mass, with peak heights proportional to the isotopic abundances (given in the pie chart). For tin there are 10 peaks, 3 of which are very small and hard to see (peaks at masses 112, 114, and 115), the rest are easily seen:

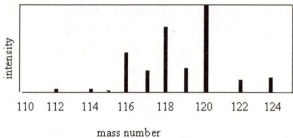

2.59   The upper left superscript in an isotopic symbol represents the sum of protons and neutrons. The number in a symbol such as Co–60 represents the mass number. The number of protons can be deduced from the elemental symbol, with the help of a periodic table:
(a) 43 protons, 56 neutrons, 43 electrons; (b) 26 p, 26 n, 26 e; (c) 54 p, 79 n, 54 e; and
(d) 53 p, 78 n, 53e

2.61   "Same number of atoms" also means "same number of moles," so work with moles:

$n_{Li} = n_{Pt}$ and $n = \dfrac{m}{MM}$

$$m_{Li} = 5.75 \text{ g Pt} \left(\frac{1 \text{ mol Pt}}{195.08 \text{ g Pt}} \right) \left(\frac{1 \text{ mol Li}}{1 \text{ mol Pt}} \right) \left(\frac{6.94 \text{ g Li}}{1 \text{ mol Li}} \right) = 0.205 \text{ g Li}$$

2.63   Charge must be conserved, and every chemical compound must be electrically neutral. Thus, two chlorine atoms gain an electron for each calcium atom that loses two electrons: (a) Ca^{2+}, Cl^-; (b) $CaCl_2$;

(c)

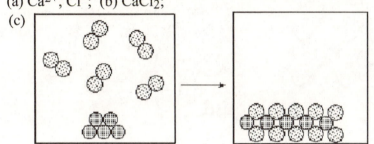

2.65   Each of these nuclei has mass number 40, so each upper left superscript is 40:
$^{40}_{17}Cl$, $^{40}_{18}Ar$, $^{40}_{19}K$, $^{40}_{20}Ca$, and $^{40}_{21}Sc$

2.67   An atomic mass spectrum contains a peak at each isotopic mass, and the peak heights are proportional to the isotopic abundances. The ratio for masses 24:25:26 of magnesium is approximately 10:1:1, so your bars should have those proportional lengths:

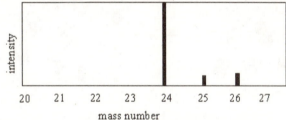

2.69   The atomic number of an element, which equals the number of protons in its nucleus, can be found from the atomic symbol and a periodic table. The right–hand superscript in an ionic symbol represents (protons – electrons):
(a) 11 protons, 10 electrons; (b) 7 p, 10 e; (c) 22 p, 18 e; and (d) 53 p, 54 e

2.71   Diatomic bromine in the liquid phase will have molecules close together with little space between them, in the gas phase there is a lot of space between molecules. Use arrows to show that equal numbers of molecules escape the liquid and are captured from the gas.

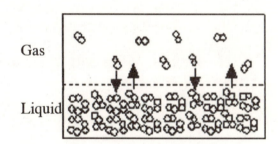

2.73   Only the elements with atomic numbers that are a multiple of 4 can have exactly 1.25 times as many neutrons as protons: $^{45}_{20}Ca$, $^{54}_{24}Cr$, $^{63}_{28}Ni$, $^{72}_{32}Ge$, $^{81}_{36}Kr$, and $^{90}_{40}Zr$

2.75   In a crystalline solid, the ions lie in a regular array, in this case cubic. In the liquid phase, the ions are still close together but the arrangement is no longer entirely regular:

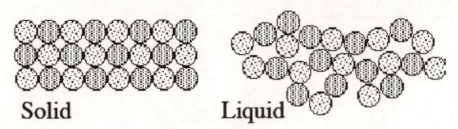

Solid   **Liquid**

2.77   The molecular picture is not correct because atoms are not conserved. Molecular pictures of reactions must show that atoms of each element are conserved. There must be equal numbers of atoms of each element on the two sides of the arrow. The question asks you to adjust the right–hand side to make the figure correct. On the left–hand side there are eight big atoms and 12 small atoms, so the right–hand side must show 2 unreacted small molecules:

3.1   Determine the chemical formula from a ball-and-stick model by counting balls of each
color and consulting the color code for the elements (Figure 1-4 of your textbook):
(a) CH_4; (b) C_2H_4; (c) C_2H_6O; (d) HBr; (e) PCl_3; (f) CH_4N_2O; and (g) C_2H_5I

3.3   Structural formulas look like ball-and-stick structures but with lines in place of sticks and
elemental symbols in place of balls:

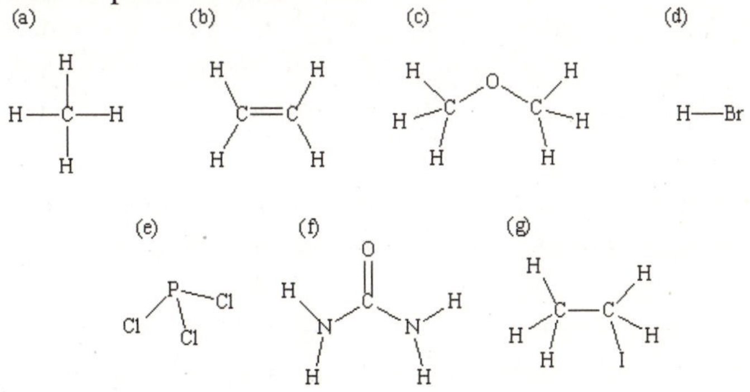

3.5   To convert a structural formula into a line structure, remove all –H connections to carbon
atoms and remove all C atom labels:

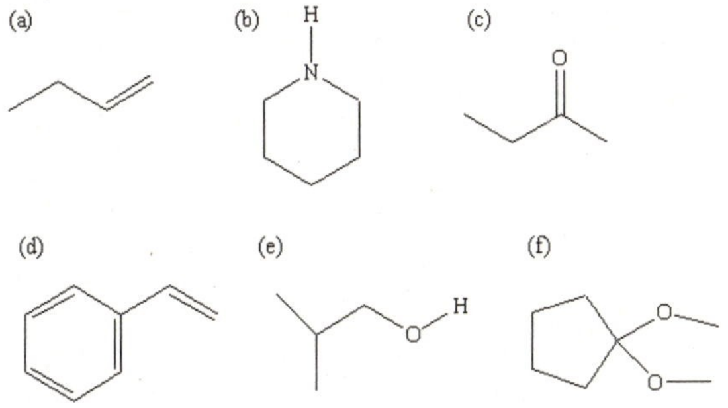

3.7   To convert a line structure into a structural formula, place a C atom at the end of each
line and at each line intersection, then add enough –H connections to give each C atom 4
connections:

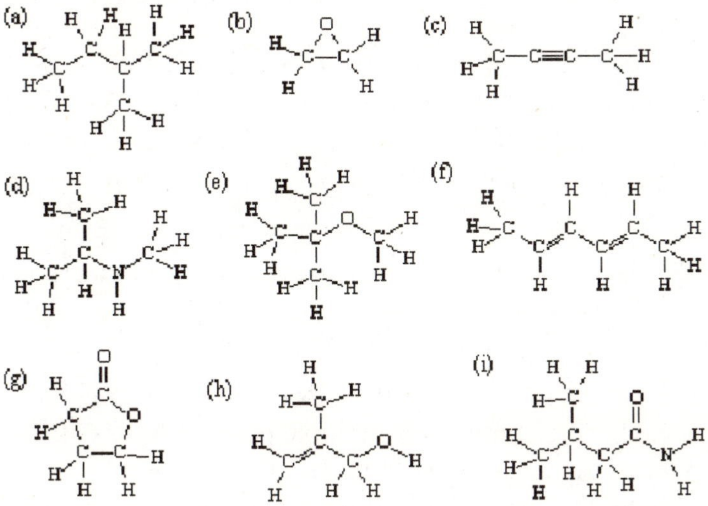

3.9   Determine a chemical formula from a space-filling model by counting atoms of each color and using the color code (Figure 1-4 of your textbook) to identify the elements. Name binary compounds using the rules given in Section 3-2 of your textbook: (a) CO_2, carbon dioxide; (b) HCl, hydrogen chloride; and (c) CCl_4, carbon tetrachloride

3.11   Determine a chemical formula from a ball-and-stick model by counting atoms of each color and using the color code (Figure 1-4 of your textbook) to identify the elements. Name binary compounds using the rules given in Section 3-2 of your textbook: (a) H_2S, hydrogen sulfide; (b) SF_4, sulfur tetrafluoride; and (c) HF, hydrogen fluoride

3.13   Identify the compound as binary or carbon-based. Then apply the rules given in Section 3.2 of your textbook: (a) CH_4; (b) HI; (c) CaH_2; (d) PCl_3; (e) N_2O_5; (f) SF_6; and (g) BF_3

3.15   Identify the compound as binary or carbon-based. Then apply the rules given in Section 3.2 of your textbook: (a) disulfur dichloride; (b) iodine heptafluoride; (c) hydrogen bromide; (d) dinitrogen trioxide; (e) silicon carbide; and (f) methanol

3.17   Ionic compounds can be identified using the decision tree shown in Figure 3-12 in your textbook. Any compound is neutral, so the formula for an ionic compound must contain equal numbers of positive and negative charges:
(a) not ionic, HF; (b) Group 2 metal (2+) and Group 17 nonmetal (1–), ionic, CaF_2;
(c) polyatomic anion (2–) and metal (3+), ionic, $Al_2(SO_4)_3$; (d) ammonium cation (1+) and Group 16 nonmeteal (2–), ionic, $(NH_4)_2S$; (e) not ionic, SO_2; and (f) not ionic, CCl_4

3.19   Identify the compound as ionic, binary but not ionic, or carbon-based. Then apply the rules given in Section 3.2 of your textbook. Ionic compounds can be identified using the decision tree shown in Figure 3-12 in your textbook:
(a) not ionic, carbon-based, dichloromethane; (b) not ionic, binary, carbon dioxide;
(c) Group 2 metal, ionic, calcium oxide; (d) polyatomic anion, ionic, potassium carbonate; (e) not ionic, binary, phosphorus tribromide; (f) not ionic, binary, hydrogen bromide; and (g) polyatomic anion, ionic, sodium hydrogen phosphate

3.21   To determine a chemical formula from the name of a compound requires knowledge of the polyatomic anions and charges. The compound must be electrically neutral.
(a) Na is a +1 cation and SO_4 is a -2 anion, Na_2SO_4; (b) K is a +1 cation and S is a -2 anion, K_2S; (c) K is a +1 cation and H_2PO_4 is a -1 anion, KH_2PO_4; (d) Co is a +2 cation, F is a -1 anion, $CoF_2 \cdot 4H_2O$; (e) Pb is a +4 metal and O is a -2 anion, PbO_2; (f) Na is a +1 cation and HCO_3 is a -1 anion, $NaHCO_3$; and (g) Li is a +1 cation and BrO_4 is a -1 anion, $LiBrO_4$

3.23   Follow the rules for naming ionic compounds (cation first, then anion), using Roman numerals for cation charge only when multiple possibilities exist:
(a) calcium chloride hexahydrate; (b) iron(II) ammonium sulfate;
(c) potassium carbonate; (d) tin(II) chloride dihydrate; (e) sodium hypochlorite;

(f) silver sulfate; (g) copper(II) sulfate; (h) potassium dihydrogen phosphate;
(i) sodium nitrate; (j) calcium sulfite; and (k) potassium permanganate

3.25   To calculate the molar mass of a compound, multiply each elemental molar mass by the
number of atoms in the formula and sum over the elements:
(a) CCl_4, MM = (12.01 g/mol C) + 4(35.45 g/mol Cl) = 153.81 g/mol
(b) K_2S, MM = 2(39.10 g/mol K) + 32.07 g/mol S = 110.27 g/mol
(c) O_3, MM = 3(16.00 g/mol O) = 48.00 g/mol
(d) LiBr, MM = 6.94 g/mol Li + 79.90 g/mol Br = 86.84 g/mol
(e) GaAs, MM = 69.72 g/mol Ga + 74.92 g/mol As = 144.64 g/mol
(f) $AgNO_3$, MM = 107.87 g/mol Ag + 14.01 g/mol N + 3(16.00 g/mol O) = 169.88 g/mol

3.27   Determine the molecular formula from the line drawing, taking into account the
"missing" carbon atoms at the ends and vertices of lines and the "missing" hydrogen
atoms attached to carbon atoms. To calculate the molar mass, multiply each elemental
molar mass by the number of atoms in the formula and sum over the elements:
tyrosine: 9 C atoms, 4 missing H atoms, 7 shown H atoms, 1 N atom, 3 O atoms;
  $C_9H_{11}NO_3$, MM = 9(12.01g/mol) + 11(1.008g/mol) + 1(14.01g/mol)
$$+ 3(16.00g/mol) = 181.19 \text{ g/mol}$$
tryptophan: 11 C atoms, 5 missing H atoms, 7 shown H atoms, 2 N atoms, 2 O atoms;
  $C_{11}H_{12}N_2O_2$, MM =11(12.01g/mol) + 12(1.008g/mol) + 2(14.01g/mol)
$$+ 2(16.00g/mol) = 204.23 \text{ g/mol}$$
glutamic acid: 5 C atoms, 9 shown H atoms, 1 N atom, 4 O atoms;
  $C_5H_9NO_4$, MM = 5(12.01g/mol) + 9(1.008g/mol) + 1(14.01g/mol)
$$+ 4 (16.00g/mol) = 147.13 \text{ g/mol}$$
lysine: 6 C atoms, 14 shown H atoms, 2 N atoms, 2 O atoms;
  $C_6H_{14}N_2O_2$, MM = 6(12.01g/mol) + 14(1.008g/mol) + 2(14.01g/mol)
$$+ 2(16.00g/mol) = 146.19 \text{ g/mol}$$

3.29   To calculate the number of atoms in a mass, convert the mass to grams, divide by molar
mass to obtain moles and multiply by Avogadro's number to obtain the number of atoms:
(a) CH_4: MM = 1(12.01 g/mol) + 4(1.008 g/mol) = 16.04 g/mol

$$\# = 5.86 \text{ mg} \left(\frac{10^{-3} \text{ g}}{1 \text{ mg}}\right)\left(\frac{1 \text{ mol}}{16.04 \text{ g}}\right)\left(\frac{6.022 \times 10^{23} \text{ molecules}}{1 \text{ mol}}\right) = 2.20 \times 10^{20} \text{ molecules}$$

(b) Phosphorus trichloride is PCl_3:
  MM = 1(30.974 g/mol) + 3(35.453 g/mol) = 137.33 g/mol

$$\# = 5.86 \text{ mg} \left(\frac{10^{-3} \text{ g}}{1 \text{ mg}}\right)\left(\frac{1 \text{ mol}}{137.33 \text{ g}}\right)\left(\frac{6.022 \times 10^{23} \text{ molecules}}{1 \text{ mol}}\right) = 2.57 \times 10^{19} \text{ molecules}$$

(c) C_2H_6O: MM = 2(12.01 g/mol) + 6(1.008 g/mol) + 1(15.999 g/mol) = 46.07 g/mol

$$\# = 5.86 \text{ mg} \left(\frac{10^{-3} \text{ g}}{1 \text{ mg}}\right)\left(\frac{1 \text{ mol}}{46.07 \text{ g}}\right)\left(\frac{6.022 \times 10^{23} \text{ molecules}}{1 \text{ mol}}\right) = 7.66 \times 10^{19} \text{ molecules}$$

(d) Uranium hexafluoride is UF_6:

$$MM = 1(238.03 \text{ g/mol}) + 6(18.998 \text{ g/mol}) = 352.02 \text{ g/mol}$$

$$\# = 5.86 \text{ mg}\left(\frac{10^{-3} \text{ g}}{1 \text{ mg}}\right)\left(\frac{1 \text{ mol}}{352.02 \text{ g}}\right)\left(\frac{6.022 \times 10^{23} \text{ molecules}}{1 \text{ mol}}\right) = 1.00 \times 10^{19} \text{ molecules}$$

3.31   To calculate the mass of some number of molecules of a substance, divide by Avogadro's number to obtain moles and multiply by molar mass to obtain grams:

(a) CH_4: $MM = 12.01 \text{ g/mol} + 4(1.008 \text{ g/mol}) = 16.04 \text{ g/mol}$

$$m = 3.75 \times 10^5 \text{ molecules}\left(\frac{1 \text{ mol}}{6.022 \times 10^{23} \text{ molecules}}\right)\left(\frac{16.04 \text{ g}}{1 \text{ mol}}\right) = 9.99 \times 10^{-18} \text{ g}$$

(b) $C_9H_{13}NO_3$:

$$MM = 9(12.01 \text{ g/mol}) + 13(1.008 \text{ g/mol}) + 14.01 \text{ g/mol} + 3(16.00 \text{ g/mol}) = 183.2 \text{ g/mol}$$

$$m = 2.5 \times 10^9 \text{ molecules}\left(\frac{1 \text{ mol}}{6.022 \times 10^{23} \text{ molecules}}\right)\left(\frac{183.2 \text{ g}}{1 \text{ mol}}\right) = 7.6 \times 10^{-13} \text{ g}$$

(c) $C_{55}H_{72}MgN_4O_5$:

$$MM = 55(12.01 \text{ g/mol}) + 72(1.008 \text{ g/mol}) + 24.305 \text{ g/mol} + 4(14.01 \text{ g/mol})$$
$$+ 5(16.00 \text{ g/mol}) = 893.5 \text{ g/mol}$$

$$m = 1 \text{ molecule}\left(\frac{1 \text{ mol}}{6.022 \times 10^{23} \text{ molecules}}\right)\left(\frac{893.5 \text{ g}}{1 \text{ mol}}\right) = 1.484 \times 10^{-21} \text{ g}$$

3.33   All parts of this question involve mass-mole-number conversions. Moles and mass in grams are related through the equation, $n = \dfrac{m}{MM}$. Use Avogadro's number to convert between number and moles. When masses are not given in grams, unit conversions must be made. The chemical formula states the number of atoms of each element per molecule of substance:

MM of vitamin A = $20(12.01 \text{ g/mol}) + 30(1.008 \text{ g/mol}) + 16.00 \text{ g/mol} = 286.44 \text{ g/mol}$

$$n = 0.75 \text{ mg}\left(\frac{10^{-3} \text{ g}}{1 \text{ mg}}\right)\left(\frac{1 \text{ mol}}{286.44 \text{ g}}\right) = 2.6 \times 10^{-6} \text{ mol of vitamin A}$$

$\#_{\text{molecules}} = nN_A = (2.6 \times 10^{-6} \text{ mol})(6.022 \times 10^{23} \text{ molecules/mol}) = 1.6 \times 10^{18} \text{ molecules}$

There are 30 H atoms for every molecule of vitamin A:

\# atoms = (atoms/molecule)(\# molecules)

$$\#_H = 1.6 \times 10^{18} \text{ molecules}\left(\frac{30 \text{ atom H}}{1 \text{ molecule}}\right) = 4.8 \times 10^{19} \text{ atoms of H}$$

in 0.75 mg of vitamin A

$$m_H = \frac{\#H}{N_A}MM = 4.8 \times 10^{19} \text{ atoms}\left(\frac{1 \text{ mol}}{6.022 \times 10^{23} \text{ atoms}}\right)\left(\frac{1.008 \text{ g}}{1 \text{ mol}}\right) = 8.0 \times 10^{-5} \text{ g}$$

3.35   To calculate mass percent composition, which involves mass ratios, work with one mole of substance. Take the ratio of the mass of each element that one mole contains to the mass of one mole (molar mass).
CaO: MM = 40.08 g/mol + 16.00 g/mol = 56.08 g/mol

$$\% \text{ Ca} = (100\%) \frac{40.08 \text{ g/mol Ca}}{56.08 \text{ g/mol CaO}} = 71.47\%$$

$$\% \text{ O} = (100\%) \frac{16.00 \text{ g/mol O}}{56.08 \text{ g/mol CaO}} = 28.53\%$$

SiO_2: MM = 28.09 g/mol + 2(16.00 g/mol) = 60.09 g/mol

$$\% \text{ Si} = (100\%) \frac{28.09 \text{ g/mol Si}}{60.09 \text{ g/mol SiO}_2} = 46.75\%$$

$$\% \text{ O} = (100\%) \frac{2(16.00 \text{ g/mol O})}{60.09 \text{ g/mol SiO}_2} = 53.25\%$$

Al_2O_3: MM = 2(26.98 g/mol) + 3(16.00 g/mol) = 101.96 g/mol;

$$\% \text{ Al} = (100\%) \frac{2(26.98 \text{ g/mol Al})}{101.96 \text{ g/mol Al}_2\text{O}_3} = 52.92\%$$

$$\% \text{ O} = (100\%) \frac{3(16.00 \text{ g/mol O})}{101.96 \text{ g/mol Al}_2\text{O}_3} = 47.08$$

Fe_2O_3: MM = 2(55.85 g/mol) + 3(16.00 g/mol) = 159.70 g/mol

$$\% \text{ Fe} = (100\%) \frac{2(55.85 \text{ g/mol Fe})}{159.70 \text{ g/mol Fe}_2\text{O}_3} = 69.94\%$$

$$\% \text{ O} = (100\%) \frac{3(16.00 \text{ g/mol O})}{159.70 \text{ g/mol Fe}_2\text{O}_3} = 30.06\%$$

3.37   When working with mass percent compositions, it is convenient to consider 100 g of the substance. Divide the mass of each element by its elemental molar mass to obtain relative amounts of each element in the substance:

$$\text{C:} \left(\frac{74.0 \text{ g}}{100 \text{ g}} \right) \left(\frac{1 \text{ mol}}{12.01 \text{ g}} \right) = 6.16 \text{ mol C/100 g}$$

$$\text{H:} \left(\frac{8.65 \text{ g}}{100 \text{ g}} \right) \left(\frac{1 \text{ mol}}{1.008 \text{ g}} \right) = 8.58 \text{ mol H/100 g}$$

$$\text{N:} \left(\frac{17.35 \text{ g}}{100 \text{ g}} \right) \left(\frac{1 \text{ mol}}{14.01 \text{ g}} \right) = 1.24 \text{ mol N/100 g}$$

To obtain the empirical formula divide each amount by the smallest among them (nitrogen with 1.24 mol/100g) and round to whole numbers. This tells us the number of each element relative to one nitrogen atom in the compound:

$$\text{C:} \left(\frac{6.16 \text{ mol C}}{100 \text{ g}} \right) \left(\frac{100 \text{ g}}{1.24 \text{ mol N}} \right) = 4.97 \text{ mol C/mol N, round to 5 C/N}$$

$$H: \left(\frac{8.58 \text{ mol H}}{100 \text{ g}}\right)\left(\frac{100 \text{ g}}{1.24 \text{ mol N}}\right) = 6.92 \text{ mol H/mol N, round to 7 H/N}$$

Empirical formula: C_5H_7N, empirical $MM = 81.12$ g/mol

To convert an empirical formula to a molecular formula, multiply the subscripts in the empirical formula by the ratio of the compound's molar mass to its empirical formula mass:

$$MM/\text{empirical } MM = \left(\frac{162 \text{ g/mol}}{81.12 \text{ g/mol}}\right) = 2; \text{ Molecular formula: } C_{10}H_{14}N_2$$

3.39   If 0.302 g of pure Hg metal was obtained from the decomposition, then the difference between that and the original cinnabar mass is the mass of sulfur:

$m_S = m_{cinnabar} - m_{Hg} = 0.350$ g - 0.302 g = 0.048 g sulfur

$$n_S = \frac{m_S}{MM_S} = 0.048 \text{ g}\left(\frac{1 \text{ mol}}{32.066 \text{ g}}\right) = 0.0015 \text{ mol}$$

$$n_{Hg} = \frac{m_{Hg}}{MM_{Hg}} = 0.302 \text{ g}\left(\frac{1 \text{ mol}}{200.59 \text{ g}}\right) = 0.0015 \text{ mol}$$

Both have the same number of moles, so the empirical formula of cinnabar is HgS.

3.41   The problem statement identifies this as an empirical formula problem. Use the combustion data to determine the mass percent composition of the burned sample, then use elemental molar masses to find the empirical formula. Assume that the only source of C in CO_2 and the only source of H in H_2O is from the unknown compound:

$$m \text{ C} = 11.8 \text{ g } CO_2\left(\frac{1 \text{ mol } CO_2}{44.01 \text{ g } CO_2}\right)\left(\frac{1 \text{ mol C}}{1 \text{ mol } CO_2}\right)\left(\frac{12.01 \text{ g C}}{1 \text{ mol C}}\right) = 3.22 \text{ g C}$$

$$\% \text{ C} = (100\%)\frac{3.22 \text{ g}}{5.00 \text{ g}} = 64.4\%$$

$$m \text{ H} = 2.42 \text{ g } H_2O\left(\frac{1 \text{ mol}}{18.02 \text{ g}}\right)\left(\frac{2 \text{ mol H}}{1 \text{ mol } H_2O}\right)\left(\frac{1.008 \text{ g}}{1 \text{ mol}}\right) = 0.271 \text{ g H}$$

$$\% \text{ H} = (100\%)\frac{0.271 \text{ g}}{5.00 \text{ g}} = 5.42\%$$

Since the unknown contains only C, H, and Fe, the % composition of Fe must be the remainder:

$\% \text{ Fe} = 100\% - (64.4\% + 5.42\%) = 30.2\%$

Now calculate the number of moles of each element in 100. g of the compound:

$$C: 64.4 \text{ g}\left(\frac{1 \text{ mol}}{12.01 \text{ g}}\right) = 5.36 \text{ mol C} \qquad H: 5.42 \text{ g}\left(\frac{1 \text{ mol}}{1.008 \text{ g}}\right) = 5.38 \text{ mol H}$$

$$Fe: 30.2 \text{ g}\left(\frac{1 \text{ mol}}{55.85 \text{ g}}\right) = 0.541 \text{ mol Fe}$$

Divide each by the smallest among them, 0.541 mol Fe:

$$\frac{5.36 \text{ mol C}}{0.541 \text{ mol Fe}} = 9.91 \text{ C/Fe, round to 10}$$

$$\frac{5.38 \text{ mol H}}{0.541 \text{ mol Fe}} = 9.94 \text{ H/Fe, round to 10}$$

The empirical formula is $C_{10}H_{10}Fe$.

3.43   The question asks you to determine the empirical and molecular formulas, for which the essential data are the combustion analysis and the approximate molar mass. Assume that the only source of C in CO_2 and H in H_2O is from the unknown compound:

$$m\text{ C} = 1.466 \text{ g CO}_2 \left(\frac{1 \text{ mol CO}_2}{44.01 \text{ g CO}_2}\right)\left(\frac{1 \text{ mol C}}{1 \text{ mol CO}_2}\right)\left(\frac{12.01 \text{ g C}}{1 \text{ mol C}}\right) = 0.400 \text{ g C}$$

$$\% \text{ C} = (100\%)\frac{0.400 \text{ g}}{0.60 \text{ g}} = 66.7\%$$

$$m\text{ H} = 0.60 \text{ g H}_2O\left(\frac{1 \text{ mol}}{18.02 \text{ g}}\right)\left(\frac{2 \text{ mol H}}{1 \text{ mol H}_2O}\right)\left(\frac{1.008 \text{ g}}{1 \text{ mol}}\right) = 0.0671 \text{ g H}$$

$$\% \text{ H} = (100\%)\frac{0.0671 \text{ g}}{0.60 \text{ g}} = 11.2\%$$

Determine the mass percent of O by subtracting the other mass percents from 100%:

$$\% \text{ O} = 100\% - (66.7\% + 11.2\%) = 22.1\%$$

Now determine the number of moles of each element in 100. g of the compound:

$$\text{C: } 66.7 \text{ g}\left(\frac{1 \text{ mol}}{12.01 \text{ g}}\right) = 5.55 \text{ mol C} \qquad \text{H: } 11.2 \text{ g}\left(\frac{1 \text{ mol}}{1.008 \text{ g}}\right) = 11.1 \text{ mol H}$$

$$\text{O: } 22.1 \text{ g}\left(\frac{1 \text{ mol}}{16.00 \text{ g}}\right) = 1.38 \text{ mol O}$$

Divide each by the smallest among them, 1.38 mol O:

$$\frac{5.55 \text{ mol C}}{1.38 \text{ mol O}} = 4.02 \text{ C/O, round to 4} \qquad \frac{11.1 \text{ mol H}}{1.38 \text{ mol O}} = 8.04 \text{ H/O, round to 8}$$

The empirical formula is C_4H_8O, with the following empirical molar mass:

$$MM = 4(12.01 \text{ g/mol}) + 8(1.008 \text{ g/mol}) + 16.00 \text{ g/mol} = 72.1 \text{ g/mol}.$$

According to the data in the problem, the compound has a molar mass around 220 g/mol:

$$MM/Emp \; MM = 220/72.1 = 3.05.$$ The molecular formula must contain three empirical formula units, so the molecular formula of the compound is $C_{12}H_{24}O_3$.

3.45   The solution process is $MgCl_2 \, (s) \rightarrow Mg^{2+} \, (aq) + 2 \, Cl^- \, (aq)$. Each mole of solid generates one mole of magnesium cations and two moles of chloride anions:

(a) Molarity is found using the equations, $M = \dfrac{n}{V}$ and $n = \dfrac{m}{MM}$

$$MM = 24.31 \text{ g/mol} + 2(35.45 \text{ g/mol}) = 95.21 \text{ g/mol}$$

$$n(Mg^{2+}) = 4.68 \text{ g MgCl}_2 \left(\frac{1 \text{ mol}}{95.21 \text{ g}} \right) \left(\frac{1 \text{ mol Mg}^{2+}}{1 \text{ mol MgCl}_2} \right) = 4.915 \times 10^{-2} \text{ moles Mg}^{2+}$$

$n(Cl^-) = 2 \, n(Mg^{2+}) = 9.831 \times 10^{-2}$ moles Cl^-

Divide by the volume in liters to obtain the concentrations:

$$V = 1.50 \times 10^2 \text{ mL} \left(\frac{10^{-3} \text{ L}}{1 \text{ mL}} \right) = 0.150 \text{ L}$$

$$M(Mg^{2+}) = \frac{4.915 \times 10^{-2} \text{ mol}}{0.150 \text{ L}} = 0.328 \text{ M}$$

$M(Cl^-) = 2 \, M(Mg^{2+}) = 0.655$ M

(Answers have three significant figures because the mass and volume are known to three significant figures.)

(b) Molecular pictures of solutions must illustrate the relative numbers of ions of each type present in the solution. Your picture must show twice as many chloride ions as magnesium ions. There are many more molecules of water than of either of the ions:

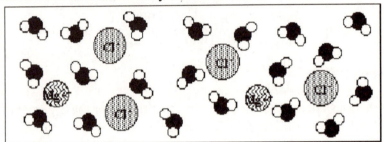

3.47 The solution process is KOH (s) → K⁺ (aq) + OH⁻ (aq). Each mole of solid generates one mole of each ion:

(a) Molarity is found using the equations, $M = \dfrac{n}{V}$ and $n = \dfrac{m}{MM}$

$MM = 39.098 + 15.999 + 1.0079 = 56.11$ g/mol

$$M = \frac{n}{V} = \left(\frac{4.75 \text{ g}}{275 \text{ mL}} \right) \left(\frac{1 \text{ mol}}{56.11 \text{ g}} \right) \left(\frac{10^3 \text{ mL}}{1 \text{ L}} \right) = 0.308 \text{ M} = [K^+] = [OH^-]$$

(b) In a dilution, the number of moles of solute remains constant while volume increases, so $M_f V_f = M_i V_i$. The starting volume is 25.00 mL and the final volume is 100.00 mL:

$$M_f = \frac{M_i V_i}{V_f} = \frac{(0.308 \text{ M})(25.00 \text{ mL})}{100. \text{ mL}} = 0.0770 \text{ M} = [K^+] = [OH^-]$$

(c) Molecular pictures of solutions must illustrate the relative numbers of ions of each type present in the solution. Because of the four-fold dilution, the solution in (b) is 1/4 as concentrated as the solution in (a). We omit solvent molecules for clarity, but remember that there are many more molecules of water than of either of the ions:

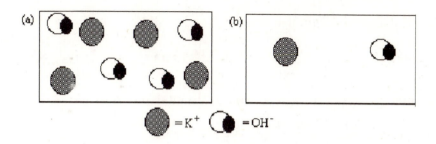

$= K^+$ $= OH^-$

3.49   This is a dilution type problem. Rearrange the dilution equation to give an expression for the initial volume:

$$M_iV_i = M_fV_f \qquad V_i = \frac{M_fV_f}{M_i}$$

$$V_i = \frac{(0.125 \text{ M})(0.500 \text{ L})}{12.1 \text{ M}} = 0.00517 \text{ L or } 5.17 \text{ mL}$$

3.51   In each of the following, remember that concentration is $M = \dfrac{n}{V}$. Begin each part by identifying the major ionic species present in solution:

(a) Na_2CO_3 contains a Group 1 metal ion, Na^+, and a polyatomic anion, CO_3^{2-}, which are the major ionic species. Determine the number of moles in 4.55 g of Na_2CO_3 and use stoichiometric ratios to determine the number of moles of each ion:

$MM(Na_2CO_3) = 2(23.99\text{g/mol}) + 12.01 \text{ g/mol} + 3(16.00\text{g/mol}) = 105.99 \text{ g/mol}$

$$n(Na_2CO_3) = 4.55 \text{ g} \left(\frac{1 \text{ mol}}{105.99 \text{ g}} \right) = 0.0429 \text{ mol } Na_2CO_3$$

$$n(CO_3^{2-}) = 0.0429 \text{ mol } Na_2CO_3 \left(\frac{1 \text{ mol } CO_3^{2-}}{1 \text{ mol } Na_2CO_3} \right) = 0.0429 \text{ mol } CO_3^{2-}$$

There are two moles of Na^+ per mole of Na_2CO_3:

$$n(Na^+) = 0.0429 \text{ mol } Na_2CO_3 \left(\frac{2 \text{ mol } Na^+}{1 \text{ mol } Na_2CO_3} \right) = 0.0858 \text{ mol } Na^+$$

Now determine the molarities by dividing the moles by the volume of the solution:

$$V = 245 \text{ mL} \left(\frac{10^{-3} \text{ L}}{1 \text{ mL}} \right) = 0.245 \text{ L}$$

$$[CO_3^{2-}] = \left(\frac{0.0429 \text{ mol}}{0.245 \text{ L}} \right) = 0.175 \text{ M}$$

$$[Na^+] = 2 [CO_3^{2-}] = 0.350 \text{ M}$$

(b) NH_4Cl contains a polyatomic ion and is therefore an ionic compound with two major ionic species, NH_4^+ and Cl^-. Determine the number of moles in 27.45 mg of the salt. The stoichiometric ratio is 1:1, so the number of moles of each ion is equal to the number of moles of the salt.·

$MM(NH_4Cl) = 14.01 \text{g/mol} + 4(1.008 \text{ g/mol}) + 35.45 \text{ g/mol} = 53.49 \text{ g/mol}$

$$n(NH_4Cl) = 27.45 \text{ mg} \left(\frac{10^{-3} \text{ g}}{1 \text{ mg}} \right) \left(\frac{1 \text{ mol}}{53.49 \text{ g}} \right) = 5.132 \times 10^{-4} \text{ mol} = n(NH_4^+) = n(Cl^-)$$

Determine the molarities of each ion by dividing moles by the volume of the solution:

$$[NH_4^+] = [Cl^-] = \frac{5.132 \times 10^{-4} \text{ mol}}{1.55 \times 10^{-2} \text{ L}} = 0.0331 \text{ M}$$

(c) Potassium sulfate (K_2SO_4) contains a Group 1 metal ion, K^+, and a polyatomic anion, SO_4^{2-}, which are the major ionic species. Determine the number of moles in 1.85 kg of the salt and use stoichiometric ratios to determine the number of moles of each ion:

$MM(K_2SO_4) = 2(39.01 \text{ g/mol}) + 32.066 \text{ g/mol} + 4(16.00 \text{ g/mol}) = 174.26 \text{ g/mol}$

$$n(K_2SO_4) = 1.85 \text{ kg} \left(\frac{10^3 \text{ g}}{1 \text{ kg}} \right) \left(\frac{1 \text{ mol}}{174.26 \text{ g}} \right) = 10.6 \text{ mol } K_2SO_4$$

$$n(SO_4^{2-}) = 10.6 \text{ mol } K_2SO_4 \left(\frac{1 \text{ mol } SO_4^{2-}}{1 \text{ mol } K_2SO_4} \right) = 10.6 \text{ mol } SO_4^{2-}$$

$n(K^+) = 2\,n(SO_4^{2-}) = 21.2 \text{ mol } K^+$

Obtain the molarities by dividing the moles of each ion by the volume of the solution:

$$[SO_4^{2-}] = \frac{10.6 \text{ mol}}{5.75 \times 10^3 \text{ L}} = 1.84 \times 10^{-3} \text{ M}$$

Similarly to part (a), K^+ has twice the concentration as the anion:

$$[K^+] = \frac{21.2 \text{ mol}}{5.75 \times 10^3 \text{ L}} = 3.69 \times 10^{-3} \text{ M}$$

3.53   Chemical names follow the standard rules for chemical nomenclature:
(a) carbon dioxide; (b) potassium nitrate; (c) sodium chloride;
(d) sodium hydrogen carbonate; (e) sodium carbonate; (f) sodium hydroxide;
(g) calcium oxide; and (h) magnesium hydroxide

3.55   To calculate mass percent composition, which involves mass ratios, work with one mole of substance. Take the ratio of the mass of each element that one mole contains to the mass of one mole (molar mass):
Fe_2SiO_4: $MM = 2(55.85 \text{ g/mol}) + 28.09 \text{ g/mol} + 4(16.00 \text{ g/mol}) = 203.79 \text{ g/mol}$

$$\% \text{ Fe} = (100\%) \frac{2(55.85 \text{ g/mol})}{203.79 \text{ g/mol}} = 54.81\%$$

$$\% \text{ Si} = (100\%) \frac{28.09 \text{ g/mol}}{203.79 \text{ g/mol}} = 13.78\%$$

$$\% \text{ O} = (100\%) \frac{4(16.00 \text{ g/mol})}{203.79 \text{ g/mol}} = 31.40\%$$

$NaAlSi_3O_8$:

$MM = 22.99 \text{ g/mol} + 26.98 \text{ g/mol} + 3(28.09 \text{ g/mol}) + 8(16.00 \text{ g/mol}) = 262.24 \text{ g/mol}$

$$\% \ Na = (100\%)\frac{22.99 \text{ g/mol}}{262.24 \text{ g/mol}} = 8.767\%$$

$$\% \ Al = (100\%)\frac{26.98 \text{ g/mol}}{262.24 \text{ g/mol}} = 10.29\%$$

$$\% \ Si = (100\%)\frac{3(28.09 \text{ g/mol})}{262.24 \text{ g/mol}} = 32.13\%$$

$$\% \ O = (100\%)\frac{8(16.00 \text{ g/mol})}{262.24 \text{ g/mol}} = 48.81\%$$

$Al_2Si_2O_5(OH)_4$:

$MM = 2(26.98 \text{ g/mol}) + 2(28.09 \text{ g/mol}) + 5(16.00 \text{ g/mol})$
$\qquad + 4(16.00 \text{ g/mol} + 1.008 \text{ g/mol}) = 258.17 \text{ g/mol}$

$$\% \ Al = (100\%)\frac{2(26.98 \text{ g/mol})}{258.17 \text{ g/mol}} = 20.90\%$$

$$\% \ Si = (100\%)\frac{2(28.09 \text{ g/mol})}{258.17 \text{ g/mol}} = 21.76\%$$

$$\% \ O = (100\%)\frac{9(16.00 \text{ g/mol})}{258.17 \text{ g/mol}} = 55.78\%$$

$$\% \ H = (100\%)\frac{4(1.008 \text{ g/mol})}{258.17 \text{ g/mol}} = 1.56\%$$

$MgSi_4O_{10}(OH)_8$:

$MM = 24.31 \text{ g/mol} + 4(28.09 \text{ g/mol}) + 10(16.00 \text{ g/mol})$
$\qquad + 8(16.00 \text{ g/mol} + 1.008 \text{ g/mol}) = 432.73 \text{ g/mol}$

$$\% \ Mg = (100\%)\frac{24.31 \text{ g/mol}}{432.73 \text{ g/mol}} = 5.62\%$$

$$\% \ Si = (100\%)\frac{4(28.09 \text{ g/mol})}{432.73 \text{ g/mol}} = 25.97\%$$

$$\% \ O = (100\%)\frac{18(16.00 \text{ g/mol})}{432.73 \text{ g/mol}} = 66.55\%$$

$$\% \ H = (100\%)\frac{8(1.008 \text{ g/mol})}{432.73 \text{ g/mol}} = 1.86\%$$

3.57   This is a mass-mole-mass problem, for which molar masses must be calculated:
NH_4NO_3:

$MM = 2(14.01 \text{ g/mol}) + 4(1.008 \text{ g/mol}) + 3(16.00 \text{ g/mol}) = 80.05 \text{ g/mol}$

$$1.00 \text{ kg N}\left(\frac{1 \text{ mol N}}{14.01 \text{ g N}}\right)\left(\frac{1 \text{ mol NH}_4\text{NO}_3}{2 \text{ mol N}}\right)\left(\frac{80.05 \text{ g NH}_4\text{NO}_3}{1 \text{ mol NH}_4\text{NO}_3}\right) = 2.86 \text{ kg}$$

$(NH_4)_2SO_4$:

$MM = 2(14.01 \text{ g/mol}) + 8(1.008 \text{ g/mol}) + 32.07 \text{ g/mol} + 4(16.00 \text{ g/mol}) = 132.15 \text{ g/mol}$

$$1.00 \text{ kg N} \left(\frac{1 \text{ mol N}}{14.01 \text{ g N}} \right) \left(\frac{1 \text{ mol } (NH_4)_2 SO_4}{2 \text{ mol N}} \right) \left(\frac{132.15 \text{ g } (NH_4)_2 SO_4}{1 \text{ mol } (NH_4)_2 SO_4} \right) = 4.72 \text{ kg}$$

$(NH_2)_2CO$:

$MM = 2(14.01 \text{ g/mol}) + 4(1.008 \text{ g/mol}) + 12.01 \text{ g/mol} + 16.00 \text{ g/mol} = 60.06 \text{ g/mol}$

$$1.00 \text{ kg N} \left(\frac{1 \text{ mol N}}{14.01 \text{ g N}} \right) \left(\frac{1 \text{ mol } (NH_2)_2 CO}{2 \text{ mol N}} \right) \left(\frac{60.06 \text{ g } (NH_2)_2 CO}{1 \text{ mol } (NH_2)_2 CO} \right) = 2.14 \text{ kg}$$

$(NH_4)_2HPO_4$:

$MM = 2(14.01 \text{ g/mol}) + 9(1.008 \text{ g/mol}) + 30.97 \text{ g/mol} + 4(16.00 \text{ g/mol}) \ 132.06 \text{ g/mol}$

$$1.00 \text{ kg N} \left(\frac{1 \text{ mol N}}{14.01 \text{ g N}} \right) \left(\frac{1 \text{ mol } (NH_4)_2 HPO_4}{2 \text{ mol N}} \right) \left(\frac{132.06 \text{ g } (NH_4)_2 HPO_4}{1 \text{ mol } (NH_4)_2 HPO_4} \right) = 4.71 \text{ kg}$$

3.59   To obtain a chemical formula, convert a line structure into a structural formula by placing a C atom at the end of each line and at each line intersection. Then add enough –H connections to give each C atom 4 connections. From the structural formula, determine the chemical formula by counting and calculate MM from numbers of atoms and atomic molar masses:

abscisic acid   indole acetic acid   zeatin

(a) abscisic acid, $C_{15}H_{20}O_4$:

$MM = 15(12.01 \text{ g/mol}) + 20(1.008 \text{ g/mol}) + 4(16.00 \text{ g/mol}) = 264.31 \text{ g/mol}$

(b) indole acetic acid, $C_{10}H_9NO_2$:

$MM = 10(12.01 \text{ g/mol}) + 9(1.008 \text{ g/mol}) + 14.01 \text{ g/mol} + 2(16.00 \text{ g/mol})$
$$= 175.18 \text{ g/mol}$$

(c) zeatin, $C_{10}H_{13}N_5O$:

$MM = 10(12.01 \text{ g/mol}) + 13(1.008 \text{ g/mol}) + 5(14.01 \text{ g/mol}) + 16.00 \text{ g/mol}$
$$= 219.25 \text{ g/mol}$$

3.61   When salts dissolve in water, they dissociate into cations and anions. Molecular substances retain their structure in solution, and H_2O is always present as a species in aqueous solutions: (a) H_2O, NH_4^+, SO_4^{2-}; (b) H_2O, CO_2; (c) H_2O, Na^+, F^-; (d) H_2O, K^+, CO_3^{2-}; (e) H_2O, Na^+, HSO_4^-; and (f) H_2O, Cl_2

3.63   To write correct chemical formulas, you must know the compositions and charges of polyatomic ions. The overall formula of a compound must be electrically neutral: (a) $NaNO_2$ and $NaNO_3$; (b) K_2CO_3 and $KHCO_3$; (c) FeO and Fe_2O_3; and (d) I_2 and I^-

3.65   The chemical formula of a substance provides all the information needed to compute its molar characteristics:

(a) $MM = 2(26.98 \text{ g/mol}) + 3[32.07 \text{ g/mol} + 4(16.00 \text{ g/mol})] = 342.17 \text{ g/mol}$

(b) $n = \dfrac{m}{MM} = 25.0 \text{ g}\left(\dfrac{1 \text{ mol}}{342.17 \text{ g}}\right) = 7.31 \times 10^{-2} \text{ mol}$

(c) To determine the percent composition, work with one mole of substance. Take the ratio of the mass of each element that one mole contains to the mass of one mole (molar mass):

$$\% \text{ Al} = (100\%)\dfrac{2(26.98 \text{ g/mol})}{342.17 \text{ g/mol}} = 15.77\%$$

$$\% \text{ S} = (100\%)\dfrac{3(32.07 \text{ g/mol})}{342.17 \text{ g/mol}} = 28.12\%$$

$$\% \text{ O} = (100\%)\dfrac{12(16.00 \text{ g/mol})}{342.17 \text{ g/mol}} = 56.11\%$$

(d) $1.00 \text{ mol O}\left(\dfrac{1 \text{ mol Al}_2(SO_4)_3}{12 \text{ mol O}}\right)\left(\dfrac{342.17 \text{ g}}{1 \text{ mol}}\right) = 28.5 \text{ g}$

3.67   Convert from mass to moles to determine molarity:

$$M = \dfrac{n}{V} = \dfrac{m}{MM\ V}$$

$$M = 8.3 \text{ mg/L}\left(\dfrac{10^{-3} \text{ g}}{1 \text{ mg}}\right)\left(\dfrac{1 \text{ mol}}{32.00 \text{ g}}\right) = 2.6 \times 10^{-4} \text{ mol/L}$$

3.69   Use nomenclature rules to determine a name from a chemical formula:
(a) ammonium chloride; (b) xenon tetrafluoride; (c) iron(III) oxide; (d) sulfur dioxide; and (e) potassium perchlorate

3.71   The question asks about mole-mass-number quantities. Although there is much interesting biomedical information provided, the only relevant data are the chemical formula of verapamil, $C_{27}H_{38}O_4N_2$, and the amount of verapamil per tablet, 120.0 mg:

(a) $MM = 27(12.01 \text{ g/mol}) + 38(1.008 \text{ g/mol}) + 4(16.00 \text{ g/mol})$
$$+ 2(14.01 \text{ g/mol}) = 454.59 \text{ g/mol}$$

(b) $n = \dfrac{m}{MM} = 120.0 \text{ mg}\left(\dfrac{10^{-3} \text{ g}}{1 \text{ mg}}\right)\left(\dfrac{1 \text{ mol}}{454.59 \text{ g}}\right) = 2.640 \times 10^{-4} \text{ mol}$

(c) # atoms $= n\ N_A = 2.640 \times 10^{-4} \text{ mol}\left(\dfrac{6.022 \times 10^{23} \text{ molec. verap.}}{1 \text{ mol}}\right)\left(\dfrac{2 \text{ atoms N}}{1 \text{ molec. verap.}}\right)$

atoms $= 3.179 \times 10^{20}$ atoms N

3.73   This question asks about solution concentration under two different sets of conditions:
(a) Use density and molar mass to determine molarity:

$MM = 2(1.01 \text{ g/mol H}) + 32.07 \text{ g/mol S} + 4(16.00 \text{ g/mol O}) = 98.09 \text{ g/mol H}_2SO_4$

$$1.75 \text{ g/mL}\left(\frac{1000 \text{ mL}}{1 \text{ L}}\right)\left(\frac{80.\%}{100\%}\right)\left(\frac{1 \text{ mol}}{98.09 \text{ g}}\right) = 14 \text{ M}$$

(b) This is a dilution problem, so rearrange $M_iV_i = M_fV_f$ to give $V_i = \dfrac{M_fV_f}{M_i}$:

$$V_i = \frac{(0.65 \text{ M})(2.50 \text{ L})}{14 \text{ M}} = 0.11 \text{ L or } 1.1 \times 10^2 \text{ mL}$$

3.75   Moles and number of ions are connected by Avogadro's number, $\# = nN_A$. Use information about ionic charges to relate overall concentration to the amounts of each individual ion:

(a) Magnesium is in Group 2 and forms cations with 2+ charge; nitrate anions have 1– charge, so the chemical formula is $Mg(NO_3)_2$. Use $n = MV$ to determine the number of moles:

$n = (1.25 \text{ M})(55.6 \text{ mL})(10^{-3} \text{ L/mL}) = 6.95 \times 10^{-2} \text{ mol Mg}(NO_3)_2$

$n\,(Mg^{2+}) = 6.95 \times 10^{-2} \text{ mol and } n\,(NO_3^-) = 2(6.95 \times 10^{-2} \text{ mol}) = 1.39 \times 10^{-1} \text{ mol}$

$\#\,(Mg^{2+}) = (6.95 \times 10^{-2} \text{ mol})(6.022 \times 10^{23}/\text{mol}) = 4.19 \times 10^{22} \text{ ions}$

$\#\,(NO_3^-) = 2(4.19 \times 10^{22} \text{ ions}) = 8.37 \times 10^{22} \text{ ions}$

(b) Potassium is in Group 1 and forms cations with 1+ charge; sulfate anions have 2– charge, so the chemical formula is K_2SO_4. We are told the number of formula units, so the volume is irrelevant:

$\#\,(K^+) = (2 \text{ K/formula unit})(9.03 \times 10^{21} \text{ formula units}) = 1.81 \times 10^{22} \text{ ions}$

$\#\,(SO_4^{2-}) = 9.03 \times 10^{21} \text{ ions}$

$$n\,(K^+) = \frac{\#}{N_A} = \frac{1.81 \times 10^{22} \text{ ions}}{6.022 \times 10^{23} \text{ ions/mol}} = 3.00 \times 10^{-2} \text{ mol}$$

$n\,(SO_4^{2-}) = (1/2)\,n\,(K^+) = 1.50 \times 10^{-2} \text{ mol}$

(c) Sodium is in Group 1 and forms cations with 1+ charge; phosphate anions have 3– charge, so the chemical formula is Na_3PO_4. We need to do a mass-mole conversion and then use the volume-molarity relationship:

$MM = 3(22.990 \text{ g/mol}) + 30.974 \text{ g/mol} + 4(15.999 \text{ g/mol}) = 163.9 \text{ g/mol}$

$m = (13.5 \text{ g/L})(3.55 \text{ mL})(10^{-3} \text{ L/mL}) = 4.79 \times 10^{-2} \text{ g}$

$$n(Na_3PO_4) = \frac{m}{MM} = \frac{4.79 \times 10^{-2} \text{ g}}{163.9 \text{ g/mol}} = 2.92 \times 10^{-4} \text{ mol}$$

$n\,(PO_4^{3-}) = n(Na_3PO_4) = 2.92 \times 10^{-4} \text{ mol}$

$n\,(Na^+) = 3\,n(Na_3PO_4) = 3(2.92 \times 10^{-4} \text{ mol}) = 8.77 \times 10^{-4} \text{ mol}$

$\#\,(PO_4^{3-}) = (2.92 \times 10^{-4} \text{ mol})(6.022 \times 10^{23}/\text{mol}) = 1.76 \times 10^{20} \text{ ions}$

$\#\,(Na^+) = 3(1.76 \times 10^{19} \text{ ions}) = 5.28 \times 10^{20} \text{ ions}$

3.77 A molar concentration is moles in one liter of liquid. First use density to calculate mass/L, then use MM to convert to mol/L:

$$mass/L = \left(\frac{1.00\text{ g}}{1\text{ mL}}\right)\left(\frac{10^3\text{ mL}}{1\text{ L}}\right) = 1.00 \times 10^3\text{ g/L}$$

$MM = 2(1.008\text{ g/mol}) + 15.999\text{ g/mol} = 18.015\text{ g/mol}$

$$molar\ concentration = \left(\frac{1.00 \times 10^3\text{ g}}{1\text{ L}}\right)\left(\frac{1\text{ mol}}{18.015\text{ g}}\right) = 55.5\text{ mol/L}$$

3.79 The mass information lets us calculate mass percentages directly:

$$Mass\ \%\ F = 100\%\left(\frac{2.93\text{ g}}{3.75\text{ g}}\right) = 78.1\%$$

Mass % S = 100% - 78.1% = 21.9%

Consider 100. g of sample, and divide by elemental molar masses to get moles:

$$n\ (F) = \frac{78.1\text{ g}}{18.998\text{ g/mol}} = 4.11\text{ mol, and } n\ (S) = \frac{21.9\text{ g}}{32.066\text{ g/mol}} = 0.682\text{ mol}$$

Divide by the smaller amount to find the F:S ratio: $\dfrac{4.11\text{ mol}}{0.682\text{ mol}} = 6.02$, rounds to 6

The empirical formula is SF_6, with

empirical $MM = (32.066\text{ g/mol}) + 6\ (18.998\text{ g/mol}) = 146.066\text{ g/mol}$

Knowing that the molar mass is less than 200 g/mol, we deduce that the molecular formula must be the same as the empirical formula.

3.81 To determine which is the larger amount, we must do appropriate conversions.

(a) $m = n\ MM$:

    $(0.35\text{ mol }H_2O)(18.02\text{ g/mol}) = 6.3\text{ g}$

    $(0.25\text{ mol }H_2O_2)(34.0\text{ g/mol}) = 8.5\text{ g}$

    The hydrogen peroxide sample has the larger mass.

(b) We could calculate number of atoms, but it is sufficient to calculate number of moles of atoms in each sample:

    $(0.88\text{ mol }CO)(2\text{ atoms/molecule}) = 1.76\text{ mol atoms}$

    $(0.66\text{ mol }CO_2)(3\text{ atoms/molecule}) = 1.98\text{ mol atoms}$

    The carbon dioxide sample contains more atoms.

(c) Use density to convert from volume to mass; then do a mass-mole conversion to get moles. Methanol is CH_3OH and ethanol is C_2H_5OH:

    $(2.5\text{ mL methanol})(0.791\text{ g/mL}) = 1.98\text{ g}$

    $MM = 12.01\text{ g/mol} + 4(1.008\text{ g/mol}) + 15.999\text{ g/mol} = 32.04\text{ g/mol}$

    $$n = \frac{m}{MM} = \frac{1.98\text{ g}}{32.04\text{ g/mol}} = 0.0618\text{ mol}$$

    $(3.5\text{ mL ethanol})(0.789\text{ g/mL}) = 2.76\text{ g}$

    $MM = 2(12.01\text{ g/mol}) + 6(1.008\text{ g/mol}) + 15.999\text{ g/mol} = 46.07\text{ g/mol}$

$$n = \frac{m}{MM} = \frac{2.76 \text{ g}}{46.07 \text{ g/mol}} = 0.0599 \text{ mol}$$

The methanol sample contains more moles.

3.83 This problem poses a set of mole-mass-number conversion questions. Sevin is $C_{12}H_{11}NO_2$:

$MM = 12(12.01 \text{ g/mol}) + 11(1.008 \text{ g/mol}) + 14.01 \text{ g/mol} + 2(16.00 \text{ g/mol}) = 201.22 \text{ g/mol}$

(a) $n_C = \frac{m}{MM}\left(\frac{\text{mol element}}{\text{mol compound}}\right) = 8.3 \text{ g}\left(\frac{1 \text{ mol}}{201.22 \text{ g}}\right)\left(\frac{12 \text{ mol C}}{1 \text{ mol Sevin}}\right) = 0.49 \text{ mol C}$

(b) $m_O = 4.5 \text{ g}\left(\frac{1 \text{ mol}}{201.22 \text{ g}}\right)\left(\frac{2 \text{ mol O}}{1 \text{ mol Sevin}}\right)\left(\frac{16.00 \text{ g}}{1 \text{ mol}}\right) = 0.72 \text{ g O}$

(c) $n = 75 \text{ mL}\left(\frac{1.00 \text{ g}}{1 \text{ mL}}\right)\left(\frac{0.010 \text{ g}}{100 \text{ g}}\right)\left(\frac{1 \text{ mol}}{201.22 \text{ g}}\right) = 3.7 \times 10^{-5} \text{ mol Sevin}$

$\# = n N_A = (3.7 \times 10^{-5} \text{ mol})(6.022 \times 10^{23} \text{ molecules/mol}) = 2.2 \times 10^{19} \text{ molecules}$

(d) 15 gallons of spray requires 15 mL of insecticide. The insecticide contains

3.7×10^{-5} mol per 75 mL, so $n = 15 \text{ mL}\left(\frac{3.7 \times 10^{-5} \text{ mol}}{75 \text{ mL}}\right) = 7.4 \times 10^{-6} \text{ mol Sevin}$

3.85 Moles and mass in grams are related through the equation, $n = \frac{m}{MM}$. When masses are not given in grams, unit conversions must be made (1 lb = 453.6 g):

$MM (H_3PO_4) = 3(1.008 \text{ g/mol}) + 30.97 \text{ g/mol} + 4(16.00 \text{g/mol}) = 97.99 \text{ g/mol}$

$n = 2.619 \times 10^8 \text{ lb}\left(\frac{453.6 \text{ g}}{1 \text{ lb}}\right)\left(\frac{1 \text{ mol}}{97.99 \text{ g}}\right) = 1.212 \times 10^9 \text{ mol } H_3PO_4$

The chemical formula shows that there is one mole of P in every mole of H_3PO_4. The problem states that 15% of the annual production of H_3PO_4 comes from elemental P:

$n(P) = 1.212 \times 10^9 \text{ mol}\left(\frac{15\%}{100\%}\right) = 1.8 \times 10^8 \text{ mol}$

$m(P) = n\,MM = 1.8 \times 10^8 \text{ mol}\left(\frac{30.97 \text{ g}}{1 \text{ mol}}\right)\left(\frac{10^{-3} \text{ kg}}{1 \text{ g}}\right) = 5.6 \times 10^6 \text{ kg of P consumed}$

3.87 To determine a molecular formula, mass percent composition and an approximate molar mass must be known. The data given are insufficient to calculate mass percent composition. To calculate the mass percentages of carbon and hydrogen, the mass of compound that was burned must be known. Unless the percentages of C and H together total 100%, further information about the other elements present in the compound is also needed.

3.89   This problem asks for mass-mole calculations on turquoise.

(a) $MM_{turquoise}$= 63.55 g/mol Cu + 6(26.98 g/mol Al) + 4(30.97 g/mol P)

$$+ 28(16.00 \text{ g/mol O}) + 16(1.01 \text{ g/mol H}) = 813.47 \text{ g/mol turquoise}$$

$$7.25 \text{ g turquoise}\left(\frac{1 \text{ mol}}{813.47 \text{ g}}\right)\left(\frac{6 \text{ mol Al}}{1 \text{mol turquoise}}\right)\left(\frac{26.98 \text{ g}}{1 \text{ mol}}\right) = 1.44 \text{ g Al}$$

(b) there are 4 phosphate ions per every 28 O atoms present:

$$5.50 \times 10^{-3} \text{g O}\left(\frac{1 \text{mol}}{16.00 \text{ g}}\right)\left(\frac{4 \text{ mol PO}_4^{3-}}{28 \text{ mol O}}\right)\left(\frac{6.022 \times 10^{23} \text{ ions}}{1 \text{ mol}}\right) = 2.96 \times 10^{19} \text{ ions}$$

(c) charge on OH = 1−, charge on PO_4 = 3−, charge on Al = 3+

Thus, the charge on Cu must balance 8(−1) + 4(−3) + 6(+3) = −2

charge on Cu = 2+

3.91   (a) To find the mass percent composition of each element find the mass of each element in one mole of the compound and divide that by the mass of one mole of the compound.

The masses of the elements in 1 mol of $C_{10}H_{16}N_5O_{13}P_3$ follow:

$$10 \text{ mol C}\left(\frac{12.011 \text{ g}}{1 \text{ mol}}\right) = 120.11 \text{ g} \qquad 16 \text{ mol H}\left(\frac{1.008 \text{ g}}{1 \text{ mol}}\right) = 16.13 \text{ g}$$

$$5 \text{ mol N}\left(\frac{14.007 \text{ g}}{1 \text{ mol}}\right) = 70.035 \text{ g} \qquad 13 \text{ mol O}\left(\frac{15.999 \text{ g}}{1 \text{ mol}}\right) = 207.987 \text{ g}$$

$$3 \text{ mol P}\left(\frac{30.974 \text{ g}}{1 \text{ mol}}\right) = 92.922 \text{ g}$$

Mass of 1 mol of $C_{10}H_{16}N_5O_{13}P_3$ = 507.18 g

$$\text{Mass percentage} = (100\%)\,\frac{\text{mass of element}}{\text{mass of compound}}$$

$$\text{C: } (100\%)\frac{120.11 \text{ g}}{507.18 \text{ g}} = 23.682\% \text{ C} \qquad \text{H: } (100\%)\frac{16.13 \text{ g}}{507.18 \text{ g}} = 3.180\% \text{ H}$$

$$\text{N: } (100\%)\frac{70.035 \text{ g}}{507.18 \text{ g}} = 13.809\% \text{ N} \qquad \text{O: } (100\%)\frac{207.987 \text{ g}}{507.18 \text{ g}} = 41.009\% \text{ O}$$

$$\text{P: } (100\%)\frac{92.922 \text{ g}}{507.18 \text{ g}} = 18.321\% \text{ P}$$

Summing the percentages gives 100%, which confirms the work.

(a) This is a mass-number conversion. First convert from mass ATP to moles ATP, then to molecules ATP, and finally to atoms of P:

$$n = \frac{m}{MM} = \left(\frac{1.75 \text{ } \mu g}{507.18 \text{ g/mol}}\right)\left(\frac{10^{-6} g}{1 \text{ } \mu g}\right) = 3.45 \times 10^{-9} \text{ mol ATP}$$

$$\# = nN_A = (3.45 \times 10^{-9} \text{ mol ATP})(6.022 \times 10^{23}/\text{mol}) = 2.078 \times 10^{15} \text{ molecules ATP}$$

$$\# \text{ atoms} = \# \text{ molecules}\left(\frac{\# \text{ atoms}}{1 \text{ molecule}}\right) = 3(2.078 \times 10^{15}) = 6.23 \times 10^{15} \text{ atoms P}$$

(b) This is a mole-mass conversion:

$$m = n\,MM = 3.0\ \text{pmol} \left(\frac{10^{-12}\ \text{mol}}{1\ \text{pmol}} \right) (507.18\ \text{g/mol}) = 1.5 \times 10^{-9}\ \text{g or 1.5 ng}$$

(c) To make this comparison, we could work with moles. A quicker way is using atom ratios. One molecule of ATP contains 16 H atoms and 5 N atoms, so the mass ratio of amounts of ATP containing equal numbers of atoms of these elements is 5 to 16:

$$m = \left(\frac{5}{16} \right) (37.5\ \text{mg}) = 11.7\ \text{mg}$$

You can verify this result by converting mg of ATP to moles, then molecules, then multiplying by 5 to get the number of N atoms. Then divide by 16 to get molecules of ATP containing the same number of H atoms and convert back to moles and mass. The mass-mole and number-mole conversions cancel, leaving only the 5:16 ratio.

3.93   Rearrange the definition of mass percentage composition to compute the molar mass:

$$\% \text{Co} = (100\%)\ \frac{(\text{atoms Co/molecule})(MM_{\text{Co}})}{MM_{\text{molecule}}}$$

$$MM_{\text{molecule}} = \frac{100\%}{\% \text{Co}}\,(\text{atoms Co/molecule})MM_{\text{Co}}$$

$$MM_{\text{molecule}} = \left(\frac{100\%}{4.34\%} \right)\left(\frac{1\ \text{atom Co}}{1\ \text{molecule}} \right)\left(\frac{58.93\ \text{g}}{1\ \text{mol}} \right) = 1.36 \times 10^3\ \text{g/mol}$$

3.95   This problem poses a set of mole-mass-number conversion questions. Malathion is $C_{10}H_{19}O_6PS_2$: $MM = 10(12.01\text{g/mol}) + 19(1.008\text{g/mol}) + 6(16.00\text{g/mol})$

$$+ 30.97\text{g/mol} + 2(32.07\text{g/mol}) = 330.36\ \text{g/mol}$$

(a) $\# = 6.5\ \text{g} \left(\frac{1\ \text{mol}}{330.36\ \text{g}} \right)\left(\frac{2\ \text{mol S}}{1\ \text{mol Malathion}} \right)\left(\frac{6.022 \times 10^{23}\ \text{atoms}}{1\ \text{mol}} \right) = 2.4 \times 10^{22}\ \text{atoms}$

(b) $m_O = 17.8\ \text{g} \left(\frac{1\ \text{mol}}{330.36\ \text{g}} \right)\left(\frac{6\ \text{mol O}}{1\ \text{mol Malathion}} \right)\left(\frac{16.00\ \text{g}}{1\ \text{mol}} \right) = 5.17\ \text{g}$

(c) $M = \dfrac{n}{V} = \left(\dfrac{1.00\ \text{g}}{1\ \text{mL}} \right)\left(\dfrac{10^3\ \text{mL}}{1\ \text{L}} \right)\left(\dfrac{50\ \text{g}}{100\ \text{g}} \right)\left(\dfrac{1\ \text{mol}}{330.36\ \text{g}} \right) = 1.5\ \text{M}$

(d) The original solution has its concentration reduced by a factor of 10^4:

$$n = MV = 1.5\ \text{M} \left(\frac{1}{10^4} \right)\left(\frac{10^{-3}\ \text{L}}{1\ \text{mL}} \right)(25\ \text{mL}) = 3.8 \times 10^{-6}\ \text{mol}$$

3.97   To prepare a solution of low concentration from a solution of higher concentration, a dilution is made using pure solvent. In this problem, we know the final concentrations and volume and the initial concentrations, so we need to calculate the initial volumes:

$$M_f V_f = M_i V_i,\ \text{so}\ V_f = \frac{M_f V_f}{M_i}$$

$$V_f(\text{sodium acetate}) = \frac{(0.30\ M)(1.5\ L)}{5.0\ M} = 0.090\ L$$

$$V_f(\text{acetic acid}) = \frac{(0.15\ M)(1.5\ L)}{5.0\ M} = 0.045\ L$$

To prepare the solution, the worker should mix 0.090 L of sodium acetate solution, 0.045 L of acetic acid solution, and enough water to make the final volume 1.5 L.

3.99 $CuSO_4$:

$$MM = 63.546\ g/mol + 32.066\ g/mol + 4(15.999\ g/mol) = 159.608\ g/mol$$

$CuSO_4 \bullet 5H_2O$:

$$MM = 159.608\ g/mol + 10(1.008\ g/mol) + 5(15.999\ g/mol) = 249.683\ g/mol$$

$$n = 125\ g\left(\frac{1\ mol}{159.608\ g}\right) = 0.783\ \text{mol of anhydrous cupper(II)sulfate}$$

$$m = 0.783\ mol\left(\frac{249.683\ g}{1\ mol}\right) = 196\ \text{g of hydrated cupper(II)sulfate}$$

4.1   A balanced chemical equation must have equal numbers of atoms of each element on each side of the arrow. Balance each element in turn, beginning with those that appear in only one reactant and product, by adjusting stoichiometric coefficients. Generally, H and O are balanced last. In each case, we start by determining the number of atoms on each side of the chemical equation:

(a) $NH_4NO_3 \rightarrow N_2O + H_2O$

$2N + 3O + 4H \rightarrow 2N + 2O + 2H$

There are two nitrogen atoms in the reactant and two in the products, thus nitrogen is already balanced. Balance H by making the stoichiometric coefficient of water 2:

$NH_4NO_3 \rightarrow N_2O + \mathbf{2}\ H_2O$

$2N + 3O + 4H \rightarrow 2N + 3O + 4H$ (all elements balanced)

(b) $P_4O_{10} + H_2O \rightarrow H_3PO_4$

$4P + 11O + 2H \rightarrow P + 4O + 3H$

Start by balancing P, which occurs in only one reactant and one product. There are 4 P on the reactant side and 1 P on the product side. Hence, balance P by giving H_3PO_4 a coefficient of 4:

$P_4O_{10} + H_2O \rightarrow \mathbf{4}\ H_3PO_4$

$4P + 11O + 2H \rightarrow 4P + 16O + 12H$

Next balance H by giving H_2O a coefficient of 6, which also balances O:

$P_4O_{10} + \mathbf{6}\ H_2O \rightarrow 4\ H_3PO_4$

$4P + 16O + 12H \rightarrow 4P + 16O + 12H$ (all elements balanced)

(c) $HIO_3 \rightarrow I_2O_5 + H_2O$

$I + H + 3O \rightarrow 2I + 2H + 6O$

Notice that there are half as many of each atom on the reactant side as on the product side. Therefore, this equation can be balanced by simply increasing the HIO_3 coefficient from 1 to 2.

$\mathbf{2}\ HIO_3 \rightarrow I_2O_5 + H_2O$

$2I + 2H + 6O \rightarrow 2I + 2H + 6O$ (all elements balanced)

(d) $As + Cl_2 \rightarrow AsCl_5$

$1As + 2Cl \rightarrow 1As + 5Cl$

As is already balanced. There are 2 Cl on the reactant side and 5 on the product. To balance Cl we need to change the Cl_2 coefficient from 1 to $\frac{5}{2}$:

$As + \frac{5}{2} Cl_2 \rightarrow AsCl_5$

$1As + 5Cl \rightarrow 1As + 5Cl$

Notice that both As and Cl are now balanced. However, we do not want fractions in a chemical equation. Therefore, multiply **all** coefficients by 2:

$\mathbf{2}(As + \frac{5}{2} Cl_2 \rightarrow AsCl_5)$

$\mathbf{2}\ As + \mathbf{5}\ Cl_2 \rightarrow \mathbf{2}\ AsCl_5$

$2\ As + 10\ Cl \rightarrow 2\ As + 10\ Cl$ (all elements balanced)

4.3   Molecular pictures must show the correct number of molecules undergoing the reaction. In 4.1d, 2 atoms of As react with 5 molecules of Cl_2 to form 2 molecules of $AsCl_5$. Remember that when drawing molecular pictures you must differentiate between the different atom types by color, labeling, or shading (here we use shading):

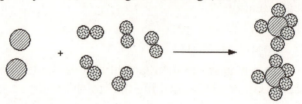

4.5   A balanced chemical equation must have equal numbers of atoms of each element on each side of the arrow. Balance each element in turn, beginning with those that appear in only one reactant and product, by adjusting stoichiometric coefficients. Generally, H and O are balanced last. When balancing an equation, start by determining the number of atoms on each side of the chemical equation:

(a) Start by determining the chemical formula of each compound.

molecular hydrogen: H_2; carbon monoxide: CO; methanol: CH_3OH

The chemical equation (unbalanced) is:

$H_2 + CO \rightarrow CH_3OH$

$2H + 1C + 1O \rightarrow 4H + 1C + 1O$

Both carbon and oxygen are balanced. There are 4H on the product side and 2H on the reactant. Change the coefficient of H_2 from 1 to 2:

2 $H_2 + CO \rightarrow CH_3OH$

$4H + 1C + 1O \rightarrow 4H + 1C + 1O$ (all elements balanced)

(a) $CaO + C \rightarrow CO + CaC_2$

$1Ca + 1O + 1C \rightarrow 1Ca + 1O + 3C$

Note that both Ca and O are balanced. Multiply the coefficient of C by 3:

$CaO + $**3** $C \rightarrow CO + CaC_2$

$1Ca + 1O + 3C \rightarrow 1Ca + 1O + 3C$ (all elements balanced)

(b) $C_2H_4 + O_2 + HCl \rightarrow C_2H_4Cl_2 + H_2O$

$2C + 5H + 2O + Cl \rightarrow 2C + 6H + 1O + 2Cl$

Carbon is already balanced. Both O and Cl are found in only one reactant and one product, so we can start with either. We choose Cl. Since there is one Cl on the reactant side and 2 on the product side, change the HCl coefficient to 2:

$C_2H_4 + O_2 + $**2** $HCl \rightarrow C_2H_4Cl_2 + H_2O$

$2C + 6H + 2O + 2Cl \rightarrow 2C + 6H + 1O + 2Cl$

This balances both Cl and H while leaving C balanced. To balance O, change the coefficient on O_2 from 1 to $\frac{1}{2}$:

$C_2H_4 + \frac{\mathbf{1}}{\mathbf{2}} O_2 + 2HCl \rightarrow C_2H_4Cl_2 + H_2O$

$2C + 6H + 1O + 2Cl \rightarrow 2C + 6H + 1O + 2Cl$

The reaction is now balanced. Multiply **all** coefficients by 2 to eliminate fractions from the chemical equation:

$$2(C_2H_4 + \frac{1}{2}O_2 + 2\,HCl \rightarrow C_2H_4Cl_2 + H_2O)$$

$$2\,C_2H_4 + O_2 + \mathbf{4}\,HCl \rightarrow \mathbf{2}\,C_2H_4Cl_2 + \mathbf{2}\,H_2O$$

4.7   Molecular pictures must show the correct number of molecules undergoing the reaction. In 4.5a, 2 molecules of hydrogen react with a molecule of CO to form one molecule of CH_3OH.

4.9   A balanced chemical equation must have equal numbers of atoms of each element on each side of the arrow. Balance each element in turn, beginning with those that appear in only one reactant and product, by adjusting stoichiometric coefficients. Start by determining the number of atoms on each side of the chemical equation.

(a) $Ca(OH)_2 + H_3PO_4 \rightarrow H_2O + Ca_3(PO_4)_2$

    $1Ca + 6O + 5H + 1P \rightarrow 3Ca + 9O + 2H + 2P$

Start by balancing Ca by changing the coefficient of $Ca(OH)_2$ from 1 to 3:

    $\mathbf{3}\,Ca(OH)_2 + H_3PO_4 \rightarrow H_2O + Ca_3(PO_4)_2$

    $3Ca + 10O + 9H + 1P \rightarrow 3Ca + 9O + 2H + 2P$

Next balance P by multiplying the H_3PO_4 coefficient by 2:

    $3\,Ca(OH)_2 + \mathbf{2}\,H_3PO_4 \rightarrow H_2O + Ca_3(PO_4)_2$

    $3Ca + 14O + 12H + 2P \rightarrow 3Ca + 9O + 2H + 2P$

Now balance H by multiplying the water coefficient by 6

    $3\,Ca(OH)_2 + 2\,H_3PO_4 \rightarrow \mathbf{6}\,H_2O + Ca_3(PO_4)_2$

    $3Ca + 14O + 12H + 2P \rightarrow 3Ca + 14O + 12H + 2P$ (all elements balanced)

(b) $Na_2O_2 + H_2O \rightarrow NaOH + H_2O_2$

    $2Na + 3O + 2H \rightarrow 1Na + 3O + 3H$

Start by balancing Na by changing the NaOH coefficient from 1 to 2:

    $Na_2O_2 + H_2O \rightarrow \mathbf{2}\,NaOH + H_2O_2$

    $2Na + 3O + 2H \rightarrow 2Na + 4O + 4H$

Since H occurs in only one reactant (O occurs in both), balance H next. Balance H by multiplying the water coefficient by 2:

    $Na_2O_2 + \mathbf{2}\,H_2O \rightarrow 2\,NaOH + H_2O_2$

    $2Na + 4O + 4H \rightarrow 2Na + 4O + 4H$ (all elements balanced)

(c) $BF_3 + H_2O \rightarrow HF + H_3BO_3$

   $1B + 3F + 2H + 1O \rightarrow 1B + 1F + 4H + 3O$

Boron is already balanced. Balance F by multiplying the HF coefficient by 3:

   $BF_3 + H_2O \rightarrow \mathbf{3}\,HF + H_3BO_3$

   $1B + 3F + 2H + 1O \rightarrow 1B + 3F + 6H + 3O$

Next balance H by multiplying the water coefficient by 3:

   $BF_3 + \mathbf{3}\,H_2O \rightarrow 3\,HF + H_3BO_3$

   $1B + 3F + 6H + 3O \rightarrow 1B + 3F + 6H + 3O$ (all elements balanced)

(c) $NH_3 + CuO \rightarrow Cu + N_2 + H_2O$.

   $1N + 3H + 1Cu + 1O \rightarrow 2N + 2H + 1Cu + 1O$

Cu is balanced. To balance H, multiply the NH_3 coefficient by 2 and H_2O by 3:

   $\mathbf{2}\,NH_3 + CuO \rightarrow Cu + N_2 + \mathbf{3}\,H_2O$

   $2N + 6H + 1Cu + 1O \rightarrow 2N + 6H + 1Cu + 3O$

This also balances N. Finally, balance O without unbalancing Cu by giving *both* CuO *and* Cu coefficients of 3:

   $2\,NH_3 + \mathbf{3}\,CuO \rightarrow \mathbf{3}\,Cu + N_2 + 3\,H_2O$

   $2N + 6H + 3Cu + 3O \rightarrow 2N + 6H + 3Cu + 3O$

4.11   We must calculate the mass of the second reactant that will completely react with 5.00 g of the first. Remember that calculations of amounts in chemistry always center on the mole. Thus, we need first to determine how many moles there are in the first reactant and then determine how many moles of the second are required to react completely:

(a) balanced equation: $2\,H_2 + CO \rightarrow CH_3OH$

$$5.00 \text{ g } H_2 \left(\frac{1 \text{ mol } H_2}{2.016 \text{ g } H_2} \right)\left(\frac{1 \text{ mol CO}}{2 \text{ mol } H_2} \right)\left(\frac{28.01 \text{ g CO}}{1 \text{ mol CO}} \right) = 34.7 \text{ g CO}$$

(b) balanced equation: $CaO + 3\,C \rightarrow CO + CaC_2$

$$5.00 \text{ g CaO} \left(\frac{1 \text{ mol CaO}}{56.08 \text{ g CaO}} \right)\left(\frac{3 \text{ mol C}}{1 \text{ mol CaO}} \right)\left(\frac{12.01 \text{ g C}}{1 \text{ mol C}} \right) = 3.21 \text{ g C}$$

(c) balanced equation: $2\,C_2H_4 + O_2 + 4\,HCl \rightarrow 2\,C_2H_4Cl_2 + 2\,H_2O$

$$5.00 \text{ g } C_2H_4 \left(\frac{1 \text{ mol } C_2H_4}{28.05 \text{ g } C_2H_4} \right)\left(\frac{1 \text{ mol } O_2}{2 \text{ mol } C_2H_4} \right)\left(\frac{32.00 \text{ g } O_2}{1 \text{ mol } O_2} \right) = 2.85 \text{ g } O_2$$

4.13   You are asked to calculate the mass of sodium iodide required to produce 1.50 kg of iodine. The balanced equation is given in the problem. Remember to convert the mass into grams before dividing by the molar mass:

$$1.50 \text{ kg } I_2 \left(\frac{1000 \text{ g}}{1 \text{ kg}} \right)\left(\frac{1 \text{ mol}}{253.8 \text{ g}} \right)\left(\frac{2 \text{ mol NaI}}{1 \text{ mol } I_2} \right)\left(\frac{149.89 \text{ g}}{1 \text{ mol}} \right) = 1.77 \times 10^3 \text{ g NaI}$$

4.15   This problem tells you how much of the reactant you have (in kg) and asks you to calculate the amount of product formed. Remember that calculations of amounts in chemistry always center on the mole:

$$1.00 \text{ kg sugar} \left(\frac{10^3 \text{ g}}{1 \text{ kg}} \right) \left(\frac{1 \text{ mol}}{180.2 \text{ g}} \right) \left(\frac{2 \text{ mol C}_2\text{H}_5\text{OH}}{1 \text{ mol sugar}} \right) \left(\frac{46.07 \text{ g}}{1 \text{ mol}} \right) = 511 \text{ g C}_2\text{H}_5\text{OH}$$

4.17   To calculate masses for a chemical reaction, balance the equation, then do appropriate mass-mole-mass conversions. The reaction is:

$$CCl_4 + HF \rightarrow CCl_2F_2 + HCl$$

$$1C + 4Cl + 1H + 1F \rightarrow 1C + 3Cl + 2F + 1H$$

Carbon is balanced. Fluorine occurs in only 1 reactant and 1 product, therefore balance it next. Fluorine can be balanced by giving HF a coefficient of 2:

$$CCl_4 + \textbf{2} \text{ HF} \rightarrow CCl_2F_2 + HCl$$

$$1C + 4Cl + 2H + 2F \rightarrow 1C + 3Cl + 2F + 1H$$

Finally, balance H and Cl by giving HCl a coefficient of 2:

$$CCl_4 + 2 \text{ HF} \rightarrow CCl_2F_2 + \textbf{2} \text{ HCl}$$

$$1C + 4Cl + 2H + 2F \rightarrow 1C + 4Cl + 2F + 2H \text{ (all elements balanced)}$$

To determine the amount of HF required to completely react with the CCl_4, convert to moles and do the appropriate conversions:

$$175 \text{ kg CCl}_4 \left(\frac{10^3 \text{ g}}{1 \text{ kg}} \right) \left(\frac{1 \text{ mol}}{153.8 \text{ g}} \right) \left(\frac{2 \text{ mol HF}}{1 \text{ mol CCl}_4} \right) \left(\frac{20.01 \text{ g}}{1 \text{ mol}} \right) \left(\frac{1 \text{ kg}}{10^3 \text{ g}} \right) = 45.5 \text{ kg HF}$$

Since this reaction is 100% efficient, all of the reactants will be converted to products. Mass of products formed:

$$175 \text{ kg CCl}_4 \left(\frac{10^3 \text{ g}}{1 \text{ kg}} \right) \left(\frac{1 \text{ mol}}{153.8 \text{ g}} \right) \left(\frac{1 \text{ mol CCl}_2F_2}{1 \text{ mol CCl}_4} \right) \left(\frac{120.9 \text{ g}}{1 \text{ mol}} \right) \left(\frac{1 \text{ kg}}{10^3 \text{ g}} \right) = 138 \text{ kg CCl}_2F_2$$

$$175 \text{ kg CCl}_4 \left(\frac{10^3 \text{ g}}{1 \text{ kg}} \right) \left(\frac{1 \text{ mol}}{153.8 \text{ g}} \right) \left(\frac{2 \text{ mol HCl}}{1 \text{ mol CCl}_4} \right) \left(\frac{36.46 \text{ g}}{1 \text{ mol}} \right) \left(\frac{1 \text{ kg}}{10^3 \text{ g}} \right) = 83.0 \text{ kg HCl}$$

4.19   In a yield problem, we compare actual amounts with theoretical amounts. Carry out standard mole-mass conversions to find the theoretical amount.
The reaction (balanced) is:

$$Ca_5(PO_4)_3F + 5 \text{ H}_2SO_4 + 10 \text{ H}_2O \rightarrow 3 \text{ H}_3PO_4 + 5 \text{ CaSO}_4 \cdot 2\text{H}_2O + HF$$

$$\% \text{ yield} = (100\%) \frac{\text{actual yield}}{\text{theoretical yield}}$$

The problem gives the actual yield (400 g), but we need to calculate the theoretical yield. Use stoichiometry to calculate the theoretical yield (FA = fluoroapatite):

$$1.00 \text{ kg FA} \left(\frac{10^3 \text{ g}}{1 \text{ kg}} \right) \left(\frac{1 \text{ mol}}{504.3 \text{ g}} \right) \left(\frac{3 \text{ mol H}_3PO_4}{1 \text{ mol FA}} \right) \left(\frac{97.99 \text{ g}}{1 \text{ mol}} \right) = 583 \text{ g H}_3PO_4$$

Now divide the actual yield by the theoretical yield to obtain the percent yield:

$$\% \text{ yield} = (100\%)\frac{\text{actual yield}}{\text{theoretical yield}} = (100\%)\frac{400 \text{ g}}{583 \text{ g}} = 68.6\%$$

4.21   In a multiple-step synthesis, the yield of each step must be multiplied to determine the overall yield. In this case, there are eight steps, each with a yield of 88%, so the overall fractional yield is $(0.88)(0.88)(0.88)(0.88)(0.88)(0.88)(0.88) = (0.88)^8 = 0.36$. Use this value and the desired amount of phenobarbital in moles to determine how many moles of toluene are required. Do the usual mole-mass conversions:
For phenobarbital, $MM = 12(12.01 \text{ g/mol}) + 12(1.01 \text{ g/mol}) + 2(14.01 \text{ g/mol})$
$$+ 3(16.00 \text{ g/mol}) = 232.3 \text{ g/mol}$$

$$\text{Moles of toluene} = 25 \text{ kg}\left(\frac{10^3 \text{ g}}{1 \text{ kg}}\right)\left(\frac{1 \text{ mol}}{232.3 \text{ g}}\right)\left(\frac{1 \text{ mol toluene}}{1 \text{ mol phenobarbital}}\right) = 107.7 \text{ mol toluene}$$

$$\text{Moles of toluene required} = \frac{107.7 \text{ mol}}{0.36} = 299 \text{ mol}$$

$$\text{Mass required} = 299 \text{ mol}\left(\frac{92.13 \text{ g}}{1 \text{ mol}}\right)\left(\frac{10^{-3} \text{ kg}}{1 \text{ g}}\right) = 28 \text{ kg}$$

The result has two significant figures because the mass and yield are only known to two significant figures.

4.23   The calculation of a yield requires a theoretical amount and an actual amount. The theoretical amount is calculated as in Problem 4.17:

$$175 \text{ kg CCl}_4\left(\frac{10^3 \text{ g}}{1 \text{ kg}}\right)\left(\frac{1 \text{ mol}}{153.8 \text{ g}}\right)\left(\frac{1 \text{ mol CCl}_2\text{F}_2}{1 \text{ mol CCl}_4}\right)\left(\frac{120.9 \text{ g}}{1 \text{ mol}}\right)\left(\frac{10^{-3} \text{ kg}}{1 \text{ g}}\right) = 137.6 \text{ kg CCl}_2\text{F}_2$$

(Carry an additional significant figure until the calculation is complete.)
The actual amount (given) is 105 kg

$$\% \text{ yield} = (100\%)\frac{105 \text{ kg}}{137.6 \text{ kg}} = 76.3\%$$

To find how much of each reactant is required, divide the desired amount by the percent yield to obtain the theoretical yield and then do the usual stoichiometric calculations:

$$\text{Theoretical yield} = 155 \text{ kg}\left(\frac{100\%}{76.3\%}\right) = 203 \text{ kg CCl}_2\text{F}_2$$

$$203 \text{ kg CCl}_2\text{F}_2\left(\frac{10^3 \text{ g}}{1 \text{ kg}}\right)\left(\frac{1 \text{ mol}}{120.9 \text{ g}}\right)\left(\frac{1 \text{ mol CCl}_4}{1 \text{ mol CCl}_2\text{F}_2}\right)\left(\frac{153.8 \text{ g}}{1 \text{ mol}}\right)\left(\frac{10^{-3} \text{ kg}}{1 \text{ g}}\right) = 258 \text{ kg CCl}_4$$

$$203 \text{ kg CCl}_2\text{F}_2\left(\frac{10^3 \text{ g}}{1 \text{ kg}}\right)\left(\frac{1 \text{ mol}}{120.9 \text{ g}}\right)\left(\frac{2 \text{ mol HF}}{1 \text{ mol CCl}_2\text{F}_2}\right)\left(\frac{20.01 \text{ g}}{1 \text{ mol}}\right)\left(\frac{10^{-3} \text{ kg}}{1 \text{ g}}\right) = 67.2 \text{ kg HF}$$

4.25   To determine which ingredient will run out first, calculate how many cheeseburgers could be made with each ingredient:

INGREDIENT	INVENTORY	AMOUNT/BURGER	# BURGERS
Roll	(12 dozen)(12/dozen)	1	144
Beef	40 lb	2(1/4 lb)	80
Cheese	(2 pkg)(65/pkg)	1	130
Tomato	40	1/4	160
Lettuce	(1 kg)(10^3 g/kg)	15 g	66.7

The lettuce will run out first, after 66 burgers have been made.

4.27   The problem gives information about the amounts of both starting materials, so this is a limiting reactant situation. We must calculate the number of moles of each species, construct a table of amounts, and use the results to determine the mass of the product formed. Starting amounts are in kilograms (1 kg = 10^3 g), so it will be convenient to work with 10^3 mol amounts. The balanced equation is given in the problem.
Begin by calculating the initial amounts:

$$75.0 \text{ kg N}_2\left(\frac{10^3 \text{ g}}{1 \text{ kg}}\right)\left(\frac{1 \text{ mol}}{28.02 \text{ g}}\right) = 2.68 \times 10^3 \text{ mol}$$

$$75.0 \text{ kg H}_2\left(\frac{10^3 \text{ g}}{1 \text{ kg}}\right)\left(\frac{1 \text{ mol}}{2.02 \text{ g}}\right) = 37.1 \times 10^3 \text{ mol}$$

Next, using the balanced chemical equation, construct an amounts table:

Reaction	N_2 +	$3 H_2$ →	$2 NH_3$
Initial amount (10^3 mol)	2.68	37.1	0
10^3 mol/coeff	2.68 (LR)	12.4	
Change in amount (10^3 mol)	-2.68	-8.04	+5.36
Final amount (10^3 mol)	0	4.36	5.36

Do a mole-mass conversion to determine the mass of ammonia that could be produced:

$$5.36 \times 10^3 \text{ mol}\left(\frac{17.04 \text{ g}}{1 \text{ mol}}\right)\left(\frac{10^{-3} \text{ kg}}{1 \text{ g}}\right) = 91.3 \text{ kg}$$

4.29   The problem gives information about the amounts of starting materials, so this is a limiting reactant situation. We must calculate the number of moles of each species, construct a table of amounts, and use the results to determine the final product masses. Starting amounts are in metric tons (1 metric ton = 10^6 g), so it will be convenient to work with 10^6 mol amounts. See the answers to problem 4.5 for the balanced equations:
(a) Calculate the initial amounts:

$$1000 \text{ kg CO}\left(\frac{10^3 \text{ g}}{1 \text{ kg}}\right)\left(\frac{1 \text{ mol}}{28.01 \text{ g}}\right) = 0.0357 \times 10^6 \text{ mol CO}$$

$$1000 \text{ kg H}_2\left(\frac{10^3 \text{ g}}{1 \text{ kg}}\right)\left(\frac{1 \text{ mol}}{2.02 \text{ g}}\right) = 0.495 \times 10^6 \text{ mol H}_2$$

Now set up an amounts table:

Reaction	2 H$_2$ +	CO →	CH$_3$OH
Initial amount (10^6 mol)	0.495	0.0357	0
10^6 mol/coeff	0.248	0.0357 (LR)	
Change (10^6 mol)	-2(0.0357)	-0.0357	+0.0357
Final amount (10^6 mol)	0.424	0.00	0.0357

The mass that could be produced is:

$$0.0357 \times 10^6 \text{ mol} \left(\frac{32.04 \text{ g}}{1 \text{ mol}}\right)\left(\frac{1 \text{ ton}}{10^6 \text{ g}}\right) = 1.14 \text{ metric ton CH}_3\text{OH}$$

(b) Calculate the initial amounts:

$$1000 \text{ kg CaO}\left(\frac{10^3 \text{ g}}{1 \text{ kg}}\right)\left(\frac{1 \text{ mol}}{56.08 \text{ g}}\right) = 0.0178 \times 10^6 \text{ mol CaO}$$

$$1000 \text{ kg C}\left(\frac{10^3 \text{ g}}{1 \text{ kg}}\right)\left(\frac{1 \text{ mol}}{12.01 \text{ g}}\right) = 0.0833 \times 10^6 \text{ mol C}$$

Now set up an amounts table:

Reaction	CaO +	3 C →	CO +	CaC$_2$
Initial amount (10^6 mol)	0.0178	0.0833	0	0
10^6 mol/coeff	0.0178 (LR)	0.0278		
Change (10^6 mol)	-0.0178	-3(0.0178)	+0.0178	+0.0178
Final amount (10^6 mol)	0	0.0299	+0.0178	+0.0178

The masses that could be produced are:

$$\text{CO: } 0.0178 \times 10^6 \text{ mol} \left(\frac{28.01 \text{ g}}{1 \text{ mol}}\right)\left(\frac{1 \text{ ton}}{10^6 \text{ g}}\right) = 0.499 \text{ metric ton}$$

$$\text{CaC}_2\text{: } 0.0178 \times 10^6 \text{ mol} \left(\frac{64.10 \text{ g}}{1 \text{ mol}}\right)\left(\frac{1 \text{ ton}}{10^6 \text{ g}}\right) = 1.14 \text{ metric ton}$$

(c) Calculate the initial amounts:

$$1000 \text{ kg C}_2\text{H}_4 \left(\frac{10^3 \text{ g}}{1 \text{ kg}}\right)\left(\frac{1 \text{ mol}}{28.05 \text{ g}}\right) = 0.0357 \times 10^6 \text{ mol C}_2\text{H}_4$$

$$1000 \text{ kg O}_2 \left(\frac{10^3 \text{ g}}{1 \text{ kg}}\right)\left(\frac{1 \text{ mol}}{32.00 \text{ g}}\right) = 0.0313 \times 10^6 \text{ mol O}_2$$

$$1000 \text{ kg HCl} \left(\frac{10^3 \text{ g}}{1 \text{ kg}}\right)\left(\frac{1 \text{ mol}}{36.46 \text{ g}}\right) = 0.0274 \times 10^6 \text{ mol HCl}$$

Now set up an amounts table:

Reaction	2 C$_2$H$_4$ +	O$_2$ +	4 HCl →	2 C$_2$H$_4$Cl$_2$ +	2 H$_2$O
Initial amount (10^6 mol)	0.0357	0.0313	0.0274	0	0
10^6 mol/coeff	0.0179	0.0313	0.00685 (LR)		
Change (10^6 mol)	$-\frac{2}{4}(0.0274)$	$-\frac{1}{4}(0.0274)$	-0.0274	$+\frac{2}{4}(0.0274)$	$+\frac{2}{4}(0.0274)$
Final amount (10^6 mol)	0.0220	0.0245	0	0.0137	0.0137

The masses that could be produced are:

$$C_2H_4Cl_2: 0.0137 \times 10^6 \text{ mol} \left(\frac{98.96 \text{ g}}{1 \text{ mol}}\right)\left(\frac{1 \text{ ton}}{10^6 \text{ g}}\right) = 1.36 \text{ metric ton}$$

$$H_2O: 0.0137 \times 10^6 \text{ mol} \left(\frac{18.02 \text{ g}}{1 \text{ mol}}\right)\left(\frac{1 \text{ ton}}{10^6 \text{ g}}\right) = 0.247 \text{ metric ton}$$

4.31 The problem gives information about the amounts of both starting materials, so this is a limiting reactant situation. We must calculate the number of moles of each species, construct a table of amounts, and use the results to determine the masses of the product formed and the remaining reactant.

Begin by determining the balanced chemical equation:

$$P_4 + O_2 \rightarrow P_4O_{10}$$

P is already balanced. To balance O, give O$_2$ a coefficient of 5:

$$P_4 + 5O_2 \rightarrow P_4O_{10}$$

Next, calculate the initial amounts:

$$3.75 \text{ g P}_4\left(\frac{1 \text{ mol}}{123.9 \text{ g}}\right) = 0.0303 \text{ mol} \qquad 6.55 \text{ g O}_2\left(\frac{1 \text{ mol}}{32.00 \text{ g}}\right) = 0.205 \text{ mol}$$

Construct an amounts table:

Reaction	P$_4$ +	5 O$_2$ →	P$_4$O$_{10}$
Initial amount (mol)	0.0303	0.205	0
mol/coeff	0.0303 (LR)	0.0410	
Change (mol)	-0.0303	-5(0.0303)	+0.0303
Final amoung (mol)	0	0.0535	0.0303

Now obtain the mass of P$_4$O$_{10}$ produced, using the information from the amounts table and the molar mass: $MM = 4(30.97 \text{ g/mol}) + 10(16.00 \text{ g/mol}) = 283.9 \text{ g/mol}$

The mass that could be produced is $0.0303 \text{ mol}\left(\frac{283.9 \text{ g}}{1 \text{ mol}}\right) = 8.60 \text{ g P}_4\text{O}_{10}$

Finally, determine the mass of O$_2$ left over:

There would be $0.0535 \text{ mol}\left(\frac{32.00 \text{ g}}{1 \text{ mol}}\right) = 1.7 \text{ g of O}_2$ left unreacted (2 significant figures because the final moles of O$_2$ has only two significant figures, although 3 are shown to avoid roundoff errors)

4.33   Identify species based on the type of substance present: Ions for salts and strong
acids/bases, molecules for other substances. H_2O is always a major species in aqueous
solution:
(a) Salt: NH_4^+, Cl^-, H_2O; (b) Salt: Fe^{2+}, ClO_4^-, H_2O; (c) Salt: Na^+, SO_4^{2-}, H_2O;
(d) Not a salt: Br_2, H_2O; (e) Salt: K^+, Br^-, H_2O

4.35   Identify species based on the type of substance present: Ions for salts and strong
acids/bases, molecules for other substances. H_2O is always a major species in aqueous
solution:
(a) Salt: K^+, HPO_4^-, H_2O; (b) Weak acid: CH_3CO_2H, H_2O;
(c) Salt: Na^+, $CH_3CO_2^-$, H_2O; (d) Weak base: NH_3, H_2O; (e) Salt: NH_4^+, Cl^-, H_2O

4.37   To determine whether a precipitate will form, consider the solubility guidelines. If any
combination of cations and anions is insoluble, that salt will precipitate:
(a) $AgNO_3$: All nitrates are soluble, so a precipitate forms only if the silver salt is
    insoluble: (a) AgCl precipitates; (b, c, and d) No precipitate; (e) AgBr precipitates
(b) Na_2CO_3: All sodium salts are soluble, so a precipitate forms only if the carbonate salt
    is insoluble: (a) No precipitate; (b) $FeCO_3$ precipitates; (c, d, and e) No precipitate
(c) $Ba(OH)_2$: Both ions form insoluble salts unless their partners convey solubility. (a, d,
    and e) No precipitate; (b) $Fe(OH)_2$ precipitates; (c) $BaSO_4$ precipitates

4.39   A net ionic equation shows which ions combine to give new products. Spectator ions do
not appear in the net equation, and charges must be balanced. Start by determining what
ions are present in solution. Then use the Solubility Guidelines from your text to
determine what precipitates can form.
(a) The ions present are: Ag^+, NO_3^-, K^+, and OH^-. The possible combinations are:
    $AgNO_3$, AgOH, KOH, and KNO_3. Guideline 2 predicts that salts of potassium and
    nitrate are soluble. Since AgOH is not soluble by the guidelines, it must be an
    insoluble salt: Ag^+ (aq) + OH^- (aq) → AgOH (s)   spectator ions: NO_3^-, K^+
(b) The ions present are: Fe^{3+}, ClO_4^-, NH_4^+, and $C_2O_4^{2-}$. The possible combinations are:
    $Fe(ClO_4)_3$, $Fe_2(C_2O_4)_3$, NH_4ClO_4, $(NH_4)_2C_2O_4$. Guidelines 1 and 2 predict that all
    ammonium salts and all perchlorate salts are soluble. $Fe(C_2O_4)_3$ is not covered in
    guidelines 1,2, or 5 and thus is insoluble by guideline 3:
        $2 Fe^{3+}$ (aq) + $3 C_2O_4^{2-}$ (aq)→ $Fe_2(C_2O_4)_3$ (s)   spectator ions: ClO_4^-, NH_4^+
(c) The ions present are: Pb^{2-}, Br^-, Na^+, and NO_3^-. By guidelines 1 and 2, all salts of
    sodium and nitrate are soluble. $PbBr_2$ is covered in guideline 4 as an insoluble salt:
        Pb^{2-} (aq) + $2Br^-$ (aq) →$PbBr_2$ (s)   spectator ions: NO_3^-, Na^+
(d) The ions present are: Ni^{2+}, OH^-, K^+, and SO_4^{2-}. By guidelines 1 and 3, all salts of
    potassium and most of sulfate are soluble. Thus, the only possible precipitate is
    $Ni(OH)_2$, which is insoluble by guideline 3:
        Ni^{2+} (aq)+ $2 OH^-$ (aq) → $Ni(OH)_2$ (s)   spectator ions: K^+, SO_4^{2-}

4.41 The problem gives information about the amounts of both starting materials, so this is a limiting reactant situation. We must calculate the number of moles of each species, construct a table of amounts, and use the results to determine the final product mass. Start by determining the balanced net ionic reaction using the solubility guidelines.

The ions present are Ag^+, NO_3^-, K^+, and CO_3^{2-}. Guidelines 1 and 2 state that all salts containing potassium and nitrate are soluble. Ag_2CO_3 is not covered in guidelines 1 or 2, and thus by guideline 3 is insoluble:

$$2 \, Ag^+ \, (aq) + CO_3^{2-} \, (aq) \rightarrow Ag_2CO_3 \, (s)$$

Calculations of initial amounts:

For Ag^+, 55.0 mL = 0.0550 L *and* 5.00×10^{-2} M = 0.0500 mol Ag^+/L

$$0.0550 \, L \left(\frac{0.0500 \, mol}{1 \, L} \right) = 2.75 \times 10^{-3} \, mol \, Ag^+$$

For CO_3^{2-}, 95.0 mL = 0.0950 L *and* 3.50×10^{-2} M = 0.0350 mol CO_3^{2-}/L

$$0.0950 \, L \left(\frac{0.0350 \, mol}{1 \, L} \right) = 3.33 \times 10^{-3} \, mol \, CO_3^{2-}$$

Next, set up an amounts table:

Reaction	2 Ag$^+$ +	CO$_3^{2-}$ $\rightarrow$	Ag$_2$CO$_3$ (s)
Initial amount (10^{-3} mol)	2.75	3.33	0
10^{-3} Mol/coeff	2.75/2 = 1.38 (LR)	3.33	
Change (10^{-3} mol)	-2.75	-2.75/2	+2.75/2
Final amount (10^{-3} mol)	0	1.96	1.38

Do a mole-mass conversion to calculate the mass of product:

$$1.38 \times 10^{-3} \, mol \left(\frac{275.8 \, g}{1 \, mol} \right) = 0.381 \, g$$

The ions remaining in solution are the excess CO_3^{2-} and the spectator ions, NO_3^- and K^+

4.43 All these are acid-base reactions, where H^+ transfers from the acid to the base:

(a) HCl is a strong acid that forms H_3O^+ and Cl^- ions in solution; $Ca(OH)_2$ is soluble and a strong base.
   Reaction: $H_3O^+ \, (aq) + OH^- \, (aq) \rightarrow 2 \, H_2O \, (l)$ spectator ions: Ca^{2+}, Cl^-

(b) H_3PO_4 is a weak acid; LiOH is a strong base that forms OH^- and Li^+ ions in solution.
   Reaction: $H_3PO_4 \, (aq) + 3 \, OH^- \, (aq) \rightarrow 3 \, H_2O \, (l) + PO_4^{3-} \, (aq)$ spectator ion: Li^+

(c) NH_3 is a weak base, HNO_3 is a strong acid and forms H_3O^+ and NO_3^- ions.
   Reaction: $NH_3 \, (aq) + H_3O^+ \, (aq) \rightarrow H_2O \, (l) + NH_4^+ \, (aq)$ spectator ion: NO_3^-

(d) CH_3CO_2H is a weak acid; KOH is a strong base that forms OH^- and K^+ ions in solution.
   Reaction: $CH_3CO_2H \, (aq) + OH^- \, (aq) \rightarrow H_2O \, (l) + CH_3CO_2^- \, (aq)$ spectator ion: K^+

4.45   This problem describes an acid-base reaction. We are asked to determine the mass of base required to completely neutralize the acid. The starting materials are HCl, a strong acid, and $Al(OH)_3$, an insoluble base. In addition to water, the major species are

$$H_3O^+, Cl^-, Al(OH)_3$$

The presence of hydronium ions and a base together as major species will result in an acid-base reaction. The balanced reaction is

$$Al(OH)_3 \, (s) + 3 \, H_3O^+ \, (aq) \rightarrow Al^{3+} \, (aq) + 6 \, H_2O \, (l)$$

The problem gives information about the amount of acid present. This is a mole-mass conversion problem:

$$\text{For } H_3O^+: 0.155 \, L \left(\frac{0.175 \text{ mol}}{1 \, L} \right) = 2.71 \times 10^{-2} \text{ mol } H_3O^+$$

$MM \, [Al(OH)_3] = 26.98 \text{ g/mol} + 3(16.00 \text{ g/mol}) + 3(1.008 \text{ g/mol}) = 78.00 \text{ g/mol}$

$$2.71 \times 10^{-2} \text{ mol } H_3O^+ \left(\frac{1 \text{ mol } Al(OH)_3}{3 \text{ mol } H_3O^+} \right) \left(\frac{78.00 \text{ g}}{1 \text{ mol}} \right) = 0.705 \text{ g } Al(OH)_3$$

4.47   This problem describes an acid-base reaction. We are asked to determine the final concentrations of all the major ions in a final solution. Begin by analyzing the chemistry. The starting materials are HCl (aq), a strong acid, and $Ba(OH)_2$ (aq), a strong base. In addition to water, the major species in solution are:

$$H_3O^+, Cl^-, Ba^{2+}, OH^-$$

The presence of hydronium and hydroxide ions together as major species in solution always results in neutralization:

$$H_3O^+ + OH^- \rightarrow 2 \, H_2O$$

The problem gives information about the amounts of both starting materials, so this is a limiting reactant situation. We must calculate the number of moles of each species, construct a table of amounts, and use the results to determine the final solution concentrations:

Calculations of initial amounts:

For HCl (aq), 100.0 mL = 0.1000 L *and* 5.00×10^{-2} M = 0.0500 mol HCl/L

(0.100 L)(0.0500 mol/L) = 0.00500 mol H_3O^+ and 0.00500 mol Cl^-

For $Ba(OH)_2$ (aq), 150.0 mL = 0.1500 L *and* 2.00×10^{-2} M = 0.0200 mol $Ba(OH)_2$/L

(0.150 L)(0.0200 mol/L) = 0.00300 mol Ba^{2+}

(0.150 L)(0.0200 mol/L)(2 mol OH^-/mol $Ba(OH)_2$) = 0.00600 mol OH^-

Now we set up a table of amounts. The acid reaction involves only H_3O^+ and OH^-. The amounts of Ba^{2+} and Cl^- do not change, so we may omit these species from the table. Moreover, the water generated in the proton transfer reaction joins the bulk solvent, so we may ignore the water in our calculations.

The balanced equation shows that the starting materials react in a 1:1 mole ratio, so we can identify the limiting reactant by inspection; the limiting reactant is H_3O^+.

Here is the complete table of amounts:

Reaction	H_3O^+ +	OH^- →	$2 H_2O$
Initial amount (mol)	0.00500	0.00600	solvent
Change (mol)	–0.00500	–0.00500	solvent
Final amount (mol)	0	0.00100	solvent

Before calculating the final concentrations, organize the final amounts:

$H_3O^+ = 0$ mol; $OH^- = 0.00100$ mol; $Cl^- = 0.00500$ mol; $Ba^{2+} = 0.00300$ mol

The final concentrations are obtained by dividing the number of moles of each species by the final volume of the solution.

$$V_{final} = 100.0 \text{ mL} + 150.0 \text{ mL} = 250.0 \text{ mL} = 0.2500 \text{ L}$$

Here are the final concentrations:

$$[H_3O^+] = 0 \qquad [OH^-] = \frac{0.00100 \text{ mol}}{0.2500 \text{ L}} = 0.00400 \text{ M}$$

$$[Cl^-] = \frac{0.00500 \text{ mol}}{0.2500 \text{ L}} = 0.0200 \text{ M} \qquad [Ba^{2+}] = \frac{0.00300 \text{ mol}}{0.2500 \text{ L}} = 0.0120 \text{ M}$$

4.49   This problem describes an acid-base titration. We are asked to determine the concentration of base used to neutralize the acid. The starting materials are HCl, a strong acid, and KOH, a strong base. In addition to water, the major species are:

$$H_3O^+, Cl^-, K^+, OH^-$$

The presence of hydronium ions and hydroxide ions together as major species will result in neutralization:

$$OH^- + H_3O^+ \rightarrow 2 H_2O \; (l)$$

Calculation of the amount of acid added:

$$27.35 \text{ mL} \left(\frac{10^{-3} \text{ L}}{1 \text{ mL}} \right) \left(\frac{0.1206 \text{ mol}}{1 \text{ L}} \right) = 3.298 \times 10^{-3} \text{ mol } H_3O^+$$

Since this is a completed titration, the moles of acid added to the solution will equal the moles of base in the solution.

$$n(H_3O^+) = 3.298 \times 10^{-3} \text{ mol} = n(OH^-)$$

The concentration of the KOH solution is obtained by dividing the number of moles of base by the volume of the KOH solution.

$$V_{solution} = 0.00500 \text{ L}$$

$$[OH^-] = \frac{3.298 \times 10^{-3} \text{ mol}}{0.00500 \text{ L}} = 0.660 \text{ M}$$

4.51   This problem describes an acid-base titration. We are asked to determine the concentration of base used to neutralize the acid. The starting materials are NaOH, a strong base, and $KHC_8H_4O_4$, the salt of a weak acid. In addition to water, the major species are Na^+, OH^-, K^+, and $HC_8H_4O_4^-$

The presence of hydroxide ions and $HC_8H_4O_4^-$ ions together as major species will result in an acid base reaction: $OH^- + HC_8H_4O_4^- \rightarrow C_8H_4O_4^{2-} + H_2O$

Calculate the amount of $KHC_8H_4O_4$:

$$MM = 39.10 \text{ g/mol} + 8(12.01 \text{ g/mol}) + 5(1.01 \text{ g/mol}) + 4(16.00 \text{ g/mol}) = 204.23 \text{ g/mol}$$

$$n = 0.6634 \text{ g} \left(\frac{1 \text{ mol}}{204.23 \text{ g}} \right) = 0.003248 \text{ mol of } HC_8H_4O_4^-$$

Since this is a completed titration, the moles of base added to the solution will equal the moles of acid in the solution.

$$n(HC_8H_4O_4^-) = 0.003248 \text{ mol} = n(OH^-)$$

Obtain the concentration of the NaOH solution by dividing the number of moles of base by the volume of the NaOH solution.

$$V_{solution} = 0.03655 \text{ L} \qquad and \qquad [OH^-] = \frac{0.003248 \text{ mol}}{0.03655 \text{ L}} = 0.08886 \text{ M}$$

4.53   A redox reaction between a metal and a cation occurs when the metal is higher in the activity series than the metal of the cation. The activity series appears as Table 4-3 of your textbook.

(a and b) Cu is below H_2 and Mg in the activity series, so no reaction occurs.

(c) Cu is above Ag in the activity series, so a displacement reaction occurs:

$$Cu_{(s)} + 2 Ag^+_{(aq)} \rightarrow Cu^{2+}_{(aq)} + 2 Ag_{(s)}$$

(d) K is above H_2 in the activity series, so a displacement reaction occurs:

$$2 K_{(s)} + 2 H_2O_{(l)} \rightarrow 2 K^+_{(aq)} + H_2_{(g)} + 2 OH^-_{(aq)}$$

4.55   Formation reactions are easily balanced once the chemical formula of the oxide is known. Oxygen carries a charge of 2 –, and the compound must be electrically neutral.

(a) Sr is in Group 2, so it forms a 2+ cation, and its oxide is SrO: $2 Sr + O_2 \rightarrow 2 SrO$

(b) Cr(III) has a 3+ charge, and its oxide is Cr_2O_3: $4 Cr + 3 O_2 \rightarrow 2 Cr_2O_3$

(c) Sn(IV) has a 4+ charge, and its oxide is SnO_2: $Sn + O_2 \rightarrow SnO_2$

4.57   This problem describes a redox reaction. We are asked to determine the mass H_2 that will form from the reaction. The starting materials are HCl (aq), a strong acid, and Al metal. In addition to water, the major species in solution are Al^{3+}, H_3O^+, and Cl^-.

Use Table 4-3 to identify the half-reactions:

$$Al \rightarrow Al^{3+} + 3e^- \qquad\qquad H_3O^+ + e^- \rightarrow \frac{1}{2} H_2 + H_2O$$

To balance the number of electrons, multiply the second equation by 3 and add the two equations: $3 H_3O^+ + Al \rightarrow Al^{3+} + \frac{3}{2} H_2 + 3 H_2O$

Multiply all coefficients by 2 to eliminate the fraction and obtain the balanced reaction:

$$6 H_3O^+ + 2 Al \rightarrow 2 Al^{3+} + 3 H_2 + 6 H_2O$$

The problem gives information about the amounts of both starting materials, so this is a limiting reactant situation. We must calculate the number of moles of each species, construct a table of amounts, and use the results to determine the final amounts. Initial amounts (We carry an additional significant figure to avoid round off error.):

$$V_{HCl} = 8.00 \text{ mL} = 0.00800 \text{ L} \qquad n_{HCl} = 0.00800 \text{ L} \left(\frac{6.00 \text{ mol}}{1 \text{ L}} \right) = 4.800 \times 10^{-2} \text{ mol}$$

$$n_{Al} = 0.355 \text{ g} \left(\frac{1 \text{ mol}}{26.98 \text{ g}} \right) = 1.316 \times 10^{-2} \text{ mol}$$

Because the problem asks only about the mass of H_2 formed, we do not need to do calculations for the other products.

Reaction	$6 H_3O^+$ +	$2 Al$ →	$3 H_2$
Initial amount (10^{-2} mol)	4.800	1.316	0
10^{-2} mol/coeff	0.800	0.658 (LR)	
Change (10^{-2} mol)	-6/2(1.316)	-1.316	+3/2 (1.316)
Final amount (10^{-2} mol)	0.852	0	1.974

Do a mole-mass conversion to calculate the mass of H_2 formed:

$$m = 1.974 \times 10^{-2} \text{ mol} \left(\frac{2.016 \text{ g}}{1 \text{ mol}} \right) = 3.98 \times 10^{-2} \text{ g } H_2$$

4.59   The first two parts of this problem ask about an ionic reaction. Identify the species present in the solution and use the products to find the net ionic reaction and the spectator ions. The species present are H_2O, NH_4^+, HCO_3^-, Na^+, and Cl^-. The solid product is $NaHCO_3$:

(a) The net ionic reaction is $Na^+ (aq) + HCO_3^- (aq) \rightarrow NaHCO_3 (s)$

(b) NH_4^+ and Cl^- are spectator ions

(c) The second two parts of this problem involve stoichiometric calculations. This is a limiting reagent problem as well as a yield problem, requiring a table of amounts to determine the theoretical yield.
Begin by determining the initial amounts of the reactants:

$$n_{Na^+} = 5.00 \times 10^2 \text{ L} \left(\frac{6.00 \text{ mol}}{1 \text{ L}} \right) = 3.00 \times 10^3 \text{ mol}$$

$$n_{HCO_3^-} = 5.00 \times 10^2 \text{ L} \left(\frac{1.50 \text{ mol}}{1 \text{ L}} \right) = 7.50 \times 10^2 \text{ mol}$$

The balanced equation shows that the starting materials react in a 1:1 mole ratio, so we can identify the limiting reactant by inspection; the limiting reactant is HCO_3^-.
Here is the complete table of amounts:

Reaction	Na^+ +	HCO_3^- →	$NaHCO_3$
Initial amount (10^2 mol)	30.0	7.50	0
Change (10^2 mol)	-7.50	-7.50	+7.50
Final amount (10^2 mol)	22.5	0	7.50

The theoretical yield of $NaHCO_3$ is 7.50×10^2 mol

Now determine the actual yield from the amount of product formed (35.0 kg):

MM (NaHCO$_3$) = 22.99 g/mol + 1.01 g/mol + 12.01 g/mol + 3(16.00 g/mol)

$$= 84.01 \text{ g/mol}$$

The actual yield is $35.0 \text{ kg} \left(\dfrac{10^3 \text{ g}}{1 \text{ kg}} \right) \left(\dfrac{1 \text{ mol}}{84.01 \text{ g}} \right) = 4.17 \times 10^2 \text{ mol}$

Obtain the percent yield by dividing the actual yield by the theoretical yield:

The percent yield is $(100\%) \dfrac{4.17 \times 10^2 \text{ mol}}{7.50 \times 10^2 \text{ mol}} = 55.6\%$

(The yield is low because NaHCO$_3$ is relatively soluble.)

(d) To find concentrations in the final solution, first determine how many moles of each ion remain in solution, then divide by the final volume of the solution:

$V_{\text{final}} = 5.00 \times 10^2 \text{ L} + 5.00 \times 10^2 \text{ L} = 1.000 \times 10^3 \text{ L}$

The amounts of the two spectator ions are unaffected by the reaction:

$[NH_4^+] = \dfrac{7.50 \times 10^2 \text{ mol}}{1.000 \times 10^3 \text{ L}} = 0.750 \text{ M}$ *and* $[Cl^-] = \dfrac{30.0 \times 10^2 \text{ mol}}{1.000 \times 10^3 \text{ L}} = 3.00 \text{ M}$

The amounts of the reactant ions are reduced by the amount that precipitates (Note the final amounts from the amounts table are the theoretical yields; we want to use the actual amounts.):

$[Na^+] = \dfrac{30.0 \times 10^2 \text{ mol} - 4.17 \times 10^2 \text{ mol}}{1.000 \times 10^3 \text{ L}} = 2.58 \text{ M}$

$[HCO_3^-] = \dfrac{7.50 \times 10^2 \text{ mol} - 4.17 \times 10^2 \text{ mol}}{1.000 \times 10^3 \text{ L}} = 0.333 \text{ M}$

4.61   This problem describes a redox reaction. We are asked to determine the percent yield for the reaction. The starting materials are Xe and F$_2$, and the product is XeF$_4$:

$$Xe + 2 F_2 \rightarrow XeF_4$$

The problem gives information about the amount of one of the starting materials (the other is in excess), so this is a simple mass-mole-mass problem. We must calculate the theoretical yield, determine the percent yield, and use the actual yield to determine the final amount of Xe left over.

Calculations of theoretical yield:

MM (XeF$_4$) = 131.3 g/mol + 4(19.0 g/mol) = 207.3 g/mol

The theoretical yield is determined from the startinig amount of Xe:

$n = 5.00 \text{ g} \left(\dfrac{1 \text{ mol}}{131.3 \text{ g}} \right) \left(\dfrac{1 \text{ mol XeF}_4}{1 \text{ mol Xe}} \right) = 3.81 \times 10^{-2} \text{ mol XeF}_4$

The actual yield of XeF$_4$ is $4.00 \text{ g} \left(\dfrac{1 \text{ mol}}{207.3 \text{ g}} \right) = 1.93 \times 10^{-2} \text{ mol}$

Now determine the percent yields using the actual yield and the theoretical yield:

$$\text{The percent yield is } (100\%)\frac{1.93 \times 10^{-2} \text{ mol}}{3.81 \times 10^{-2} \text{ mol}} = 50.7\%$$

The mass of Xe left unreacted is the difference between the starting mass and the amount converted into XeF_4:

$$m = 5.00 \text{ g} - (1.93 \times 10^{-2} \text{ mol})\left(\frac{131.3 \text{ g}}{1 \text{ mol}}\right) = 2.47 \text{ g}$$

4.63   This problem describes a redox reaction. We are asked to determine the mass of iron(III) oxide that will form from the reaction. Begin by analyzing the chemistry. The starting materials are FeS_2 (pyrite) and excess O_2; the products are iron(III) oxide and SO_2. The reaction (unbalanced) is:

$$FeS_2 + O_2 \rightarrow Fe_2O_3 + SO_2$$
$$1Fe + 2S + 2O \rightarrow 2Fe + 1S + 5O$$

Balance the Fe by giving FeS_2 a coefficient of 2:

$$\mathbf{2} \text{ } FeS_2 + O_2 \rightarrow Fe_2O_3 + SO_2$$
$$2Fe + 4S + 2O \rightarrow 2Fe + 1S + 5O$$

Next, balance the sulfur by changing the coefficient on SO_2 from 1 to 4:

$$2 \text{ } FeS_2 + O_2 \rightarrow Fe_2O_3 + \mathbf{4} \text{ } SO_2$$
$$2Fe + 4S + 2O \rightarrow 2Fe + 4S + 11O$$

Finally, balance oxygen by giving the O_2 a coefficient of 11/2 and eliminate the fraction by multiplying by 2:

$$4 \text{ } FeS_2 + 11 \text{ } O_2 \rightarrow 2 \text{ } Fe_2O_3 + 8 \text{ } SO_2$$

Determine the amount of pyrite by multiplying the percent composition by the mass of the ore:

$$m_{\text{pyrite}} = 175 \text{ ton}\left(\frac{55\%}{100\%}\right) = 96.25 \text{ ton}$$

Use mass-mole-mass calculations to obtain the theoretical mass of iron(III) oxide formed:

$$m_{\text{oxide}} = 96.25 \text{ ton}\left(\frac{1 \text{ mol}}{120.2 \text{ g}}\right)\left(\frac{2 \text{ mol Fe}_2O_3}{4 \text{ mol FeS}_2}\right)\left(\frac{159.7 \text{ g}}{1 \text{ mol}}\right) = 63.9 \text{ ton}$$

Multiply by the percent efficiency to determine the actual mass recovered:

$$\text{Mass recovered} = 63.9 \text{ ton}\left(\frac{85\%}{100\%}\right) = 54 \text{ tons}$$

4.65   When determining reaction products, first identify the types of substances present to determine what kind of reaction can occur.
(a) Oxygen can oxidize metals to metal oxides:
$$2 \text{ Al } (s) + 3 \text{ } O_2 \text{ } (g) \rightarrow 2 \text{ } Al_2O_3 \text{ } (s) \text{ (redox reaction)}$$
(b) Oxygen can oxidize hydrocarbons to carbon dioxide and water when heated:
$$C_3H_8 \text{ } (g) + 5 \text{ } O_2 \text{ } (g) \rightarrow 3 \text{ } CO_2 \text{ } (g) + 4 \text{ } H_2O \text{ } (g) \text{ (redox reaction)}$$

(c) Mg displaces hydrogen gas from strong acids such as HBr:

$Mg\ (s) + 2\ H_3O^+\ (aq) \rightarrow Mg^{2+}\ (aq) + H_2\ (g) + 2\ H_2O\ (l)$ (redox reaction)

(d) NaOH is a strong base, HCl is a strong acid:

$OH^-\ (aq) + H_3O^+\ (aq) \rightarrow 2H_2O\ (l)$ (acid-base reaction)

(e) Both compounds are salts, so a precipitate will form if any combination is insoluble:

$Pb^{2+}\ (aq) + CO_3^{2-}\ (aq) \rightarrow PbCO_3\ (s)$ (precipitation reaction)

(f) $Ca(OH)_2$ is a strong base, H_2SO_4 is a strong acid for the first H:

$OH^-\ (aq) + H_3O^+\ (aq) \rightarrow 2\ H_2O\ (l)$ (acid-base reaction)

4.67   The formation reactions for oxides are easily balanced by inspection. The number of electrons lost by the metal atom can be determined by assigning two additional electrons to each oxygen atom in the oxide:

$4\ Ru + 3\ O_2 \rightarrow 2\ Ru_2O_3$: Three O atoms gain six electrons, so each Ru loses three

$Ru + O_2 \rightarrow RuO_2$: Two O atoms gain four electrons, so each Ru loses four

$2\ Ru + 3\ O_2 \rightarrow 2\ RuO_3$: Three O atoms gain six electrons, so each Ru loses six

$Ru + 2\ O_2 \rightarrow RuO_4$: Four O atoms gain eight electrons, so each Ru loses eight

4.69   Examine the molecular picture to see what chemical reaction occurs. The reactants are atomic X and atomic Y and the product of the reaction is YX_2 (there are atoms of X left over). Thus, the reaction is: $Y + 2\ X \rightarrow YX_2$

4.71   A balanced chemical equation has equal numbers of atoms of each element on each side of the arrow. Adjust stoichiometric coefficients to balance each element in turn, beginning with those that appear in only one reactant and product:

(a) $H_2 + NO \rightarrow NH_3 + H_2O$

$2H + 1N + 1O \rightarrow 5H + 1N + 1O$

N and O are already balanced. Give H_2 a coefficient of 5/2 to balance H, then multiply through by 2 to clear fractions:

$5\ H_2 + 2\ NO \rightarrow 2\ NH_3 + 2\ H_2O$

(b) $CO + NO \rightarrow N_2 + CO_2$

$1C + 1N + 2O \rightarrow 1C + 2N + 2O$

C is already balanced. Give N_2 a coefficient of 1/2 to balance N, then multiply through by 2 to clear fractions:

$2\ CO + 2\ NO \rightarrow N_2 + 2\ CO_2$

(c) $NH_3 + O_2 \rightarrow N_2O + H_2O$

$3H + 1N + 2O \rightarrow 2H + 2N + 2O$

Give NH_3 a coefficient of 2 to balance N

$2\ NH_3 + O_2 \rightarrow N_2O + H_2O$

$6H + 2N + 2O \rightarrow 2H + 2N + 2O$

Both N and O are now balanced. To balance H, give H_2O a coefficient of 3:

$2 NH_3 + O_2 \rightarrow N_2O + 3 H_2O$

$6H + 2N + 2O \rightarrow 6H + 2N + 4O$

This balances H, but now O is unbalanced. To balance O, give O_2 a coefficient of 2, which gives the balanced reaction:

$2 NH_3 + 2 O_2 \rightarrow N_2O + 3 H_2O$

$6H + 2N + 4O \rightarrow 6H + 2N + 4O$

(d) The reaction (unbalanced) is:

$NO + NH_3 \rightarrow N_2 + H_2O$

$3H + 2N + 1O \rightarrow 2H + 2N + 1O$

N appears in three reagents, so balance it last. Balance H by giving NH_3 a coefficient of 2 and H_2O a coefficient of 3:

$NO + \mathbf{2} NH_3 \rightarrow N_2 + \mathbf{3} H_2O$

$6H + 3N + 1O \rightarrow 6H + 2N + 3O$

Next balance O by giving NO a coefficient of 3:

$3 NO + 2 NH_3 \rightarrow N_2 + 3 H_2O$

$6H + 5N + 3O \rightarrow 6H + 2N + 3O$

Now there are 5 N on the left, so give N_2 a coefficient of 5/2 to balance N and multiply through by 2 to clear fractions and obtain the balanced reaction:

$6 NO + 4 NH_3 \rightarrow 5 N_2 + 6 H_2O$

(e) The reaction (unbalanced) is:

$H_2O + NO \rightarrow O_2 + NH_3$

$2H + 1N + 2O \rightarrow 3H + 1N + 2O$

N is already balanced. Balance H by giving NH_3 a coefficient of 2 and H_2O a coefficient of 3:

$\mathbf{3} H_2O + NO \rightarrow O_2 + \mathbf{2} NH_3$

$6H + 1N + 4O \rightarrow 6H + 2N + 2O$

Next give NO a coefficient of 2 to balance N:

$3 H_2O + \mathbf{2} NO \rightarrow O_2 + 2 NH_3$

$6H + 2N + 5O \rightarrow 6H + 2N + 2O$

Now there are 5 O on the left, so give O_2 a coefficient of 5/2 to balance O and multiply through by 2 to clear fractions:

$6 H_2O + 4 NO \rightarrow 5 O_2 + 4 NH_3$

(f) The reaction (unbalanced) is:

$H_2 + O_2 \rightarrow H_2O$

$2H + 2O \rightarrow 2H + 1O$

Give O_2 a coefficient of 1/2 to balance O, then multiply through by 2 to clear fractions:

$2 H_2 + O_2 \rightarrow 2 H_2O$

4.73   A convenient synthesis of ionic salts starts with solutions of soluble salts of the cation and anion which, when mixed, form a precipitate of the desired product. Sodium salts are inexpensive soluble sources of anions and chlorides or nitrates are inexpensive soluble sources of cations. Do mole-mass conversions to compute the required masses.

(a) Mix solutions of Na_3PO_4 and $FeCl_3$. The net ionic reaction is:

$$Fe^{3+} (aq) + PO_4^{3-} (aq) \rightarrow FePO_4 (aq)$$

$$n_{FePO_4} = \frac{m}{MM} = 2.50 \text{ kg} \left(\frac{10^3 \text{ g}}{1 \text{ kg}} \right) \left(\frac{1 \text{ mol}}{150.82 \text{ g}} \right) = 16.6 \text{ mol}$$

$$n_{FeCl_3} = n_{Na_3PO_4} = n_{FePO_4}$$

$$m_{FeCl_3} = n \, MM = 16.6 \text{ mol} \left(\frac{162.21 \text{ g}}{1 \text{ mol}} \right) \left(\frac{10^{-3} \text{ kg}}{1 \text{ g}} \right) = 2.69 \text{ kg}$$

$$m_{Na_3PO_4} = n \, MM = 16.6 \text{ mol} \left(\frac{163.94 \text{ g}}{1 \text{ mol}} \right) \left(\frac{10^{-3} \text{ kg}}{1 \text{ g}} \right) = 2.72 \text{ kg}$$

(b) Mix solutions of NaOH and $ZnCl_2$. The net ionic reaction is:

$$Zn^{2+} (aq) + 2 \, OH^- (aq) \rightarrow Zn(OH)_2 (s)$$

$$n_{Zn(OH)_2} = \frac{m}{MM} = 2.50 \text{ kg} \left(\frac{10^3 \text{ g}}{1 \text{ kg}} \right) \left(\frac{1 \text{ mol}}{99.41 \text{ g}} \right) = 25.1 \text{ mol}$$

$$n_{ZnCl_2} = n_{Zn(OH)_2} \quad and \quad n_{NaOH} = 2 \, n_{Zn(OH)_2}$$

$$m_{ZnCl_2} = n \, MM$$

$$= 25.1 \text{ mol Zn(OH)}_2 \left(\frac{1 \text{ mol ZnCl}_2}{1 \text{ mol Zn(OH)}_2} \right) \left(\frac{136.3 \text{ g}}{1 \text{ mol}} \right) \left(\frac{10^{-3} \text{ kg}}{1 \text{ g}} \right) = 3.42 \text{ kg}$$

$$m_{NaOH} = n \, MM$$

$$= 25.1 \text{ mol Zn(OH)}_2 \left(\frac{2 \text{ mol NaOH}}{1 \text{ mol Zn(OH)}_2} \right) \left(\frac{40.01 \text{ g}}{1 \text{ mol}} \right) \left(\frac{10^{-3} \text{ kg}}{1 \text{ g}} \right) = 2.01 \text{ kg}$$

(c) Mix solutions of Na_2CO_3 and $NiCl_2$. The net ionic reaction is

$$Ni^{2+} (aq) + CO_3^{2-} (aq) \rightarrow NiCO_3 (s)$$

$$n_{NiCO_3} = \frac{m}{MM} = 2.50 \text{ kg} \left(\frac{10^3 \text{ g}}{1 \text{ kg}} \right) \left(\frac{1 \text{ mol}}{118.7 \text{ g}} \right) = 21.1 \text{ mol}$$

$$n_{NiCl_2} = n_{Na_2CO_3} = n_{NiCO_3}$$

$$m_{NiCl_2} = n \, MM = 21.1 \text{ mol} \left(\frac{129.59 \text{ g}}{1 \text{ mol}} \right) \left(\frac{10^{-3} \text{ kg}}{1 \text{ g}} \right) = 2.73 \text{ kg}$$

$$m_{Na_2CO_3} = n \, MM = 21.1 \text{ mol} \left(\frac{105.99 \text{ g}}{1 \text{ mol}} \right) \left(\frac{10^{-3} \text{ kg}}{1 \text{ g}} \right) = 2.24 \text{ kg}$$

4.75   Molecular pictures and limiting reactant calculations require a balanced chemical equation for the reaction under consideration. For this process, the balanced equation is

$$N_2 + 3\,H_2 \rightarrow 2\,NH_3$$

(a) There are 15 H_2 molecules and 6 N_2 molecules in this picture. Divide each number of molecules by the corresponding coefficient to see which reactant is limiting:

$\dfrac{15}{3} = 5$ for H_2, $\dfrac{6}{1} = 6$ for N_2. The smaller number identifies the limiting reactant, H_2.

(b) The new molecular picture must show that all the H_2 has been consumed and NH_3 has been produced, but the number of atoms of each element must be the same as before: 30 atoms of H and 12 atoms of N. Fifteen H_2 molecules react with five N_2 molecules to produce 10 molecules of NH_3:

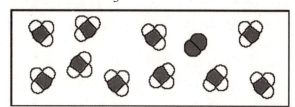

(c) Construct an amounts table for the stoichiometric calculations:

Reaction	N_2	+	$3\,H_2 \rightarrow$	$2\,NH_3$
Initial amount (mol)	6		15	0
Change (mol)	$-\dfrac{15}{3}$		-15	$+\dfrac{2}{3}(15)$
Final amount (mol)	1		0	10

Use the results from the amounts table to calculate the desired masses:

$$1\ \text{mol } N_2 \left(\frac{28.02\ \text{g}}{1\ \text{mol}} \right) = 28\ \text{g } N_2 \text{ remaining}$$

$$10\ \text{mol } NH_3 \left(\frac{17.03\ \text{g}}{1\ \text{mol}} \right) = 170\ \text{g } NH_3 \text{ produced}$$

4.77   This problem describes a combustion reaction. Begin by analyzing the chemistry. The problem asks about the amount of a product that forms from a given amount of reactant. First balance the chemical equation. The reaction is:

$$C_2H_5OH + O_2 \rightarrow CO_2 + H_2O$$
$$2C + 6H + 3O \rightarrow 1C + 2H + 3O$$

Give CO_2 a coefficient of 2 and H_2O a coefficient of 3 to balance C and H:

$$C_2H_5OH + O_2 \rightarrow \mathbf{2}\,CO_2 + \mathbf{3}\,H_2O$$
$$2C + 6H + 3O \rightarrow 2C + 6H + 7O$$

There are seven O on the product side, so give O_2 a coefficient of 3 to balance O:

$$C_2H_5OH + 3\,O_2 \rightarrow 2\,CO_2 + 3\,H_2O$$

Use the balanced equation to do the appropriate mass-mole-number calculations:

(a) $n_{H_2O} = 4.6\ g \left(\dfrac{1\ mol}{46.07\ g}\right)\left(\dfrac{3\ mol\ H_2O}{1\ mol\ C_2H_5OH}\right) = 0.30\ mol\ H_2O$

(b) $\# = n\ N_A = 0.30\ mol\left(\dfrac{6.022 \times 10^{23}\ molecules}{1\ mol}\right) = 1.8 \times 10^{23}\ molecules$

(c) $m = n\ MM = 0.30\ mol\left(\dfrac{18.02\ g}{1\ mol}\right) = 5.4\ g$

4.79 Identify species based on the type of substance present: Ions for salts and strong acids/bases, molecules for other substances. H_2O is always a major species in aqueous solution. Reactions that can occur include precipitation, proton transfer (acid-base), and redox:

(a) NH_3 is a weak base, so the species present are NH_3 and H_2O; HCl is a strong acid, so the species present are H_2O, H_3O^+ and Cl^-; an acid-base reaction occurs:

$$NH_3\ (aq) + H_3O^+\ (aq) \rightarrow NH_4^+\ (aq) + H_2O\ (l)$$

(b) Both substances are salts, so the species present are the appropriate cations and anions: H_2O, Ca^{2+}, Cl^-, Na^+, and SO_4^{2-}; sulfates are generally insoluble, so a precipitation reaction occurs:

$$Ca^{2+}\ (aq) + SO_4^{2-}\ (aq) \rightarrow CaSO_4\ (s)$$

(c) KOH is a strong base, so the species present are H_2O, K^+, and OH^-; HBr is a strong acid, so the species present are H_2O, H_3O^+, and Br^-; an acid-base reaction occurs:

$$H_3O^+\ (aq) + OH^-\ (aq) \rightarrow 2\ H_2O\ (l)$$

(d) HNO_2 is a weak acid, so the species present are H_2O and HNO_2; KOH is a strong base, so the species present are H_2O, K^+, and OH^-; an acid-base reaction occurs:

$$HNO_2\ (aq) + OH^-\ (aq) \rightarrow NO_2^-\ (aq) + H_2O\ (l)$$

4.81 There is much information about propylene oxide provided in this problem, but all that is needed for the calculations are the balanced chemical equation, the amount of starting material, and molar masses. The stoichiometry of this reaction is 1:1 in all reagents:
MM (propylene oxide) = 3(12.01 g/mol) + 6(1.01 g/mol) + 16.00 g/mol = 58.1 g/mol
MM (propene) = 3(12.01 g/mol) + 6(1.01 g/mol) = 42.1 g/mol
MM (hydroperoxide) = 4(12.01 g/mol) + 10(1.01 g/mol) + 2(16.00 g/mol) = 90.1 g/mol

(a) $m_{propylene\ oxide} = 75\ kg\left(\dfrac{1\ kmol}{90.1\ kg}\right)\left(\dfrac{1\ kmol\ propylene\ oxide}{1\ kmol\ hydroperoxide}\right)\left(\dfrac{58.1\ kg}{1\ kmol}\right) = 48\ kg$

(b) $m_{propene} = 75\ kg\left(\dfrac{1\ kmol}{90.1\ kg}\right)\left(\dfrac{1\ kmol\ propene}{1\ kmol\ hydroperoxide}\right)\left(\dfrac{42.1\ kg}{1\ kmol}\right) = 35\ kg$

4.83   A molecular picture must show the stoichiometry of the reaction and conserve atoms of each element. The net ionic reaction is $H_3O^+ + OH^- \rightarrow 2H_2O$. The starting solutions have equal overall reactant concentrations, $[H_3O^+] = [OH^-]$. Your picture should show this, for example using 3 OH^- ions and 3 H_3O^+ ions. Omit solvent water molecules and spectator ions for clarity:

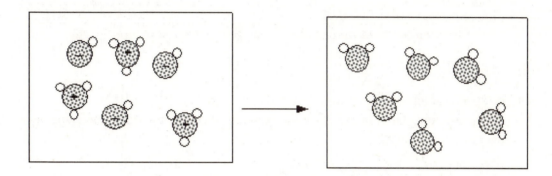

4.85   A molecular picture must show the stoichiometry of the reaction and conserve atoms of each element. The balanced reaction is $2 Mg + O_2 \rightarrow 2 MgO$. The desired picture starts with six Mg atoms and four O_2 molecules. The six Mg atoms react with three O_2 molecules to produce six MgO molecules, leaving one O_2 molecule unreacted. Your figure should show the Mg-containing species as solids:

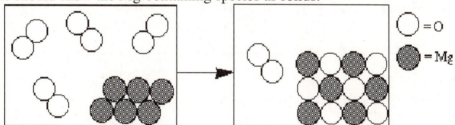

4.87   (a) The reaction is:

$$CaCO_3 + H_3O^+ \rightarrow Ca^{2+} + CO_2 + H_2O$$

$$1Ca + 1C + 4O + 3H \rightarrow 1Ca + 1C + 3O + 2H$$

Carbon and calcium are already balanced. Give H_3O^+ a coefficient of 2 to balance the charges and H_2O a coefficient of 3 to balance H. Then O is also balanced (5 on each side): $CaCO_3 + 2 H_3O^+ \rightarrow Ca^{2+} + CO_2 + 3 H_2O$

(b and c)

The problem gives information about the amounts of both starting materials, so this is a limiting reactant situation. We must calculate the number of moles of each species, construct a table of amounts, and use the results to determine the final amounts. Calculations of initial amounts:

$CaCO_3$: $MM = 40.08$ g/mol $+ 12.01$ g/mol $+ 3(16.00$ g/mol$) = 100.09$ g/mol

$$n = 5.0 \text{ g} \left(\frac{1 \text{ mol}}{100.09 \text{ g}} \right) = 0.050 \text{ mol } CaCO_3$$

H_3O^+: $n = 0.50$ L $(0.10$ M$) = 0.050$ mol H_3O^+

Use the balanced equation and the initial amounts to construct an amounts table:

Reaction	$CaCO_3$ +	$2 H_3O^+$ →	Ca^{2+} +	CO_2
Initial amount (mol)	0.050	0.050	0	0
mol/coeff	0.050	0.025 (LR)		
Change (mol)	–0.025	–0.050	+0.025	+0.025
Final amount (mol)	0.025	0	0.025	0.025

Use the final amounts from the table to obtain the desired results:

(b) $m = n\ MM = 0.025$ mol $\left(\dfrac{44.01\ g}{1\ mol}\right) = 1.1$ g CO_2 formed

(c) The spectator ion concentration remains unchanged: $[Cl^-] = 0.10$ M

Use the amounts table to calculate the concentrations of the other ionic species (Ca^{2+}) present in the solution:

$$[Ca^{2+}] = \dfrac{0.025\ mol}{0.500\ L} = 0.050\ M$$

4.89   To predict reaction products, first determine what species are present if an aqueous solution is involved, then identify the types of reaction each reactant can undergo:

(a) Species are Ca, H_3O^+, Cl^-, and H_2O. Ca is a metal that displaces H_2 from acidic solutions (redox): $Ca\ (s) + 2\ H_3O^+\ (aq) \rightarrow Ca^{2+}\ (aq) + H_2\ (g) + 2\ H_2O\ (l)$

(b) Li is a Group 1 metal, which forms an oxide, Li_2O (redox):

$$4\ Li + O_2 \rightarrow 2\ Li_2O$$

(c) Species are H_3O^+, Br^-, NH_3, and H_2O. The strong acid reacts with the weak base (acid/base): $NH_3\ (aq) + H_3O^+\ (aq) \rightarrow NH_4^+\ (aq) + H_2O\ (l)$

(d) C_3H_8 is a hydrocarbon that burns in air (redox, all gases):

$$C_3H_8 + 5\ O_2 \rightarrow 3\ CO_2 + 4\ H_2O$$

4.91   Balance each element in turn by adjusting coefficients, leaving O, which appears in several products, to the end. Because of the odd number of atoms of C and H, it is convenient to give the starting material a coefficient of 2 at the outset:

$$2\ C_3H_5N_3O_9\ (l) \rightarrow N_2\ (g) + CO_2\ (g) + H_2O\ (g) + O_2\ (g)$$
$$6C + 10H + 6N + 18O \rightarrow 1C + 2H + 2N + 5O$$

Give N_2 a coefficient of 3 to balance N, CO_2 a coefficient of 6 to balance C, and H_2O a coefficient of 5 to balance H:

$$2\ C_3H_5N_3O_9\ (l) \rightarrow \mathbf{3}\ N_2\ (g) + \mathbf{6}\ CO_2\ (g) + \mathbf{5}\ H_2O\ (g) + O_2\ (g)$$
$$6C + 10H + 6N + 18O \rightarrow 6C + 10H + 6N + 19O$$

Now there are 18 O atoms on the left and 17 in CO_2 and H_2O on the right, so give O_2 a coefficient of 1/2 to balance O and multiply through by 2 to clear fractions:

$$4\ C_3H_5N_3O_9\ (l) \rightarrow 6\ N_2\ (g) + 12\ CO_2\ (g) + 10\ H_2O\ (g) + O_2\ (g)$$

4.93 The problem asks about yields and provides information about the amounts of both starting materials, so it is both a limiting reagent and a yield problem. Use a table of amounts to determine the theoretical yield.
Calculations of initial amounts:

$$N_2: 84.0 \text{ kg} \left(\frac{1 \text{ g}}{10^{-3} \text{ kg}} \right) \left(\frac{1 \text{ mol}}{28.02 \text{ g}} \right) = 3.00 \times 10^3 \text{ mol}$$

$$H_2: 24.0 \text{ kg} \left(\frac{1 \text{ g}}{10^{-3} \text{ kg}} \right) \left(\frac{1 \text{ mol}}{2.016 \text{ g}} \right) = 1.19 \times 10^4 \text{ mol}$$

Reaction	N_2 +	$3 H_2$ →	$2 NH_3$
Initial amount (10^3 mol)	3.00	11.9	0
mol/coeff	3.00 (LR)	3.97	
Change (10^3 mol)	-3.00	-9.00	+6.00
Final amount (10^3 mol)	0	2.9	6.00

Use the table to obtain the theoretical yield of NH_3:

$$6.00 \times 10^3 \text{ mol} \left(\frac{17.04 \text{ g}}{1 \text{ mol}} \right) \left(\frac{10^{-3} \text{ kg}}{1 \text{ g}} \right) = 102 \text{ kg}$$

The theoretical yield is 102 kg, and the actual yield is 68 kg:

$$\% \text{ yield} = (100\%) \frac{68 \text{ kg}}{102 \text{ kg}} = 67\%$$

Multiply the theoretical change by the percent yield to get the actual change:
    Actual change = 67%(6.00 x 10^3 mol) = 4.00 x 10^3 mol
Repeat the table of amounts using the actual change rather than the theoretical change:

Reaction	N_2 +	$3 H_2$ →	$2 NH_3$
Amount (10^3 mol)	3.00	11.9	0
Change (10^3 mol)	-2.00	-6.00	+4.00
Final (10^3 mol)	1.00	5.9	4.00

Do mole-mass conversions to determine masses of leftover reactants:

$$N_2: 1.00 \times 10^3 \text{ mol} \left(\frac{28.02 \text{ g}}{1 \text{ mol}} \right) \left(\frac{1 \text{ kg}}{10^3 \text{ g}} \right) = 28 \text{ kg}$$

$$H_2: 5.9 \times 10^3 \text{ mol} \left(\frac{2.016 \text{ g}}{1 \text{ mol}} \right) \left(\frac{1 \text{ kg}}{10^3 \text{ g}} \right) = 12 \text{ kg}$$

4.95 The quantity asked for is a volume ratio, which can be related to a mass ratio using densities and to a mole ratio using molar masses. First balance the chemical reaction:
    $N_2H_4 \text{ (l)} + H_2O_2 \text{ (l)} \rightarrow N_2 \text{ (g)} + H_2O \text{ (g)}$
    $2N + 6H + 2O \rightarrow 2N + 2H + O$

N is already balanced; give H_2O a coefficient of 2 to balance O:

$$N_2H_4 \, (l) + H_2O_2 \, (l) \rightarrow N_2 \, (g) + 2 \, H_2O \, (g)$$

$$2N + 6H + 2O \rightarrow 2N + 4H + 2O$$

Now there are six H on the left and four on the right; multiply the coefficients of both H_2O_2 and H_2O by 2 to give 8 on each side:

$$N_2H_4 \, (l) + 2 \, H_2O_2 \, (l) \rightarrow N_2 \, (g) + 4 \, H_2O \, (g)$$

The two reactants should be present in a mole ratio of 1 N_2H_4: 2 H_2O_2.

Use $m = n \, MM$ and $V = \dfrac{m}{\rho}$, giving $V = \dfrac{n \, MM}{\rho}$:

Vol. ratio = $1 \text{ mol } N_2H_4 \left(\dfrac{32.05 \text{ g}}{1 \text{ mol}} \right) \left(\dfrac{1 \text{ mL}}{1.44 \text{ g}} \right) : 2 \text{ mol } H_2O_2 \left(\dfrac{34.02 \text{ g}}{1 \text{ mol}} \right) \left(\dfrac{1 \text{ mL}}{1.01 \text{ g}} \right)$

Vol. ratio = 22.26:67.37 or 1:3.03

4.97   This problem describes an acid-base titration. We are asked to determine the mass of acid in the tablet. $n_{base} = 0.1045 \text{ M}(0.02445 \text{ L}) = 0.002555 \text{ mol base}$

At the stoichiometric point: $n_{acid} = n_{base} = 0.002555 \text{ mol}$

Convert to mass: $m_{acid} = n \, MM = 0.002555 \text{ mol} \left(\dfrac{176.13 \text{ g}}{1 \text{ mol}} \right) \left(\dfrac{10^3 \text{ mg}}{1 \text{ g}} \right) = 450.0 \text{ mg}$

This mass differs from that of a pure tablet (500.0 mg), so the tablet is not pure.
The mass of the impurities is the difference of a pure tablet and that of the actual tablet:

500.0 mg - 450.0 mg = 50.0 mg impurities

$$\text{Mass\%} = (100\%) \dfrac{50.0 \text{ mg}}{500.0 \text{ mg}} = 10.0\% \text{ impurities}$$

4.99   This problem asks about a mixture of two salts, $CaSO_4$ and $Ca(H_2PO_4)_2$:

(a) The 2:1 mole ratio indicates the coefficients for the balanced equation. There is no water present during this reaction, so we need not ask about species present:

$$2 \, H_2SO_4 + Ca_3(PO_4)_2 \rightarrow 2 \, CaSO_4 + Ca(H_2PO_4)_2$$

(b) Because the mixture has fixed stoichiometric proportions of 2:1, it has a combined molar mass of 2 $MM(CaSO_4) + MM(Ca(H_2PO_4)_2)$:

$MM_{combination} = 2(136.1 \text{ g/mol}) + 234.1 \text{ g/mol} = 506.3 \text{ g/mol}$

$n_{combination} = 50.0 \text{ kg} \left(\dfrac{10^3 \text{ g}}{1 \text{ kg}} \right) \left(\dfrac{1 \text{ mol}}{506.3 \text{ g}} \right) = 98.76 \text{ mol}$

$n_{acid} = 2 \, n_{combination}$ $n_{Ca_3(PO_4)_2} = n_{combination}$

$m_{acid} = 98.76 \text{ mol} \left(\dfrac{2 \text{ mol } H_2SO_4}{1 \text{ mol combination}} \right) \left(\dfrac{98.09 \text{ g}}{1 \text{ mol}} \right) \left(\dfrac{10^{-3} \text{ kg}}{1 \text{ g}} \right) = 19.4 \text{ kg}$

$m_{Ca_3(PO_4)_2} = 98.76 \text{ mol} \left(\dfrac{1 \text{ mol } Ca_3(PO_4)_2}{1 \text{ mol combination}} \right) \left(\dfrac{310.2 \text{ g}}{1 \text{ mol}} \right) \left(\dfrac{10^{-3} \text{ kg}}{1 \text{ g}} \right) = 30.6 \text{ kg}$

(c) Each mole of $Ca_3(PO_4)_2$ yields 2 mol of phosphate ion, so the amount of phosphate ion available is 2 $n_{Ca_3(PO_4)_2}$ = 2(98.76 mol) = 198 mol

4.101   Mole calculations for a chemical reaction require mole ratios, which require a balanced chemical equation. Balance the equation, then do mass-mole-mass conversions. The reaction is:

$$C_5H_{12}O + O_2 \rightarrow H_2O + CO_2$$

$$5C + 12H + 3O \rightarrow 1C + 2H + 3O$$

Balance C by giving CO_2 a coefficient of 5 and H by giving H_2O a coefficient of 6:

$$C_5H_{12}O + O_2 \rightarrow \mathbf{6}\ H_2O + \mathbf{5}\ CO_2$$

$$5C + 12H + 3O \rightarrow 5C + 12H + 16O$$

This gives 16 O atoms on the right. Give O_2 a coefficient of 15/2 so there are 16 O atoms on the left, then multiply through by 2 to clear fractions:

$$2\ C_5H_{12}O + 15\ O_2 \rightarrow 12\ H_2O + 10\ CO_2$$

Use the density equation to calculate the mass of the compound: $m = \rho V$

$$3.15\ \text{mL}\ C_5H_{12}O \left(\frac{0.740\ \text{g}}{1\ \text{mL}} \right) \left(\frac{1\ \text{mol}}{88.15\ \text{g}} \right) \left(\frac{10\ \text{mol}\ CO_2}{2\ \text{mol}\ C_5H_{12}O} \right) = 0.132\ \text{mol}\ CO_2$$

4.103   First identify the chemical reactions. Here, the metal in the sample is dissolved in acid, and AgCl is then precipitated from the solution. The precipitate contains all the silver that was present in the metal sample, so we can calculate the composition of the metal from the amount of silver present in the precipitate.

$$m_{Ag} = (m_{AgCl}) \left(\frac{MM_{Ag}}{MM_{AgCl}} \right) = 0.156\ \text{g} \left(\frac{107.9\ \text{g/mol}}{143.4\ \text{g/mol}} \right) = 0.1174\ \text{g; (carry an extra digit to}$$

avoid round-off error)

$$\%\ Ag = (100\%) \left(\frac{m_{Ag}}{m_{total}} \right) = (100\%) \frac{0.1174\ \text{g}}{0.135\ \text{g}} = 87.0\%$$

$$\%\ Cu = 100\% - 87.0\% = 13.0\%$$

4.105   The problem asks about the amount of reactant needed to react completely with a given amount of another reactant. The reaction is $CO_2\ (g) + 2\ KOH\ (s) \rightarrow K_2CO_3\ (s) + H_2O\ (l)$ Use mole amounts and stoichiometric ratios:

$$n_{CO_2} = (5\ \text{astronaut})(6\ \text{days}) \left(\frac{1.0\ \text{kg}}{(1\,\text{astronaut})\ (1\,\text{day})} \right) \left(\frac{10^3\ \text{g}}{1\ \text{kg}} \right) \left(\frac{1\ \text{mol}}{44.0\ \text{g}} \right) = 6.8 \times 10^2\ \text{mol}$$

$$m_{KOH} = 6.8 \times 10^2\ \text{mol}\ CO_2 \left(\frac{2\ \text{mol}\ KOH}{1\ \text{mol}\ CO_2} \right) \left(\frac{56.1\ \text{g}}{1\ \text{mol}} \right) \left(\frac{1\ \text{kg}}{10^3\ \text{g}} \right) = 76\ \text{kg}\ KOH$$

The more expensive LiOH is used on the space shuttle because it is 43 kg lighter.

5.1   A pinhole in the top of the tube would let air leak into the space until the internal pressure matched the external, atmospheric pressure. The barometer would then indicate zero pressure, because the height of the mercury column is determined by the pressure *difference* between inside and outside, and this difference would be zero in the presence of a pinhole.

5.3   When air is pumped into a flat tire, the tire expands and becomes hard. Both these observations are the result of the pressure exerted by the gas molecules making up air.

5.5   Useful pressure conversion factors are 1 atm = 1.01325×10^5 Pa, 1 atm = 760 torr, and 1 bar = 10^5 pascals (Pa)

(a) $455 \text{ torr} \left(\dfrac{1 \text{ atm}}{760 \text{ torr}} \right) \left(\dfrac{1.01325 \times 10^5 \text{ Pa}}{1 \text{ atm}} \right) = 6.07 \times 10^4 \text{ Pa} = 0.607 \text{ bar}$

(b) $2.45 \text{ atm} \left(\dfrac{1.01325 \times 10^5 \text{ Pa}}{1 \text{ atm}} \right) = 2.48 \times 10^5 \text{ Pa} = 2.48 \text{ bar}$

(c) $0.46 \text{ torr} \left(\dfrac{1 \text{ atm}}{760 \text{ torr}} \right) \left(\dfrac{1.01325 \times 10^5 \text{ Pa}}{1 \text{ atm}} \right) = 61 \text{ Pa} = 6.1 \times 10^{-4} \text{ bar}$

(d) $1.33 \times 10^{-3} \text{ atm} \left(\dfrac{1.01325 \times 10^5 \text{ Pa}}{1 \text{ atm}} \right) = 1.35 \times 10^2 \text{ Pa} = 1.35 \times 10^{-3} \text{ bar}$

5.7   The first part of this question asks about number of moles. Rearrange the ideal gas equation to solve for n, then substitute the appropriate values and do the calculation:
$$n = \frac{PV}{RT} = \frac{(5.00 \text{ atm})(20.0 \text{ L})}{(0.08206 \text{ L atm/mol K})(298 \text{ K})} = 4.09 \text{ mol}$$
The second part of the question asks about the volume if the pressure was different. We can calculate this volume using n and the ideal gas equation, but we can also do the calculation by using proportionalities.
$$P_i V_i = P_f V_f, \text{ from which } V_i = \frac{P_f V_f}{P_i} = \frac{(5.00 \text{ atm})(20.0 \text{ L})}{(1.00 \text{ atm})} = 100. \text{ L}$$

5.9   When some variables are held fixed, rearrange the ideal gas equation, $PV = nRT$, to collect fixed values on the right.

(a) n, R, V are constant: $\dfrac{P}{T} = \dfrac{nR}{V} = \text{constant}; \quad \dfrac{P_i}{T_i} = \dfrac{P_f}{T_f}$

(b) $V = \dfrac{nRT}{P}$

(c) n, R, T are constant: $PV = nRT = \text{constant}; \quad P_i V_i = P_f V_f$

5.11   When some conditions change but others remain fixed, rearrange the ideal gas equation so the constant terms are grouped on the right. In this problem, n and P are fixed:

$$\frac{V}{T} = \frac{nR}{P} = \text{constant, so } \frac{V_i}{T_i} = \frac{V_f}{T_f}$$

$V_i = 0.255$ L

Convert temperature to kelvins:

$T_i = 25 + 273.15 = 298$ K $T_f = -15 + 273.15 = 258$ K

$$V_f = \frac{V_i T_f}{T_i} = \frac{(0.255 \text{ L})(258 \text{ K})}{(298 \text{ K})} = 0.221 \text{ L}$$

5.13   The equation is valid only for a gas for which n and T are fixed.
(a) n and T are fixed, so the equation is valid. (b) n can change, so the equation is not valid. (c) T changes, so the equation is not valid. (d) The equation is not valid for liquids.

5.15   Molecular speed is related to temperature by the equations, $u = \sqrt{\dfrac{2E_{kinetic}}{m}}$, and

$\bar{E}_{kinetic} = \dfrac{3RT}{2N_A}$, so $u_{avg} = \sqrt{\dfrac{3RT}{mN_A}}$. Thus, molecular speed increases with the square

root of the absolute temperature, so molecules will reach the detector sooner at higher temperature, because they have greater speed.

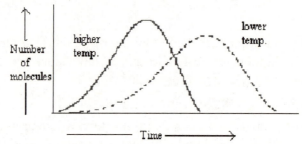

5.17   Molecular speed is related to temperature by two equations:

$$u = \sqrt{\frac{2E_{kinetic}}{m}}$$

$\bar{E}_{kinetic} = \dfrac{3RT}{2N_A}$, so $\bar{u} = \sqrt{\dfrac{3RT}{mN_A}}$

Each distribution of molecular speeds will have the same general shape, with the position of the peak

depending on $\sqrt{\dfrac{T}{m}}$:

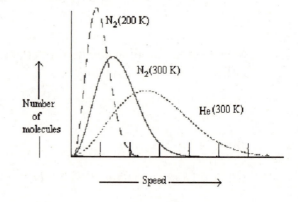

5.19   Average molecular kinetic energy is directly proportional to temperature and is independent of molecular mass: $\bar{E}_{kinetic} = \dfrac{3RT}{2N_A}$. N_2 and He at 300 K have identical kinetic energy distributions, while the distribution for N_2 at 200 K is shifted to lower energy: The shape of the curve is slightly different for energies than for speeds.

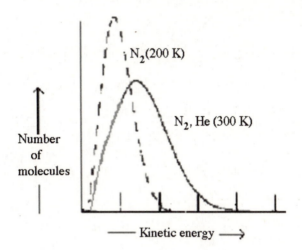

5.21   The most probable speed at any temperature is related to the most probable kinetic energy by the equation $E_{kinetic} = \dfrac{mu^2}{2}$, or $u_{mp} = \sqrt{\dfrac{2E_{kinetic,mp}}{m}}$. According to your textbook, the most probable kinetic energy at 300 K is 4.13×10^{-21} J/molecule. Because energy is proportional to temperature, the most probable kinetic energy at any other temperature can be calculated using proportions: $\dfrac{E_{kinetic,2}}{E_{kinetic,1}} = \dfrac{T_2}{T_1}$. Average kinetic energy, which is not the same as most probable kinetic energy because the distribution of energies is asymmetric, is calculated using $E_{kinetic,\ molar} = \dfrac{3RT}{2}$ (T in kelvins; 1 J = 1 kg m^2 s^{-2}):

(a) He: $m = \left(\dfrac{4.00\ g}{1\ mol}\right)\left(\dfrac{10^{-3}\ kg}{1\ g}\right)\left(\dfrac{1\ mol}{6.022 \times 10^{23}\ atoms}\right) = 6.64 \times 10^{-27}$ kg

627 °C + 273 = 900 K, $E_{kinetic,\ mp} = 4.13 \times 10^{-21}$ J $\left(\dfrac{900\ K}{300\ K}\right) = 1.24 \times 10^{-20}$ J

$u_{mp} = \sqrt{\dfrac{2(1.24 \times 10^{-20}\ J)}{6.64 \times 10^{-27}\ kg}} = 1.93 \times 10^3$ m/s

$E_{kinetic,\ molar} = \left(\dfrac{3}{2}\right)\left(\dfrac{8.314J}{1\ mol\ K}\right)(900K) = 1.12 \times 10^4$ J/mol

(b) O_2: $m = \left(\dfrac{32.00 \text{ g}}{1 \text{ mol}}\right)\left(\dfrac{10^{-3} \text{ kg}}{1 \text{ g}}\right)\left(\dfrac{1 \text{ mol}}{6.022 \times 10^{23} \text{ molecules}}\right) = 5.31 \times 10^{-26} \text{ kg}$

27 °C + 273 = 300 K, $E_{\text{kinetic, mp}} = 4.13 \times 10^{-21}$ J

$u_{mp} = \sqrt{\dfrac{2(4.13 \times 10^{-21} \text{ J})}{5.31 \times 10^{-26} \text{ kg}}} = 3.94 \times 10^2$ m/s

$E_{\text{kinetic, molar}} = \left(\dfrac{3}{2}\right)\left(\dfrac{8.314 \text{ J}}{1 \text{ mol K}}\right)(300 \text{ K}) = 3.74 \times 10^3$ J/mol

(c) SF_6: $MM = [32.066 \text{ g/mol} + 6(18.998 \text{ g/mol})] = 146.1$ g/mol

$m = \left(\dfrac{146.1 \text{ g}}{1 \text{ mol}}\right)\left(\dfrac{10^{-3} \text{ kg}}{1 \text{ g}}\right)\left(\dfrac{1 \text{ mol}}{6.022 \times 10^{23} \text{ atoms}}\right) = 2.43 \times 10^{-25}$ kg

627 °C + 273 = 900 K, $E_{\text{kinetic, mp}} = 4.13 \times 10^{-21}$ J $\left(\dfrac{900 \text{ K}}{300 \text{ K}}\right) = 1.24 \times 10^{-20}$ J

$u_{mp} = \sqrt{\dfrac{2(1.24 \times 10^{-20} \text{ J})}{2.43 \times 10^{-25} \text{ kg}}} = 3.19 \times 10^2$ m/s

$E_{\text{kinetic, molar}} = \left(\dfrac{3}{2}\right)\left(\dfrac{8.314 \text{ J}}{1 \text{ mol K}}\right)(900 \text{K}) = 1.12 \times 10^4$ J/mol

5.23   The ideal gas is defined by the conditions that molecular volumes and intermolecular forces both are negligible.
(a) At very high pressure, molecules are very close together, so their volumes are significant compared to the volume of their container; because the first condition is not met, the gas is not ideal.
(b) At very low temperature, molecules move very slowly, so the forces between molecules, even though small, are sufficient to influence molecular motion; because the second condition is not met, the gas is not ideal.

5.25   (a) If the piston is stationary and there is no friction, the forces and pressures on each side must be equal, so the internal pressure is also 1 atm. This pressure is generated by gas molecules colliding with the face of the piston.
(b) When temperature doubles, the increase in average molecular speed results in a doubling of the pressure. The piston will move outward, causing the concentration of gas molecules to decrease. This will lead to a lower frequency of collisions with the wall, reducing the pressure. The piston will stop when the internal pressure is once again 1 atm.

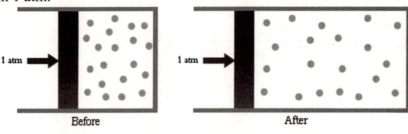

5.27   As a gas cools, its molecules move more slowly, so they impart smaller impulses on the walls of their container. This reduces the internal pressure, so the balloon collapses until the increase in gas density inside the balloon brings the internal pressure back up to the external pressure of 1.000 atm.

5.29   The ideal gas equation, $PV = nRT$, can be used to calculate moles using P-V-T data.

Molar mass is related to moles through $n = \dfrac{m}{MM}$:

$$n = \frac{PV}{RT} = \frac{m}{MM} \qquad\qquad MM = \frac{mRT}{PV}$$

Begin by converting the initial data into the units of R:

$T = 25.0 + 273.15 = 298.2 \text{ K}$

$$P = 262 \text{ torr} \left(\frac{1 \text{ atm}}{760 \text{ torr}} \right) = 0.345 \text{ atm}$$

Use the modified ideal gas equation to obtain the molar mass:

$$MM = \frac{(2.55 \text{ g})(8.206 \times 10^{-2} \frac{\text{L atm}}{\text{mol K}})(298.2 \text{ K})}{(0.345 \text{ atm})(1.50 \text{ L})} = 121 \text{ g/mol}$$

The compound contains only C (12.0 g/mol), F (19.0 g/mol), and Cl (35.5 g/mol). The formula can be determined by trial and error. The combination of 1 C, 2 F, and 2 Cl has
$MM = 12.0 \text{ g/mol} + 2(19.0 \text{ g/mol}) + 2(35.5 \text{ g/mol}) = 121 \text{ g/mol}$
This matches the experimental value. The formula is CF_2Cl_2.

5.31   Density can be calculated from the ideal gas equation and mole-mass conversions:

$$n = \frac{m}{MM} = \frac{PV}{RT} \qquad\qquad \frac{m}{V} = \frac{P \, MM}{RT}$$

$$P = 755 \text{ torr} \left(\frac{1 \text{ atm}}{760 \text{ torr}} \right) = 0.993 \text{ atm}$$

$$\frac{m}{V} = \frac{P \, MM}{RT} = \frac{(0.993 \text{ atm})(146.05 \text{ g/mol})}{(8.206 \times 10^{-2} \frac{\text{L atm}}{\text{mol K}})(27 + 273) \text{ K}} = 5.89 \text{ g/L}$$

5.33   Molecular speed can be calculated from temperature and molar mass:

$$\bar{E}_{\text{kinetic}} = \frac{3RT}{2N_A} \qquad\qquad \bar{E}_{\text{kinetic}} = \frac{mu_{\text{rms}}^2}{2}$$

$$u_{\text{rms}}^2 = \frac{3RT}{mN_A} = \frac{3RT}{MM} \qquad\qquad u_{\text{rms}} = \left(\frac{3RT}{MM} \right)^{1/2}$$

$T = 27 \ ^o\text{C} + 273 = 300. \text{ K}; MM = 32.06 \text{ g/mol} + 6(18.998 \text{ g/mol}) = 146.0 \text{ g/mol}$
The mass must be converted to kg/mol for mass units to cancel (1 J = 1 kg m²/s²):
$MM = (146.0 \text{ g/mol})(10^{-3} \text{ kg/g}) = 1.460 \times 10^{-1} \text{ kg/mol}$

$$u_{\text{rms}} = \left(\frac{3(8.314 \text{ J mol}^{-1} \text{ K}^{-1})(1 \text{ kg m}^2\text{s}^{-2}/\text{J})(300. \text{ K})}{1.460 \times 10^{-1} \text{ kg mol}^{-1}} \right)^{1/2} = 226 \text{ m/s}$$

5.35   Rates of diffusion and effusion depend on the molar masses of the gases because average speeds determine these rates, and these are given by Equation 5-4: $u_{rms} = \left(\dfrac{3\,RT}{MM} \right)^{1/2}$

For two gases at the same temperature, the one with the smaller molar mass has the faster molecular speed and will diffuse faster. Thus CH_4 ($MM = 16$ g/mol) diffuses faster through the atmosphere than C_2H_6 ($MM = 30$ g/mol).

5.37   According to Dalton's law of partial pressures, each gas exerts a pressure equal to the total pressure times its mole fraction. Mole fraction can be found from concentration in ppm:

$$X = \left(\dfrac{ppm}{10^6} \right)$$

$$p_{NO_2} = 758.4 \text{ torr} \left(\dfrac{1 \text{ atm}}{760 \text{ torr}} \right) \left(\dfrac{0.78 \text{ molecules } NO_2}{10^6 \text{ molecules of air}} \right) = 7.8 \times 10^{-7} \text{ atm}$$

5.39   According to Dalton's law of partial pressures, each gas exerts a pressure equal to its mole fraction times the total pressure: $p = X P_{tot}$. Standard atmospheric conditions correspond to $P_{tot} = 1$ atm.

$$p_{N_2} = (0.7808)(1 \text{ atm}) \left(\dfrac{760 \text{ torr}}{1 \text{ atm}} \right) = 593.4 \text{ torr}$$

$$p_{O_2} = (0.2095)(1 \text{ atm}) \left(\dfrac{760 \text{ torr}}{1 \text{ atm}} \right) = 159.2 \text{ torr}$$

$$p_{Ar} = (9.34 \times 10^{-3})(1 \text{ atm}) \left(\dfrac{760 \text{ torr}}{1 \text{ atm}} \right) = 7.10 \text{ torr}$$

$$p_{CO_2} = (3.25 \times 10^{-4})(1 \text{ atm}) \left(\dfrac{760 \text{ torr}}{1 \text{ atm}} \right) = 2.47 \times 10^{-1} \text{ torr}$$

5.41   Molecular pictures show the relative numbers of molecules of various substances, which also represents the relative numbers of moles of those substances. The figure contains 8 He atoms and 4 O_2 molecules.

(a) and (b) There are more He atoms, so the pressure due to He and the mole fraction of He are higher.

(c) $X_{He} = \dfrac{n_{He}}{n_{tot}} = \dfrac{8}{12} = 0.67$

5.43   To calculate partial pressures from a total pressure, mole fractions are required: $p_i = X_i P_{tot}$. Mole fractions can be calculated from the analytical data for the gas sample. The equations needed to solve this problem are:

$$n = \dfrac{m}{MM} \qquad\qquad X_i = \dfrac{n_i}{n_{tot}} \qquad\qquad p_i = \dfrac{n_i RT}{V}$$

$$n_{CH_4} = 1.57 \text{ g}\left(\frac{1 \text{ mol}}{16.04 \text{ g}}\right) = 9.79 \times 10^{-2} \text{ mol}$$

$$n_{C_2H_6} = 0.41 \text{ g}\left(\frac{1 \text{ mol}}{30.07 \text{ g}}\right) = 1.36 \times 10^{-2} \text{ mol}$$

$$n_{C_3H_8} = 0.020 \text{ g}\left(\frac{1 \text{ mol}}{44.09 \text{ g}}\right) = 4.54 \times 10^{-4} \text{ mol}$$

$n_{tot} = 9.79 \times 10^{-2} \text{ mol} + 1.36 \times 10^{-2} \text{ mol} + 4.54 \times 10^{-4} \text{ mol} = 1.120 \times 10^{-1} \text{ mol}$

Determine the mole fraction by dividing the moles of each component by the total number of moles:

$$X_{CH_4} = \frac{9.79 \times 10^{-2} \text{ mol}}{1.120 \times 10^{-1} \text{ mol}} = 0.874 \qquad X_{C_2H_6} = \frac{1.36 \times 10^{-2} \text{ mol}}{1.120 \times 10^{-1} \text{ mol}} = 0.121$$

$$X_{C_3H_8} = \frac{4.54 \times 10^{-4} \text{ mol}}{1.120 \times 10^{-1} \text{ mol}} = 4.06 \times 10^{-3}$$

Multiply the mole fractions by the total pressure to obtain the partial pressure (round the final values for ethane and propane to two significant figures, because their masses are only known to two significant figures):

$$p_{CH_4} = (0.874)(2.35 \text{ atm}) = 2.05 \text{ atm}$$

$$p_{C_2H_6} = (0.121)(2.35 \text{ atm}) = 0.28 \text{ atm}$$

$$p_{C_3H_8} = (4.06 \times 10^{-3})(2.35 \text{ atm}) = 9.5 \times 10^{-3} \text{ atm}$$

5.45   Stoichiometric calculations require moles and a balanced chemical equation. Balance the equation, determine the number of moles of CO_2 produced, and apply the ideal gas equation to calculate V:

The unbalanced chemical equation is

$$C_6H_{12}O_6 + O_2 \rightarrow CO_2 + H_2O$$
$$6C + 12H + 8O \rightarrow 1C + 2H + 3O$$

Give both CO_2 and H_2O a coefficient of 6 to balance C and H:

$$C_6H_{12}O_6 + O_2 \rightarrow 6CO_2 + 6 H_2O;$$
$$6C + 12H + 8O \rightarrow 6C + 12H + 18O$$

This leaves 18 O on the product side. 6 O are from glucose, so O_2 needs a coefficient of 6 to balance O and obtain the balanced chemical equation:

$$C_6H_{12}O_6 \text{ (g)} + 6 O_2 \text{ (g)} \rightarrow 6 CO_2 \text{ (g)} + 6 H_2O \text{ (l)}$$

$MM_{glucose} = 6(12.01 \text{ g/mol}) + 12(1.01 \text{ g/mol}) + 6(16.00 \text{ g/mol}) = 180.18 \text{ g/mol}$

Use mass-mole conversions to determine the moles of CO_2 formed:

$$n_{CO_2} = 4.65 \text{ g } C_6H_{12}O_6\left(\frac{1 \text{ mol}}{180.18 \text{ g}}\right)\left(\frac{6 \text{ mol } CO_2}{1 \text{ mol } C_6H_{12}O_6}\right) = 1.548 \times 10^{-1} \text{ mol}$$

$$V = \frac{nRT}{P} = \frac{(1.548 \times 10^{-1} \text{ mol})(8.206 \times 10^{-2} \frac{\text{L atm}}{\text{mol K}})(310 \text{ K})}{1.00 \text{ atm}} = 3.94 \text{ L}$$

5.47   This is a stoichiometry problem that involves gases. We are asked to determine the final pressure of the container (which is the pressure of chlorine gas). Begin by analyzing the chemistry. The starting materials are Na metal and Cl_2, a gas. The product is NaCl.

The balanced chemical reaction is: $2 Na\,(s) + Cl_2\,(g) \rightarrow 2 NaCl\,(s)$

The problem gives information about the amounts of both starting materials, so this is a limiting reactant situation. We must calculate the number of moles of each species, construct a table of amounts, and use the results to determine the final pressure.
Calculations of initial amounts:

$$n_{Na} = 6.90 \text{ g} \left(\frac{1 \text{ mol}}{22.99 \text{ g}} \right) = 0.300 \text{ mol}$$

For chlorine gas, use the ideal gas equation to do pressure-mole conversions. The data must have the same units as those for R:

$$p_{Cl_2} = 1.25 \times 10^3 \text{ torr} \left(\frac{1 \text{ atm}}{760 \text{ torr}} \right) = 1.64 \text{ atm}$$

$$n_{Cl_2} = \frac{PV}{RT} = \frac{(1.64 \text{ atm})(3.00 \text{ L})}{(8.206 \times 10^{-2}\, \frac{\text{L atm}}{\text{mol K}})(300 \text{ K})} = 0.200 \text{ mol}$$

Divide each initial amount by its coefficient to determine the limiting reactant:

$$Cl_2: \ 0.200 \text{ mol} \qquad Na: \ \frac{0.300 \text{ mol}}{2} = 0.\,150 \text{ mol (LR)}$$

Use the initial amounts and the balanced equation to construct a table of amounts:

Reaction:	2 Na (s) +	Cl_2 (g) →	2 NaCl (s)
Initial amount (mol)	0.300	0.200	0.000
Change (mol)	–0.300	–0.150	+0.300
Final amount (mol)	0.000	0.050	0.300

Use the final amount of Cl_2 and the new temperature (47°C) to calculate the pressure:

$$P = \frac{nRT}{V} = \frac{(5.0 \times 10^{-2} \text{ mol})(8.206 \times 10^{-2}\, \frac{\text{L atm}}{\text{mol K}})(320 \text{ K})}{3.00 \text{ L}} = 0.438 \text{ atm}$$

5.49   This is a stoichiometry problem that involves gases. We are asked to determine the mass of product produced. Begin by analyzing the chemistry. The starting materials are N_2 and H_2, both gases. The product of the reaction is NH_3.

The balanced chemical reaction is: $N_2 + 3H_2 \rightarrow 2NH_3$

The problem gives information about the amounts of both starting materials, so this is a limiting reactant situation. We must calculate the number of moles of each species, construct a table of amounts, and use the results to determine the partial pressure.
Use the ideal gas equation to determine the initial amounts of each gas. The reactor initially contains the gases in 1:1 mole ratio, so

$$p_i = \frac{275 \text{ atm}}{2} = 137.5 \text{ atm for each gas}$$

$$n_i = \frac{p_i V}{RT} = \frac{(137.5 \text{ atm})(8.75 \times 10^3 \text{ L})}{(8.206 \times 10^{-2}\, \frac{\text{L atm}}{\text{mol K}})(455 + 273.15) \text{ K}} = 2.01 \times 10^4 \text{ mol}$$

Since both gases have the same initial amount, the limiting reactant will be the one with the larger stoichiometric coefficient: H_2 is limiting. Here is the complete amounts table:

Reaction:	N_2 (g) +	$3 H_2$ (g) $\rightarrow$	$2 NH_3$ (g)
Initial amount (10^4 mol)	2.01	2.01	0.00
Change (10^4 mol)	$-(1/3)(2.01)$	-2.01	$+(2/3)(2.01)$
Final amount (10^4 mol)	1.34	0.00	1.34

Obtain the mass of product formed from the final amount in the table:

$$m_{NH_3} = n\,MM = 1.34 \times 10^4 \text{ mol} \left(\frac{17.04 \text{ g}}{1 \text{ mol}} \right) = 2.28 \times 10^5 \text{ g}$$

5.51   The yield in a reaction is the ratio of the actual amount produced to the theoretical amount. The calculation in Problem 5.49 gives the theoretical amount:

$$\text{actual amount} = \text{theoretical amount} \left(\frac{\% \text{ yield}}{100\%} \right)$$

$$\text{actual amount} = 2.28 \times 10^5 \text{ g} \left(\frac{13\%}{100\%} \right) = 3.0 \times 10^4 \text{ g}$$

5.53   Table 5-3 gives the mole fractions of the components in dry air (0% humidity), from which $X_{O_2} = 0.2095$. Table 5-4 gives the vapor pressure of water at 20 °C, 17.535 torr; this is the partial pressure at 100% humidity for that temperature. To find the mole fraction of O_2 at 100% humidity, first determine the partial pressure of O_2:

$$p_{H_2O} = 17.535 \text{ torr} \left(\frac{1 \text{ atm}}{760 \text{ torr}} \right) = 2.307 \times 10^{-2} \text{ atm}$$

$$P_{(N_2 + O_2)} = 1.000 \text{ atm} - 2.307 \times 10^{-2} \text{ atm} = 0.9769 \text{ atm}$$

$$p_{O_2} = (0.2095)(0.9769 \text{ atm}) = 0.2047 \text{ atm}$$

$$X_{O_2} = \frac{p_{O_2}}{P_{\text{total}}} = \frac{0.2047 \text{ atm}}{1.000 \text{ atm}} = 0.2047 \qquad \text{change} = 0.2095 - 0.2047 = +0.0048$$

In other words, the mole fraction of O_2 in the atmosphere increases by 0.0048 when the relative humidity decreases from 100% to 0%.

5.55   The dew point is defined as the temperature at which the partial pressure of water vapor in the atmosphere matches the vapor pressure of water. Refer to Table 5-4 for vapor pressures: at 25 °C, $vp = 23.756$ torr

$$p_{H_2O} = 23.756 \text{ torr} \left(\frac{78\%}{100\%} \right) = 18.53 \text{ torr}$$

This vapor pressure corresponds to a dew point between 20 and 25 °C. Assume a linear variation in vapor pressure between these two temperatures:

$$\text{dew point} = 20.0 \text{ °C} + \left[\frac{(18.53 \text{ - } 17.535)}{(23.756 \text{ - } 17.535)} \right] (5 \text{ °C}) = 20.0 + 0.8 = 20.8 \text{ °C}$$

5.57   We must calculate the volume of dry air that contains 10 mg of neon. Use moles, mole fractions from Table 5-3, and the ideal gas equation:

$$n_{Ne} = \frac{m}{MM} = 10 \text{ mg} \left(\frac{10^{-3}\text{g}}{1 \text{ mg}}\right)\left(\frac{1 \text{ mol}}{20.180 \text{ g}}\right) = 5.0 \times 10^{-4} \text{ mol}$$

From Table 5-3, $X_{Ne} = 1.82 \times 10^{-5} = \dfrac{n_{Ne}}{n_{air}}$

Thus, the number of moles of air containing 10 mg of neon is

$$n_{air} = \frac{n_{Ne}}{X_{Ne}} = \frac{5.0 \times 10^{-4} \text{mol}}{1.82 \times 10^{-5}} = 27.5 \text{ mol}$$

The volume occupied by this amount of air can be calculated from the ideal gas equation:

$$V = \frac{nRT}{P} = \frac{(27.5 \text{ mol})(8.206 \times 10^{-2} \frac{\text{L atm}}{\text{mol K}})(298 \text{ K})}{1.00 \text{ atm}} = 6.7 \times 10^2 \text{ L}$$

(two significant figures because the moles of air should only have two significant figures)

5.59   Use the ideal gas equation to determine moles, then do mole-mass-number conversions:

$$n = \frac{PV}{RT} = \left(\frac{(10.0 \text{ atm})(725 \text{ mL})}{(8.206 \times 10^{-2} \frac{\text{L atm}}{\text{mol K}})(925 + 273) \text{ K}}\right)\left(\frac{10^{-3} \text{ L}}{1 \text{ mL}}\right) = 7.37 \times 10^{-2} \text{ mol}$$

$$m = n\, MM = 7.37 \times 10^{-2} \text{ mol}\left(\frac{83.80 \text{ g}}{1 \text{ mol}}\right) = 6.18 \text{ g}$$

$$\text{\# of atoms} = n\, N_A = 7.37 \times 10^{-2} \text{ mol}\left(\frac{6.022 \times 10^{23} \text{ atoms}}{1 \text{ mol}}\right) = 4.44 \times 10^{22} \text{ atoms}$$

5.61   Molecular pictures show the relative numbers of molecules of various substances, which also represent the relative numbers of moles of those substances. Chamber A contains 6 atoms, B contains 12 atoms, and C contains 9 atoms, for a total of 27 atoms, and the chambers have equal volumes and are at the same temperature.
(a) Chamber B contains the most atoms, so it exhibits the highest pressure
(b) The pressures in the chambers are proportional to the number of atoms:

$$P_A = P_B\left(\frac{\text{\#A}}{\text{\#B}}\right) = 1.0 \text{ atm}\left(\frac{6}{12}\right) = 0.50 \text{ atm}$$

(c) The pressure will increase in proportion to the number of atoms:

$$P(\text{new}) = P(\text{old})\left(\frac{\text{\#new}}{\text{\#old}}\right) = 1.0 \text{ atm}\left(\frac{27}{6}\right) = 4.5 \text{ atm}$$

(d) When the valves are opened, the number of atoms in each chamber equalizes:
$$27/3 = 9$$
$$P(\text{new}) = P(\text{old})\left(\frac{\text{\#new}}{\text{\#old}}\right) = 0.50 \text{ atm}\left(\frac{9}{12}\right) = 0.38 \text{ atm}$$

5.63   Much of the information in this problem is not needed, because molecular speed can be calculated from temperature and molar mass:

$$\bar{E}_{kinetic} = \frac{3RT}{2N_A} \qquad\qquad \bar{E}_{kinetic} = \frac{m\overline{u^2}}{2}$$

$$\overline{u^2} = \frac{3RT}{mN_A} = \frac{3RT}{MM} \qquad\qquad \bar{u} = \left(\frac{3\,RT}{MM}\right)^{1/2}$$

The mass must be converted to kg/mol for mass units to cancel ($1\,J = 1\,kg\,m^2/s^2$):

$$MM = 3\left(\frac{16.00\,g}{1\,mol}\right)\left(\frac{1\,kg}{1000g}\right) = 48.00 \times 10^{-3}\,kg/mol$$

$$\bar{u} = \left(\frac{3(8.314\,J\,mol^{-1}\,K^{-1})(-25+273)\,K}{48.00 \times 10^{-3}\,kg\,mol^{-1}}\right)^{1/2} = 359\,m/s$$

5.65   Use the ideal gas equation to find molar density and Avogadro's number to convert to molecular density:

$$\frac{n}{V} = \frac{P}{RT} = \frac{10^{-10}\,atm}{(8.206 \times 10^{-2}\,\frac{L\,atm}{mol\,K})(310\,K)} = 4 \times 10^{-12}\,mol/L$$

$$\#/V = N_A\left(\frac{n}{V}\right) = \left(\frac{6.022 \times 10^{23}\,molecules}{1\,mol}\right)\left(\frac{4 \times 10^{-12}\,mol}{1\,L}\right) = 2 \times 10^{12}\,molecules/L$$

5.67   under 100% humidity conditions. To determine the dew point, first calculate the partial pressure of water vapor at the existing temperature and then estimate the temperature at which this partial pressure equals the vapor pressure of water, assuming a linear variation of vapor pressure with temperature. Use data in Table 5-4:

$$p_{H_2O} = (vP)(rel.\ hum.)/(100\%)$$

(a) $p_{H_2O} = 42.175\,torr\left(\dfrac{80\%}{100\%}\right) = 33.74\,torr$

dew point $= 30.0\ ^\circ C + \left[\dfrac{(33.74 - 31.824)}{(42.175 - 31.824)}\right](5\ ^\circ C) = 30.0 + 0.9 = 30.9\ ^\circ C$

(b) $p_{H_2O} = 12.788\,torr\left(\dfrac{50\%}{100\%}\right) = 6.394\,torr$

dew point $= 0\ ^\circ C + \left[\dfrac{(6.394 - 4.579)}{(6.543 - 4.579)}\right](5\ ^\circ C) = 0.0 + 4.6 = 4.6\ ^\circ C$

(c) $p_{H_2O} = 23.756\,torr\left(\dfrac{30\%}{100\%}\right) = 7.1268\,torr$

dew point $= 5\ ^\circ C + \left[\dfrac{(7.1268 - 6.543)}{(9.209 - 6.543)}\right](5\ ^\circ C) = 5.0 + 1.1 = 6.1\ ^\circ C$

5.69   The molar mass of gaseous oxygen can be determined using the ideal gas law:

$$MM = \frac{mRT}{PV}$$

Pump out a bulb of known volume, weigh the empty bulb, fill the bulb with oxygen at a measured pressure, and weigh again. For monatomic oxygen, this experiment would give $MM = 16.00$ g/mol, whereas it would give $MM = 32.00$ g/mol for the diatomic gas.

5.71   Use Dalton's law of partial pressures to determine partial pressures, then apply the ideal gas equation to calculate masses.

$$X_{C_3H_6} = \frac{1.00}{5.00} = 0.200 \qquad\qquad p_{C_3H_6} = 0.200 \text{ atm}$$

$$X_{O_2} = 1.000 - 0.200 = 0.800 \qquad p_{O_2} = 0.800 \text{ atm}$$

$$m_{C_3H_6} = \frac{pV\,MM}{RT} = \frac{(0.200 \text{ atm})(2.00 \text{ L})(42.078 \text{ g/mol})}{(8.206 \times 10^{-2} \frac{\text{L atm}}{\text{mol K}})(23.5 + 273.15) \text{ K}} = 0.692 \text{ g}$$

$$m_{O_2} = \frac{pV\,MM}{RT} = \frac{(0.800 \text{ atm})(2.00 \text{ L})(32.00 \text{ g/mol})}{(8.206 \times 10^{-2} \frac{\text{L atm}}{\text{mol K}})(23.5 + 273.15) \text{K}} = 2.10 \text{ g}$$

5.73   All conditions except the volumes are different for the two samples.
(a) The bulb with the larger number of moles will contain more molecules:

$$n_{H_2} = \frac{PV}{RT} = \frac{(2 \text{ atm})V}{(273 \text{ K})R} = 7 \times 10^{-3} \left(\frac{V}{R}\right) \text{ mol}$$

$$n_{O_2} = \frac{PV}{RT} = \frac{(1 \text{ atm})V}{(298 \text{ K})R} = 3 \times 10^{-3} \left(\frac{V}{R}\right) \text{ mol}$$

The bulb of hydrogen contains more molecules.

(b) $m_{H_2} = (7 \times 10^{-3})(V/R) \text{ mol}\left(\frac{2.01 \text{ g}}{1 \text{ mol}}\right) = 1 \times 10^{-2}(V/R) \text{ g}$

$m_{O_2} = (3 \times 10^{-3})(V/R) \text{ mol}\left(\frac{32.00 \text{ g}}{1 \text{ mol}}\right) = 1 \times 10^{-1}(V/R) \text{ g}$

The bulb of oxygen contains more mass.

(c) The average kinetic energy of molecules depends only on the temperature of the gas, so the oxygen molecules, being at higher temperature, have greater average kinetic energy.

(d) The average molecular speed depends on $\sqrt{\dfrac{T}{m}}$ because $u_{avg} = \sqrt{\dfrac{3RT}{mN_A}}$.

The temperature ratio of H_2:O_2 is $\left(\dfrac{273}{298}\right) = 0.916$, while the mass ratio is

$\left(\dfrac{2}{32}\right) = 0.0625$. Thus, $\sqrt{\dfrac{T}{m}} = \sqrt{\dfrac{0.916}{0.0625}} = 3.82$, so the hydrogen molecules have greater average speed by this factor.

5.75   (a) The most probable kinetic energy at 300 K is 4×10^{-21} J

(b) $E_{\text{most probable}} = \dfrac{m(u_{\text{most probable}})^2}{2}$ (Remember 1 J 1 kg m^2 s^{-2})

$$u_{\text{most probable}} = \sqrt{\dfrac{2(4 \times 10^{-21}\ \text{J})(6.022 \times 10^{23}\ \text{molecules mole}^{-1})}{32.00 \times 10^{-3}\ \text{kg mol}^{-1}}} = 4 \times 10^2\ \text{m/s}$$

5.77   Your explanation should include the fact that sulfur in coal combines with oxygen from the air to form SO_2, and that when SO_2 reacts with water, sulfuric acid is an end product. Chemical equations:

$S\ (s) + O_2\ (g) \rightarrow SO_2\ (g)$

$2\ SO_2\ (g) + O_2\ (g) \rightarrow 2\ SO_3\ (g)$

$SO_3\ (g) + H_2O\ (l) \rightarrow H_2SO_4\ (aq)$

5.79   Each part of this problem represents a change of one or more conditions, for which a rearranged version of the ideal gas law can be used.

(a) T and n are fixed, so $PV = nRT = $ constant and $V_f = \dfrac{P_i V_i}{P_f}$

$$V_f = \dfrac{(525\ \text{torr})(3.00\ \text{L})}{755\ \text{torr}} = 2.09\ \text{L}$$

(b) T and n are fixed, so $PV = nRT = $ constant and $P_f = \dfrac{P_i V_i}{V_f}$

$$P_f = \dfrac{(525\ \text{torr})(3.00\ \text{L})}{2.00\ \text{L}} = 788\ \text{torr}$$

(c) V and n are fixed, so $\dfrac{P}{T} = \dfrac{nR}{V} = $ constant and $P_f = \dfrac{P_i T_f}{T_i}$

$$P_f = \dfrac{(525\ \text{torr})(50.0 + 273.15)\text{K}}{273.15\ \text{K}} = 621\ \text{torr}$$

5.81   The figure illustrates atomic density and atomic motion of a sample of helium gas. The figure contains eight atoms:

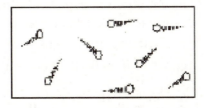

(a) Your drawing should show the same atomic density as the original figure but with longer "tails" on the atoms, indicating that they are moving at higher speeds:

(b) Your drawing should show the same lengths of "tails" as in the original figure, but there should be 2 fewer atoms:

(c) Your drawing should be the same as the original figure except that half the atoms should be replaced with diatomic molecules:

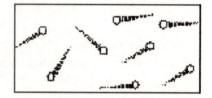

5.83   Use the ideal gas equation, $PV = nRT$, to determine which statements are correct:

(a) At constant T and V, $\dfrac{P}{n} = \dfrac{RT}{V} = $ constant. Thus, P is directly proportional to number of moles of gas. The statement is false.

(b) At constant V and n, $\dfrac{P}{T} = \dfrac{nR}{V} = $ constant. The statement is true.

(c) At fixed V and n, $\dfrac{P}{T} = \dfrac{nR}{V} = $ constant. $\dfrac{P}{T}$ is constant; PT is not. The statement is false.

5.85   This is an application of the ideal gas equation to changing gas conditions. The only constant quantity is n, so $\dfrac{PV}{T} = nR = $ constant. Therefore, $\dfrac{P_i V_i}{T_i} = \dfrac{P_f V_f}{T_f}$

$P_i = 0.969$ atm $V_i = 0.963$ L $P_f = 1.00$ atm
$T_i = 22 + 273.15 = 295$ K $T_f = 15 + 273.15 = 288$ K

$$V_f = \left(\frac{P_i V_i}{T_i}\right)\left(\frac{T_f}{P_f}\right) = \frac{(0.969 \text{ atm})(0.963 \text{ L})(288 \text{ K})}{(295 \text{ K})(1.00 \text{ atm})} = 0.911 \text{ L}$$

5.87   Concentrations in parts per billion can be converted to mole fractions, and the partial pressure can then be calculated using $p_i = X_i P_{total}$. To determine molecular concentration, first use the ideal gas equation to calculate mol, then multiply by N_A/V:

$X_{SF_6} = (1.0 \text{ ppb})(10^{-9}/\text{ppb}) = 1.0 \times 10^{-9}$ $p_{SF_6} = (1.0 \times 10^{-9})(1 \text{ atm}) = 1.0 \times 10^{-9}$ atm

$$V = 1.0 \text{ cm}^3 \left(\frac{10^{-3} \text{ L}}{1 \text{ cm}^3}\right) = 1.0 \times 10^{-3} \text{ L}$$

$$n = \frac{pV}{RT} = \frac{(1.0 \times 10^{-9} \text{ atm})(1.0 \times 10^{-3} \text{ L})}{(8.206 \times 10^{-2} \frac{\text{L atm}}{\text{mol K}})(21 + 273.15)\text{K}} = 4.1 \times 10^{-14} \text{ mol}$$

$$\#/V = 4.1 \times 10^{-14} \text{ mol} \left(\frac{6.022 \times 10^{23} \text{ molecules}}{1 \text{ mol}}\right) = 2.5 \times 10^{10} \text{ molecules/cm}^3$$

5.89   This is a *P-V-T* problem. First calculate moles of CO_2, then use the ideal gas equation to calculate the final pressure:

$$n = \frac{m}{MM} = 15.00 \text{ g}\left(\frac{1 \text{ mol}}{44.01g}\right) = 0.3408 \text{ mol}$$

$$p_{CO_2} = \frac{(0.3408 \text{ mol})(8.206 \times 10^{-2} \frac{\text{L atm}}{\text{mol K}})(273.15 \text{ K})}{0.750 \text{ L}} = 10.2 \text{ atm}$$

5.91   Density can be calculated from the ideal gas equation and mole-mass conversions:

$$n = \frac{m}{MM} = \frac{PV}{RT}, \text{ from which } \frac{m}{V} = \frac{P\,MM}{RT}$$

First, calculate the average molar masses of dry air and of moist air:

$$MM(\text{air}) = \Sigma_i\,(X_i MM_i)$$

$MM(\text{dry air}) = (0.7808)(28.01 \text{ g/mol}) + (0.2095)(32.00 \text{ g/mol})$
$$+ (9.34 \times 10^{-3})(39.948 \text{ g/mol}) = 28.95 \text{ g/mol}$$

For moist air at 298 K:

$$P_{H_2O} = 23.756 \text{ torr}\left(\frac{1 \text{ atm}}{760 \text{ torr}}\right) = 3.13 \times 10^{-2} \text{ atm} \qquad X_{H_2O} = 0.0313$$

The total mole fraction of the dry air components is reduced from 1 to $(1 - .0313) = 0.9687$. Each dry air mole fraction must be multiplied by this factor to obtain the mole fractions in moist air. Thus,

$$MM(\text{moist air}) = (3.13 \times 10^{-2})(18.01 \text{ g/mol}) + (0.9687)(28.95 \text{ g/mol}) = 28.61 \text{ g/mol}$$

$$\frac{m}{V}(\text{dry air}) = \frac{(1 \text{ atm})(28.95 \text{ g/mol})}{(8.206 \times 10^{-2} \frac{\text{L atm}}{\text{mol K}})(298 \text{ K})} = 1.18 \text{ g/L}$$

$$\frac{m}{V}(\text{moist air}) = \frac{(1 \text{ atm})(28.61 \text{ g/mol})}{(8.206 \times 10^{-2} \frac{\text{L atm}}{\text{mol K}})(298 \text{ K})} = 1.17 \text{ g/L}$$

5.93   First, determine moles of liquid oxygen in 150 L. Then calculate how many moles of dry air contain this amount of oxygen. Finally, use the ideal gas equation to calculate the volume:

$$n = \frac{\rho V}{MM} = 150 \text{ L}\left(\frac{10^3 \text{ mL}}{1 \text{ L}}\right)\left(\frac{1.14 \text{ g}}{1 \text{ mL}}\right)\left(\frac{1 \text{ mol}}{32.00 \text{ g}}\right) = 5.34 \times 10^3 \text{ mol}$$

$$n_{\text{air}} = \frac{n_{O_2}}{X_{O_2}} = \frac{5.34 \times 10^3 \text{ mol}}{0.2095} = 2.55 \times 10^4 \text{ mol}$$

$$P = 750 \text{ torr}\left(\frac{1 \text{ atm}}{760 \text{ torr}}\right) = 0.987 \text{ atm}$$

$$V = \frac{nRT}{P} = \frac{(2.55 \times 10^4 \text{ mol})(8.206 \times 10^{-2} \frac{\text{L atm}}{\text{mol K}})(298 \text{ K})}{0.987 \text{ atm}} = 6.32 \times 10^5 \text{ L}$$

5.95   In addition to a yield problem, information is given about both starting materials, so this is a limiting reactant problem that involves gases. We must calculate the number of moles of each species, construct a table of amounts, and use the results to determine the theoretical yield. Water can be ignored, because CH_4 is the product of interest. All of the reagents of interest are gases, so pressures can be used as the measures of amounts:

Reaction:	$3\ H_2\ (g)$ +	$CO\ (g)\ \rightarrow$	$CH_4\ (g)$
Initial pressure (atm)	20.0	10.0	0.0
Change (atm)	-20.0	-6.67	+6.67
Final pressure (atm)	0.0	3.33	6.67

Use the ideal gas equation and mole-mass conversion to obtain the theoretical yield in g:

$$n = \frac{PV}{RT} = \frac{(6.67\ \text{atm})(100\ \text{L})}{(8.206 \times 10^{-2}\ \text{L atm/mol K})(575\ \text{K})} = 14.1\ \text{mol}$$

Theoretical amount $= n\ MM = 14.1\ \text{mol}\left(\dfrac{16.04\ \text{g}}{1\ \text{mol}}\right) = 226\ \text{g}$

Obtain the percent yield by dividing the actual yield by the theoretical yield:

$$\%\ \text{yield} = 100\%\left(\frac{145\ \text{g}}{226\ \text{g}}\right) = 64.1\%$$

5.97   The chemical reaction is $N_2 + 3\ H_2 \rightarrow 2\ NH_3$. The molecular picture shows 4 mol of N_2 and 12 mol of H_2, which represents stoichiometric proportions. If the reaction goes to completion, all molecules will react to form 8 mol of NH_3:

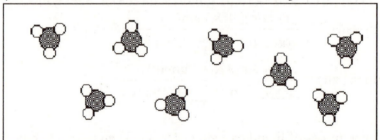

5.99   The question looks like an empirical formula problem, but the only data provided are P-V-T data for the gaseous substance, which suggests that an ideal gas law calculation can be done. Use the gas data to calculate n, then use $n = \dfrac{m}{MM}$ to determine MM and subsequently determine the formula of the compound:

$$P = 552\ \text{torr}\left(\frac{1\ \text{atm}}{760\ \text{torr}}\right) = 0.726\ \text{atm}$$

$$n = \frac{PV}{RT} = \frac{(0.726\ \text{atm})(0.100\ \text{L})}{(8.206 \times 10^{-2}\ \text{L atm/mol K})(30 + 273)\text{K}} = 2.92 \times 10^{-3}\ \text{mol}$$

$$MM = \frac{m}{n} = \frac{0.500\ \text{g}}{2.92 \times 10^{-3}\ \text{mol}} = 171\ \text{g/mol}$$

The compound contains one nickel atom ($MM = 58.69$ g/mol) and x CO molecules ($MM = 28.01$ g/mol), so $\quad$ 171 g/mol = 58.69 g/mol + (x)(28.01 g/mol)

$$x = \frac{171 \text{ g/mol} - 58.69 \text{ g/mol}}{28.01 \text{ g/mol}} = 4 \qquad \text{The formula is Ni(CO)}_4$$

5.101 At the top of Mount Everest, the atmospheric pressure is 250 mm, much lower than the pressure at sea level, 760 mm. A lower pressure corresponds to a smaller molecular density. Thus, when people say "the air is thin," they are referring to the fact that air is less dense at higher elevations.

5.103 Because Table 5-3 refers to dry air, the first step is to take into account the presence of H_2O at 50% relative humidity. The partial pressure of water must be subtracted from the total pressure to obtain the pressure due to dry air. Then $p_i = X_{\text{dry air}}\, p_{\text{dry air}}$.

From Table 5-4, $vp(25\ ^{\circ}C) = 23.756$ torr

$$p_{H_2O} = 23.756 \text{ torr}\left(\frac{1 \text{ atm}}{760 \text{ torr}}\right)\left(\frac{50\%}{100\%}\right) = 1.6 \times 10^{-2} \text{ atm}$$

$$p_{\text{dry air}} = 765 \text{ torr}\left(\frac{1 \text{ atm}}{760 \text{ torr}}\right) - 1.6 \times 10^{-2} \text{ atm} = 0.991 \text{ atm}$$

$$p_{N_2} = (0.7808)(0.991 \text{ atm}) = 0.774 \text{ atm}$$

$$p_{O_2} = (0.2095)(0.991 \text{ atm}) = 0.208 \text{ atm}$$

$$p_{Ar} = (9.34 \times 10^{-3})(0.991 \text{ atm}) = 9.26 \times 10^{-3} \text{ atm}$$

$$p_{CO_2} = (3.25 \times 10^{-4})(0.991 \text{ atm}) = 3.22 \times 10^{-4} \text{ atm}$$

$$p_{Ne} = (1.82 \times 10^{-5})(0.991 \text{ atm}) = 1.80 \times 10^{-5} \text{ atm}$$

$$p_{He} = (5.24 \times 10^{-6})(0.991 \text{ atm}) = 5.19 \times 10^{-6} \text{ atm}$$

$$p_{CH_4} = (1.4 \times 10^{-6})(0.991 \text{ atm}) = 1.4 \times 10^{-6} \text{ atm}$$

5.105 This is an empirical formula problem, with P-V-T data added to allow calculation of the molar mass. First, determine the empirical formula from mass percentages. Consider 100 g of the compound, which contains the following amounts:

$$\text{C: } 71.22 \text{ g}\left(\frac{1 \text{ mol}}{12.011 \text{ g}}\right) = 5.930 \text{ mol C} \qquad \text{H: } 14.94\left(\frac{1 \text{ mol}}{1.0079 \text{ g}}\right) = 14.82 \text{ mol H}$$

$$\text{N: } 13.84 \text{ g}\left(\frac{1 \text{ mol}}{14.007 \text{ g}}\right) = 0.9881 \text{ mol N}$$

Divide each by the smallest among them, 0.9881 mol N:

$$\text{C: } 5.930 \text{ mol C}\left(\frac{100 \text{ g}}{0.9881 \text{ mol N}}\right) = 6.001 \text{ mol C/mol N, round to 6 C/N}$$

$$\text{H: } 14.82 \text{ mol H}\left(\frac{100 \text{ g}}{0.9881 \text{ mol N}}\right) = 15.00 \text{ mol H/mol N}$$

Empirical formula: $C_6H_{15}N$, empirical $MM = 101$ g/mol

Next use the *P-V-T* data to determine the actual molar mass:

$$m = 250 \text{ mg}\left(\frac{10^{-3} \text{ g}}{1 \text{ mg}}\right) = 0.250 \text{ g} \qquad P = 435 \text{ torr}\left(\frac{1 \text{ atm}}{760 \text{ torr}}\right) = 0.572 \text{ atm}$$

$$V = 150 \text{ mL}\left(\frac{10^{-3} \text{ L}}{1 \text{ mL}}\right) = 0.150 \text{ L} \qquad T = (150 + 273)\text{K} = 423 \text{ K}$$

$$MM = \frac{mRT}{PV} = \frac{(0.250 \text{ g})(8.206 \times 10^{-2} \frac{\text{L atm}}{\text{mol K}})(423 \text{ K})}{(0.572 \text{ atm})(0.150 \text{ L})} = 101 \text{ g/mol}$$

The molar mass is the same as the empirical molar mass, so the molecular formula is the same as the empirical formula.

6.1   Conservation of energy states that energy is neither created nor destroyed. It can only be transformed from one form into another:
(a) Total energy, which is conserved, is the sum of potential, kinetic, and thermal energy. On the tree, an apple possesses some gravitational potential energy. As an apple falls, that potential energy is converted into kinetic energy of motion.
(b) When an apple hits the ground, the impact transfers energy to molecules in the earth and in the apple. As a result, there is a slight increase in temperature; the kinetic energy of the apple has been converted into thermal energy.

6.3   Speed and kinetic energy are related through the equation, $E_{kinetic} = 1/2\ mu^2$:
$$E_{kinetic} = \frac{(9.1094 \times 10^{-31} kg)(4.55 \times 10^5 m/s)^2}{2}\left(\frac{1\ J}{1\ kg\ m^2 s^{-2}}\right) = 9.43 \times 10^{-20}\ J$$

6.5   Electrical energy can be calculated using Equation 6-1: $E_{electrical} = k\ \dfrac{q_1 q_2}{r}$

When distance is expressed in picometers (1 pm = 10^{-12} m) and charges are in electronic units, the constant in the equation is $k = 2.31 \times 10^{-16}$ J pm. Attractive energies are negative because when close together, the charges have lower potential energy than when far apart. The charges on the ions in this example are +1 and –1:
$$E_{electrical} = \frac{(2.31 \times 10^{-16} J\ pm)(+1)(-1)}{320\ pm} = -7.22 \times 10^{-19}\ J$$

6.7   Energy comes in various forms: radiant, kinetic, potential, and thermal:
(a) Radiant energy is consumed and thermal energy is produced.
(b) Chemical potential energy is consumed and kinetic energy is produced.
(c) Chemical potential energy is consumed and thermal energy is produced.

6.9   Speed and kinetic energy are related through the equation, $u = \sqrt{\dfrac{2E_{kinetic}}{m}}$. The joule has units kg m^2/s^2, so mass must be expressed in kg and u has units m/s:
m (Table 2-1) = 1.6749 x 10^{-27} kg
$$u = \sqrt{\frac{2(3.75 \times 10^{-23}\ J)}{1.6749 \times 10^{-27}\ kg}} = 2.12 \times 10^2\ m/s$$

6.11   Calculate the temperature change resulting from an energy input using C values from Table 6-1 of your textbook and Equation 6-2:

$q = nC\Delta T$ $\qquad\qquad\qquad\qquad\qquad\qquad \Delta T = \dfrac{q}{nC}$

(a) $n = \dfrac{m}{MM} = (10.0 \text{ g})\left(\dfrac{1 \text{ mol}}{26.98 \text{ g}}\right) = 0.3706 \text{ mol Al}$

$\Delta T = \dfrac{25.0 \text{ J}}{(0.3706 \text{ mol})(24.35 \text{ J mol}^{-1} \text{ }^\circ\text{C}^{-1})} = 2.77 \text{ }^\circ\text{C}$

$T_f = T_i + \Delta T = 15.0 + 2.77 = 17.8 \text{ }^\circ\text{C}$

(b) $n = \dfrac{m}{MM} = (25.0 \text{ g})\left(\dfrac{1 \text{ mol}}{26.98 \text{ g}}\right) = 0.9266 \text{ mol Al}$

$\Delta T = \dfrac{25.0 \text{ J}}{(0.9266 \text{ mol})(24.35 \text{ J mol}^{-1} \text{ }^\circ\text{C}^{-1})} = 1.108 \text{ }^\circ\text{C}$

$T_f = T_i + \Delta T = 29.5 + 1.108 = 30.6 \text{ }^\circ\text{C}$

(c) $n = \dfrac{m}{MM} = (25.0 \text{ g})\left(\dfrac{1 \text{ mol}}{107.9 \text{ g}}\right) = 0.2317 \text{ mol Ag}$

$\Delta T = \dfrac{25.0 \text{ J}}{(0.2317 \text{ mol})(25.351 \text{ J mol}^{-1} \text{ }^\circ\text{C}^{-1})} = 4.256 \text{ }^\circ\text{C}$

$T_f = T_i + \Delta T = 29.5 + 4.256 = 33.8 \text{ }^\circ\text{C}$

(d) $n = \dfrac{m}{MM} = (25.0 \text{ g})\left(\dfrac{1 \text{ mol}}{18.02 \text{ g}}\right) = 1.387 \text{ mol H}_2\text{O}$

$\Delta T = \dfrac{25.0 \text{ J}}{(1.387 \text{ mol})(75.291 \text{ J mol}^{-1} \text{ }^\circ\text{C}^{-1})} = 0.2394 \text{ }^\circ\text{C}$

$T_f = T_i + \Delta T = 22.0 + 0.2394 = 22.2 \text{ }^\circ\text{C}$

6.13   Calculate an energy change accompanying a temperature change using C values from Table 6-1 of your textbook and Equation 6-2:

$q = nC\Delta T$ $\qquad\qquad \Delta T = 95.0 - 23.0 = 72.0 \text{ }^\circ\text{C}$

$n_{\text{kettle}} = 1.35 \text{ kg}\left(\dfrac{1000 \text{ g}}{1 \text{ kg}}\right)\left(\dfrac{1 \text{ mol}}{55.85 \text{ g}}\right) = 24.2 \text{ mol Fe}$

$q_{\text{kettle}} = (24.2 \text{ mol})(25.10 \text{ J mol}^{-1} \text{ }^\circ\text{C}^{-1})(72.0 \text{ C}) = 4.37 \text{ x } 10^4 \text{ J}$

$n_{\text{water}} = 2.75 \text{ kg}\left(\dfrac{1000 \text{ g}}{1 \text{ kg}}\right)\left(\dfrac{1 \text{ mol}}{18.02 \text{ g}}\right) = 153 \text{ mol water}$

$q_{\text{water}} = (153 \text{ mol})(75.291 \text{ J mol}^{-1} \text{ }^\circ\text{C}^{-1})(72.0 \text{ }^\circ\text{C}) = 8.29 \text{ x } 10^5 \text{ J}$

6.15   When two objects at different temperatures are placed in contact, energy flows from the warmer to the cooler object until the two objects are at the same temperature. Let that temperature be T. The energy lost by the warmer object equals the energy gained by the cooler object: $q_{\text{cool}} = -q_{\text{warm}}$, and $q = nC\Delta T$.

Water: $n = 37.5 \text{ g}\left(\dfrac{1 \text{ mol}}{18.02 \text{ g}}\right) = 2.081 \text{ mol water}$

$q = (2.081 \text{ mol})(75.351 \text{ J mol}^{-1} \text{ }^\circ\text{C}^{-1})(T - 20.5 \text{ }^\circ\text{C}) = (156.8 \text{ J }^\circ\text{C}^{-1})(T - 20.5 \text{ }^\circ\text{C})$

Coin: $n = (27.4 \text{ g})\left(\dfrac{1 \text{ mol}}{107.9 \text{ g}}\right) = 0.2539 \text{ mol Ag}$

$q = (0.2539 \text{ mol})(25.351 \text{ J mol}^{-1}\,{}^\circ\text{C}^{-1})(T - 100.0\,{}^\circ\text{C}) = (6.437 \text{ J }{}^\circ\text{C}^{-1})(T - 100.0\,{}^\circ\text{C})$
Substitute and solve for T:

$(156.8 \text{ J }{}^\circ\text{C}^{-1})(T - 20.5\,{}^\circ\text{C}) = -(6.437 \text{ J }{}^\circ\text{C}^{-1})(T - 100.0\,{}^\circ\text{C})$
$156.8\,T - 3214.5\,{}^\circ\text{C} = -6.437\,T + 643.7\,{}^\circ\text{C}$
$163.2\,T = 3858.2\,{}^\circ\text{C}$ $T = 23.6\,{}^\circ\text{C}$

6.17   Calculate the energy change for a temperature change using Equation 6-2 and C values from Table 6-1 of your textbook:

$q = nC\Delta T$ $\Delta T = 87.6 - 21.5 = 66.1\,{}^\circ\text{C}$

$n_{\text{water}} = \dfrac{m}{MM} = 475 \text{ mL}\left(\dfrac{1.00 \text{ g}}{1 \text{ mL}}\right)\left(\dfrac{1 \text{ mol}}{18.02 \text{ g}}\right) = 26.36 \text{ mol}$

$q_{\text{water}} = (26.36 \text{ mol})\left(\dfrac{75.291 \text{ J}}{1 \text{ mol }{}^\circ\text{C}}\right)(66.1\,{}^\circ\text{C}) = 1.31 \times 10^5 \text{ J} = 131 \text{ kJ}$

6.19   To work this problem, use data in Chemistry and Life Box. First determine the energy difference (ΔE) between chicken and beef:

Chicken: $250 \text{ g}\left(\dfrac{6.0 \text{ kJ}}{1 \text{ g}}\right) = 1500 \text{ kJ}$ Beef: $250 \text{ g}\left(\dfrac{16 \text{ kJ}}{1 \text{ g}}\right) = 4000 \text{ kJ}$

$\Delta E = 4000 \text{ kJ} - 1500 \text{ kJ} = 2500 \text{ kJ}$

According to data in Chemistry and Life Box, a 55-kg person walking 6.0 km/hr

consumes 1090 kJ/hr: $t = 2500 \text{ kJ}\left(\dfrac{1 \text{ hr}}{1090 \text{ kJ}}\right) = 2.3 \text{ hr}$

Therefore, to consume the additional energy a person must walk:

$distance = 2.3 \text{ hr}\left(\dfrac{6.0 \text{ km}}{1 \text{ hr}}\right) = 14 \text{ km}$

6.21   (a) In a combustion reaction, the products are CO_2 and H_2O:

$$C_7H_6O_2 + O_2 \rightarrow CO_2 + H_2O \text{ (unbalanced)}$$

Follow standard procedures to balance the equation. Give CO_2 a coefficient of 7 to balance C, H_2O a coefficient of 3 to balance H:

$$C_7H_6O_2 + O_2 \rightarrow 7\,CO_2 + 3\,H_2O$$
$$7C + 6H + 4O \rightarrow 7C + 6H + 17O$$

Balance O by giving O_2 a coefficient of 15/2, then multiply by 2 to clear fractions:

$$2\,C_7H_6O_2 + 15\,O_2 \rightarrow 14\,CO_2 + 6\,H_2O$$

(b) Find energy per mole from the energy released by 1.350 g using the molar mass (122.1 g/mol):

$\Delta E \text{ (kJ/mol)} = \left(\dfrac{-35.61 \text{ kJ}}{1.350 \text{ g}}\right)\left(\dfrac{122.1 \text{ g}}{1 \text{ mol}}\right) = -3.221 \times 10^3 \text{ kJ/mol } C_7H_6O_2$

(c) 15 moles O_2 is consumed for each 2 mol of benzoic acid, so the energy released per mol of O_2 is:

$$\Delta E = \left(\frac{-3.221 \times 10^3 \text{kJ}}{1 \text{ mol acid}} \right) \left(\frac{2 \text{ mol acid}}{15 \text{ mol } O_2} \right) = -4.295 \times 10^2 \text{ kJ/mol } O_2$$

6.23   In Table 6-2, we find the following bond energies for H-X bonds where X is an element in Group 16:

Bond	H – O	H – S	H - Te
Energy (kJ/mol)	460	365	240

There is a trend: moving down the column, the bond energy decreases. From this trend, we predict that the H – Se bond has an energy between H – S and H – Te, around 300 kJ/mol (the experimental value is 315 kJ), and the H – Po bond will be weaker than H – Te, perhaps around 200 kJ/mol (no experimental value is available).

6.25   To estimate the energy change in a reaction, list the number of bonds of each type in reactants and products, and subtract the sum of average bond energies for products from the sum of average bond energies for reactants, using values from Table 6-2 of your textbook:

The reaction is $N_2 + 2 O_2 \rightarrow N_2O_4$

$$N\equiv N \quad 2\, O{=}O \longrightarrow$$

Reactants

Bond	No.	Energy(kJ/mol)
N≡N	1	945
O=O	2	495

Products

Bond	No.	Energy(kJ/mol)
N–N	1	160
N–O	2	200
N=O	2	605

$\Delta E_{\text{reaction}} \cong [1 \text{ mol}(945 \text{ kJ/mol}) + 2 \text{ mol}(495 \text{ kJ/mol})]$

$\qquad\qquad - [1 \text{ mol}(160 \text{ kJ/mol}) + 2 \text{ mol}(200 \text{ kJ/mol}) + 2 \text{ mol}(605 \text{ kJ/mol})]$

$\qquad = 1935 \text{ kJ} - 1770 \text{ kJ} \cong 165 \text{ kJ}$

6.27   Figure 6-13 in your textbook shows three paths for a reaction. One is the actual path. A second is decomposition of reactants into elements and then recombination of the elements into products. A third is breakage of all bonds in the reactants into atoms of the elements, then reaction of the atoms to form all bonds in the products:

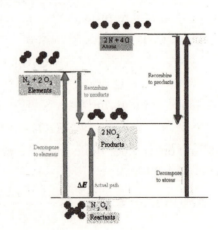

6.29   To estimate the energy change in a reaction, list the number of bonds of each type in the
       reagents, and subtract the sum of average bond energies for products from the sum of
       average bond energies for reactants, using values from Table 9-2 of your textbook:

$$2 H_2 + O_2 \rightarrow 2 H_2O$$

Reactants			**Products**		
Bond	No.	Energy(kJ/mol)	Bond	No.	Energy (kJ/mol)
H–H	2	435	O–H	4	460
O=O	1	495			

$\Delta E_{reaction} \cong$ [2 mol(435 kJ/mol) + 1 mol(495 kJ/mol)] – [4 mol(460 kJ/mol)] = – 475 kJ

$$3 H_2 + N_2 \rightarrow 2 NH_3$$

Reactants			**Products**		
Bond	No.	Energy(kJ/mol)	Bond	No.	Energy(kJ/mol)
H–H	3	435	N–H	6	390
N≡N	1	945			

$\Delta E_{reaction} \cong$ [3 mol(435 kJ/mol) + 1 mol(945 kJ/mol)] – [6 mol(390 kJ/mol)] = –90 kJ

$$3 H_2 + CO \rightarrow CH_4 + H_2O$$

Reactants			**Products**		
Bond	No.	Energy(kJ/mol)	Bond	No.	Energy(kJ/mol)
H–H	3	435	C–H	4	415
C≡O	1	1070	O–H	2	460

$\Delta E_{reaction} \cong$ [3 mol(435 kJ/mol) + 1 mol(1070 kJ/mol)]
– [4 mol(415 kJ/mol) + 2 mol(460 kJ/mol)] = 2375 kJ – 2580 kJ = –205 kJ

6.31   Calculate the heat accompanying a temperature change in a calorimeter using Equation 6-
       6 and the heat capacity C of the calorimeter: $q_{calorimeter} = C_{cal}\Delta T$.

$\Delta T = 23.8 – 21.6 = 2.2$ °C

The problem statement tells us to assume that $C_{cal} \cong C_{water}$:

$$C_{cal} \cong C_{water} = (155 \text{ mL})(1.00 \text{ g/mL}) \left(\frac{1 \text{ mol}}{18.02 \text{ g}} \right)(75.291 \text{ J mol}^{-1} \text{ °C}^{-1}) = 647.6 \text{ J °C}^{-1}$$

$q_{calorimeter} = (647.6 \text{ J °C}^{-1})(2.2 \text{ °C}) = 1.4 \times 10^3 \text{ J}$

The heat gained by the calorimeter is heat lost by the solution process, giving

$q_{solution} = - 1.4 \times 10^3 \text{ J}$

6.33   The heat capacity of a calorimeter can be found from energy and temperature data using
       Equation 6-6, $q_{calorimeter} = C_{cal} \Delta T$.

Here, the heat released by the combustion process is absorbed by the calorimeter:

$$q_{calorimeter} = -q_{glucose} = -1.7500 \text{ g} \left(\frac{-15.57 \text{ kJ}}{1 \text{ g}} \right) = 27.25 \text{ kJ}$$

$\Delta T = 23.34 – 21.45 = 1.89$ °C

$$C_{cal} = \frac{q_{calorimeter}}{\Delta T} = \frac{27.25 \text{ kJ}}{1.89 \text{ °C}} = 14.4 \text{ kJ/°C}$$

6.35   Calculate the energy released during combustion from the temperature increase and the total heat capacity of the calorimeter. Then convert to molar energy using molar mass:

$$q_{calorimeter} = C_{cal} \Delta T = (7.85 \text{ kJ/°C})(29.04 \text{ °C} - 24.65 \text{ °C}) = 34.46 \text{ kJ}$$

$$\Delta E_{reaction} = q_{reaction} = -q_{calorimeter} = -34.46 \text{ kJ}$$

$$MM = 8(12.01) + 10(1.0078) + 4(14.01) + 2(16.00) = 194.2 \text{ g/mol}$$

$$\Delta E_{molar} = \left(\frac{-34.46 \text{ kJ}}{1.35 \text{ g}} \right)\left(\frac{194.2 \text{ g}}{1 \text{ mol}} \right) = -4.96 \times 10^3 \text{ kJ/mol}$$

6.37   Expansion work can be calculated using Equation 6-8, $w_{sys} = -P_{ext}\Delta V_{sys}$. The external pressure opposing inflation of a balloon is atmospheric pressure, 1.00 atm:

$V_f = 2.5 \text{ L}, V_i = 0.0 \text{ L}$

$\Delta V = 2.5 \text{ L} - 0.0 \text{ L} = 2.5 \text{ L}$

$w_{sys} = -P_{ext}\Delta V_{sys} = -(1.00 \text{ atm})(2.5 \text{ L}) = -2.5 \text{ L atm}$

Convert to joules:

$$w_{sys} = -2.5 \text{ L atm}\left(\frac{101.325 \text{ J}}{1 \text{ L atm}} \right) = -2.5 \times 10^2 \text{ J}$$

6.39   Standard enthalpy changes are calculated from Equation 6-12 using standard enthalpies of formation, which can be found in Appendix D of your textbook:

$$\Delta H°_{reaction} = \Sigma \text{ coeff}_p \, \Delta H°_f \text{ (products)} - \Sigma \text{ coeff}_r \, \Delta H°_f \text{ (reactants)}$$

(a) $\Delta H°_{reaction} = [2 \text{ mol}(-393.5 \text{ kJ/mol}) + 2 \text{ mol}(-285.83 \text{ kJ/mol})]$
$$- [1 \text{ mol}(52.4 \text{ kJ/mol}) + 3 \text{ mol}(0 \text{ kJ/mol})] = -1411.1 \text{ kJ}$$

(b) $\Delta H°_{reaction} = 1 \text{ mol } (0 \text{ kJ/mol}) + 3 \text{ mol}(0 \text{ kJ/mol}) - 2 \text{ mol}(-45.9 \text{ kJ/mol}) = 91.8 \text{ kJ}$

(c) $\Delta H°_{reaction} = [1 \text{ mol}(-2984.0 \text{ kJ/mol}) + 5 \text{ mol}(0 \text{kJ/mol})]$
$$- [5 \text{ mol}(-277.4 \text{ kJ/mol}) + 4 \text{ mol}(0 \text{ kJ/mol})] = -1597.0 \text{ kJ}$$

(d) $\Delta H°_{reaction} = [1 \text{ mol}(-910.7 \text{ kJ/mol}) + 4 \text{ mol}(-92.3 \text{ kJ/mol})]$
$$- [1 \text{mol}(-687.0 \text{ kJ/mol}) + 2 \text{ mol}(-285.83 \text{ kJ/mol})] = -21.2 \text{ kJ}$$

6.41   Reaction energies are related to reaction enthalpies (Prob. 6.39) through Equation 6-11:

$$\Delta H_{reaction} \cong \Delta E_{reaction} + \Delta(nRT)_{gases}$$

Since temperature does not change, we can rearrange the last term:

$$\Delta E_{reaction} \cong \Delta H_{reaction} - RT\Delta(n_{gases})$$

Begin by calculating the change in the number of moles of *gases*, then use the rearranged equation to determine $\Delta E_{reaction}$:

(a) $\Delta n_{gases} = 2 - (3 + 1) = -2 \text{ mol};$

$$\Delta E_{reaction} \cong -1411.1 \text{ kJ} - (-2 \text{ mol})\left(\frac{8.314 \text{ J}}{1 \text{ mol K}} \right)(298 \text{ K})\left(\frac{10^{-3} \text{ kJ}}{1 \text{ J}} \right)$$

$$= -1411.1 + 4.96 = -1406.1 \text{ kJ}$$

(b) $\Delta n_{gases} = 3 + 1 - 2 = 2$ mol

$$\Delta E_{reaction} \cong 91.8 \text{ kJ} - (2 \text{ mol})\left(\frac{8.314 \text{ J}}{1 \text{ mol K}}\right)(298 \text{ K})\left(\frac{10^{-3} \text{ kJ}}{1 \text{ J}}\right) = 91.8 - 4.96 = 86.8 \text{ kJ}$$

(c) all compounds are solid, $\Delta n_{gases} = 0$ mol $\Delta E_{reaction} \cong -1597.0$ kJ

(d) $\Delta n_{gases} = 0$ mol $\Delta E_{reaction} \cong -21.2$ kJ

6.43   A formation reaction has elements in their standard states as reactants and 1 mol of a
single substance as the product:

(a) $3 \text{ K } (s) + \text{P } (s) + 2 \text{ O}_2 (g) \rightarrow \text{K}_3\text{PO}_4 (s)$

(b) $2 \text{ C } (grapite) + 2 \text{ H}_2 (g) + \text{O}_2 (g) \rightarrow \text{CH}_3\text{CO}_2\text{H } (l)$

(c) $3 \text{ C } (grapite) + 9/2 \text{ H}_2 (g) + 1/2 \text{ N}_2 (g) \rightarrow (\text{CH}_3)_3\text{N } (g)$

(d) $2 \text{ Al } (s) + 3/2 \text{ O}_2 (g) \rightarrow \text{Al}_2\text{O}_3 (s)$

6.45   Standard enthalpy changes are calculated from Equation 6-12 using standard enthalpies
of formation, which can be found in Appendix D of your textbook:

$$\Delta H^{\circ}_{reaction} = \Sigma \text{ coeff}_p \, \Delta H^{\circ}_f \text{ (products)} - \Sigma \text{ coeff}_r \, \Delta H^{\circ}_f \text{ (reactants)}$$

(a) $\Delta H^{\circ}_{reaction} = [4 \text{ mol}(91.3 \text{ kJ/mol}) + 6 \text{ mol}(-285.83 \text{ kJ/mol})]$
$\qquad\qquad\qquad - [4 \text{ mol}(-45.9 \text{ kJ/mol}) + 5 \text{ mol}(0 \text{ kJ/mol})] = -1166.2$ kJ

(b) $\Delta H^{\circ}_{reaction} = [2 \text{ mol}(0 \text{ kJ/mol}) + 6 \text{ mol}(-285.83 \text{ kJ/mol})]$
$\qquad\qquad\qquad - [4 \text{ mol}(-45.9 \text{ kJ/mol}) + 3 \text{ mol}(0 \text{ kJ/mol})] = -1531.4$ kJ

6.47   Figure 6-21 shows that the energy needs per capita in the computer age are
1×10^6 kJ/day. To determine the annual energy usage of the United States, multiply this
need by the total population and the number of days in a year:
Annual energy usage $= (1 \times 10^6$ kJ/person-day)(365 days/year)(281,421,906 persons)
Annual energy usage $= 1 \times 10^{17}$ kJ

6.49   Use a ruler to measure the energy consumption for each source for the year 1975 on
Figure 6-22. Here are the values:

Source	Amount (10^{18} kJ)	%
Wood	1.7	2.2%
Hydro	3.4	4.3%
Nuclear	2.1	2.7%
Coal	13.8	17.5%
Gas	20.7	26.2%
Petroleum	37.2	47.1%
Total	78.9	100%

Use these values to construct a
pie chart, which appears on the
right.

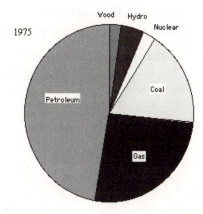

6.51   Because energy is a state function, the energy released (ΔE) when one gram of gasoline burns is the same regardless of the conditions. The energy released can be set equal to the work done (w) plus the heat transferred (q). Work is done when an automobile accelerates, but no work is done when an automobile idles:

Heat released (idle) = Heat released (acceleration) + Work done (acceleration)

Thus, more heat must be removed under idling conditions (this is part of the reason why automobiles tend to overheat in traffic jams).

6.53   The difference between $\Delta H_{reaction}$ and $\Delta E_{reaction}$ can be estimated using Equation 6-11:

$$\Delta H_{reaction} \cong \Delta E_{reaction} + \Delta(nRT)_{gases} \qquad \Delta H_{reaction} - \Delta E_{reaction} \cong \Delta(nRT)_{(gases)}$$

The reaction is $C\ (graphite) + H_2O\ (l) \rightarrow CO\ (g) + H_2\ (g)$

$\Delta n_{gases} = 1\ mol + 1\ mol - 0\ mol = 2\ mol\ gases$

$$\Delta H_{reaction} - \Delta E_{reaction} \cong (2\ mol)\left(\frac{8.314\ J}{1\ mol\ K}\right)(298\ K) = 4.96 \times 10^3\ J$$

6.55   To work a problem involving heat transfers, it is useful to set up a block diagram illustrating the process. In this problem, a copper block transfers energy to ice:

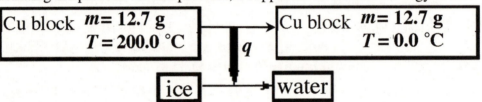

Thus, $q_{ice} = -q_{Cu}$

$q_{Cu} = n_{Cu}\,C\Delta T$ \qquad\qquad $q_{ice} = n_{ice}\Delta H_{fus}$

Substituting gives:

$$n_{ice}\,\Delta H_{fus} = -n_{Cu}\,C\Delta T$$

Here are the data needed for the calculation:

$$n_{Cu} = \frac{m}{MM} = 12.7\ g\left(\frac{1\ mol}{63.546\ g}\right) = 0.1999\ mol$$

$C = 24.435\ J/mol\ °C$

$\Delta T = (0.0\ °C - 200.0\ °C) = -200\ °C$

$\Delta H_f = 6.01\ kJ/mol = 6.01 \times 10^3\ J/mol$

Substitute and solve for the amount of ice that melts:

$$n_{ice} = \frac{-n_{Cu}C\Delta T}{\Delta H_f} = -\frac{(0.1999\ mol)(24.435\ J\ mol^{-1}\ °C^{-1})(-200\ °C)}{6.01 \times 10^3 J\ mol^{-1}} = 0.1625\ mol$$

Finally, convert to mass:

$$m_{ice} = n\ MM = 0.1625\ mol\left(\frac{18.02\ g}{1\ mol}\right) = 2.93\ g\ of\ ice\ melts$$

6.57   (a) In a combustion reaction, a substance reacts with O_2 to form CO_2 and H_2O:

$$C_6H_{12}O_6 + O_2 \rightarrow CO_2 + H_2O$$

Balance the equation using standard procedures. Give CO_2 a coefficient of 6 and H_2O a coefficient of 6 to balance C and H:

$$C_6H_{12}O_6 + O_2 \rightarrow 6\ CO_2 + 6\ H_2O$$

$$6C + 12H + 8\ O \rightarrow 6C + 12H + 18O$$

Give O_2 a coefficient of 6 to balance O:

$$C_6H_{12}O_6(s) + 6\ O_2\ (g) \rightarrow 6\ CO_2\ (g) + 6\ H_2O\ (l)$$

(b) To determine the molar heat of combustion, multiply the heat released (negative, indicating an exothermic reaction) in burning one gram by the molar mass:

$$\Delta H_{molar} = \left(\frac{-15.7\ kJ}{1\ g}\right)\left(\frac{180.15\ g}{1\ mol}\right) = -2.83 \times 10^3\ kJ/mol = \Delta H^o_{reaction}$$

(c) Use the molar heat of combustion along with Equation 6-12 and data from Appendix D to determine the heat of formation of glucose:

$$\Delta H^o_{reaction} = \sum coeff_p\ \Delta H^o_f\ (products) - \sum coeff_r\ \Delta H^o_f\ (reactants)$$

$$-2.83 \times 10^3\ kJ/mol = [6\ mol(-393.5\ kJ/mol) + 6\ mol(-285.83\ kJ/mol)]$$

$$- [6\ mol(0\ kJ/mol) + \Delta H^o_f\ (glucose)]$$

$$\Delta H^o_f\ (glucose) = -1.25 \times 10^3\ kJ/mol$$

6.59   To predict stability from bond energies, tabulate the number of bonds of each type and add up their contributions. Expand the structures in order to count bonds of each type:

Ethanol: Diethyl ether:

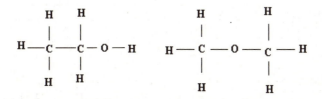

5 C–H bonds, 1 C–O bond   6 C–H bonds, 2 C–O bonds
1 O–H bond, 1 C–C bond

We can ignore the 5 C–H bonds and 1 C–O bond that the two compounds have in common, leaving O–H and C–C in ethanol vs. C–H and C–O in diethyl ether:

Ethanol: 460 kJ/mol (O–H) + 345 kJ/mol (C–C) = 805 kJ/mol

Diethyl ether: 415 kJ/mol (C–H) + 360 kJ/mol (C–O) = 775 kJ/mol

Based on average bond energies, ethanol is more stable by 30 kJ/mol.

6.61   When an airplane accelerates down a runway, its engines burn fuel, converting chemical energy into kinetic energy of motion. During takeoff, some kinetic energy is converted into gravitational potential energy. The net transformation is conversion of stored chemical energy of the airplane's fuel into kinetic energy and gravitational potential energy of the airplane (much of the chemical energy also is "wasted" as heat that is transferred to the air passing through the engines).

6.63   A molar heat capacity can be calculated from a temperature change using $q = n\,C\Delta T$:

$\Delta T = 47.6\,°C - 28.0\,°C = 19.6\,°C$ $n = 52.5\ g\left(\dfrac{1\ mol}{207.2\ g}\right) = 0.2534\ mol$

$C = \dfrac{q}{n\Delta T} = \dfrac{(100.0\ J)}{(0.2534\ mol)(19.6\,°C)} = 20.1\ J/mol\ K$

6.65   A calculation using average bond energies requires a list of all bonds in the reagents. To obtain an estimate for reaction energy per mole of material, use Equation 6-5:

$\Delta E \cong \Sigma BE_{reactants} - \Sigma BE_{products}$:

Ethane reagents:	H_3C-CH_3	$7/2\ O_2$	$2\ CO_2$	$3\ H_2O$
Bonds:	1 C–C, 6 C–H	7/2 (1 O=O)	2 (2 C=O)	3 (2 O–H)
Energies (kJ/mol):	345, 415	495	800	460

$\Delta E \cong [1(345\ kJ/mol) + 6(415\ kJ/mol) + 7/2(495\ kJ/mol)]$

$\qquad\qquad - [4(800\ kJ/mol) + 6(460\ kJ/mol)] \cong -1392\ kJ/mol$

$\Delta E(\text{per gram}) = \dfrac{\Delta E}{MM} = \left(\dfrac{-1392\ kJ}{1\ mol}\right)\left(\dfrac{1\ mol}{30.08\ g}\right) = -46.3\ kJ/g$

Ethylene reagents:	$H_2C=CH_2$	$3\ O_2$	$2\ CO_2$	$2\ H_2O$
Bonds:	1 C=C, 4 C–H	3 (1 O=O)	2 (2 C=O)	2 (2 O–H)
Energies (kJ/mol):	615, 415	495	800	460

$\Delta E \cong [1(615\ kJ/mol) + 4(415\ kJ/mol) + 3(495\ kJ/mol)] - [4(800\ kJ/mol) + 4(460\ kJ/mol)]$

$\Delta E \cong -1280\ kJ/mol$

$\Delta E(\text{per gram}) = \dfrac{\Delta E}{MM} = \left(\dfrac{-1280\ kJ}{1\ mol}\right)\left(\dfrac{1\ mol}{28.06\ g}\right) = -45.6\ kJ/g$

Acetylene reagents:	$HC\equiv CH$	$5/2\ O_2$	$2\ CO_2$	H_2O
Bonds:	1 C≡C, 2 C–H	5/2 (1 O=O)	2 (2 C=O)	2 O–H
Energies (kJ/mol):	835, 415	495	800	460

$\Delta E \cong [1(835\ kJ/mol) + 2(415\ kJ/mol) + 2.5(495\ kJ/mol)]$

$\qquad\qquad - [4(800\ kJ/mol) + 2(460\ kJ/mol)] = -1217\ kJ/mol$

$\Delta E(\text{per gram}) = \dfrac{\Delta E}{MM} = \left(\dfrac{-1217\ kJ}{1\ mol}\right)\left(\dfrac{1\ mol}{26.04\ g}\right) = -46.7\ kJ/g$

These estimates indicate that acetylene releases slightly more energy per unit mass than ethane or ethylene, but experimental values show that ethane releases the most. Remember that average bond energy calculations provide estimates, not exact values.

6.67   Kinetic energy is given by $E_{kinetic} = mu^2/2$, expressed in SI units:

$2250\ lb\left(\dfrac{0.45359\ kg}{1\ lb}\right) = 1.021\ x\ 10^3\ kg$

$\left(\dfrac{57.5\ mi}{1\ hr}\right)\left(\dfrac{1.609344\ km}{1\ mi}\right)\left(\dfrac{10^3\ m}{1\ km}\right)\left(\dfrac{1\ hr}{60\ min}\right)\left(\dfrac{1\ min}{60\ s}\right) = 25.7\ m/s$

$$E_{kinetic} = \frac{\left(1.021 \times 10^3 \text{ kg}\right)(25.7 \text{ m/s})^2}{2} = 3.37 \times 10^5 \text{ kg m}^2/\text{s}^2 = 3.37 \times 10^5 \text{ J}$$

6.69   Standard enthalpy changes are calculated from Equation 6-12 using standard enthalpies of formation, which can be found in Appendix D of your textbook:

$$\Delta H^o_{reaction} = \Sigma \text{ coeff}_p \Delta H^o_f \text{ (products)} - \Sigma \text{ coeff}_r \Delta H^o_f \text{ (reactants)}$$

(a) $\Delta H^o_{reaction}$ = 2 mol(–395.7 kJ/mol)

$$- [2 \text{ mol}(–296.8 \text{ kJ/mol}) + 1 \text{ mol}(0 \text{ kJ/mol})] = –197.8 \text{ kJ}$$

(b) $\Delta H^o_{reaction}$ = 1 mol(11.1 kJ/mol) – 2 mol(33.2 kJ/mol) = –55.3 kJ

(c) $\Delta H^o_{reaction}$ = 1 mol(–1675.7 kJ/mol) + 2 mol(0 kJ/mol)

$$- [1 \text{ mol}(–824.2 \text{ kJ/mol}) + 2 \text{ mol}(0 \text{ kJ/mol})] = –851.5 \text{ kJ}$$

6.71   To estimate the energy change in a reaction, determine Lewis structures to obtain types of bonds, list the number of bonds of each type in reactants and products, and subtract the sum of average bond energies for products from the sum of average bond energies for reactants, using values from Table 6-2 of your textbook.

$$\text{H–Cl} + \text{H}_2\text{C=CH}_2 \rightarrow \text{H}_3\text{C–CH}_2\text{Cl}$$

Reactants			Products		
Bond	No.	Energy (kJ/mol)	Bond	No.	Energy (kJ/mol)
H–Cl	1	430	C–Cl	1	330
C–H	4	415	C–H	5	415
C=C	1	615	C–C	1	345

$\Delta E_{reaction} \cong$ [1 mol(430 kJ/mol) + 4 mol(415 kJ/mol) + 1 mol(615 kJ/mol)]

$\qquad$ – [1 mol(330 kJ/mol) + 5 mol(415 kJ/mol) + 1 mol(345 kJ/mol)] = – 45 kJ

$$\text{Cl}_2 + \text{H}_2\text{C=CH}_2 \rightarrow \text{ClH}_2\text{C–CH}_2\text{Cl}$$

Reactants			Products		
Bond	No.	Energy (kJ/mol)	Bond	No.	Energy (kJ/mol)
Cl–Cl	1	240	C–Cl	2	330
C–H	4	415	C–H	4	415
C=C	1	615	C–C	1	345

$\Delta E_{reaction} \cong$ [1 mol(240 kJ/mol) + 4 mol(415 kJ/mol) + 1 mol(615 kJ/mol)]

$\qquad$ – [2 mol(330 kJ/mol) + 4 mol(415 kJ/mol) + 1 mol(345 kJ/mol)] = – 150 kJ

6.73   An enthalpy of formation can be calculated from the heat of a reaction provided all other enthalpies of formation are known. For this reaction, formation enthalpies of three of the four reagents appear in Appendix D of your textbook:

$$\Delta H^o_{reaction} = \Sigma \text{ coeff}_p \Delta H^o_f \text{ (products)} - \Sigma \text{ coeff}_r \Delta H^o_f \text{ (reactants)}$$

–3677 kJ = 1 mol(–2984.0 kJ/mol) + 3 mol(–296.8 kJ/mol)

$$- [\Delta H^o_f (\text{P}_4\text{S}_3) + 8 \text{ mol}(0 \text{ kJ/mol})]$$

$\Delta H^o_f (\text{P}_4\text{S}_3)$ = 3677 kJ –2984.0 kJ – 890.4 kJ = –197 kJ

6.75   To work a problem involving heat transfers, it is useful to set up a block diagram illustrating the process. In this problem, the unknown metal transfers energy to water:

44.0 g metal $T = 100.0\,°C$	→	44.0 g metal $T = 28.4\,°C$
	q	
80.0 g water $T = 24.8\,°C$	→	80.0 g water $T = 28.4\,°C$

Thus, $q_{metal} = -q_{water} = -(nC\Delta T)_{water}$

$$n_{water} = \frac{m}{MM} = \left(\frac{80.0\ g}{18.02\ g/mol}\right) = 4.44\ mol\ water$$

$C = 75.291\ J/mol\,°C$ $\qquad\qquad$ $\Delta T = (28.4 - 24.8) = 3.6\,°C$

$$q_{metal} = -q_{water} = -(4.44\ mol)\left(\frac{75.291\ J}{1\ mol\ °C}\right)(3.6\,°C) = -1203\ J$$

We can use $q = mC\Delta T$ and the mass of the metal to find its heat capacity:

$\Delta T = 28.4 - 100.0 = -71.6\,°C$

$$C = \frac{q}{m\Delta T} = \frac{(-1203\ J)}{(44.0\ g)(-71.6\,°C)} = 0.382\ J/g\,°C$$

This matches the heat capacity of Cu (0.385 J/g °C), so the metal is copper.

6.77   The enthalpy of a reaction can be calculated from Equation 6-12 and standard heats of formation (see Appendix D):

$$\Delta H_{reaction} = \sum coeff_p\ \Delta H°_f\ (products) - \sum coeff_r\ \Delta H°_f\ (reactants)$$

(a) The question asks for ΔH when 1 mol of hydrazine burns:

$$N_2H_4\ (l) + 1/2\ N_2O_4\ (g) \rightarrow 3/2\ N_2\ (g) + 2\ H_2O\ (g)$$

$\Delta H_{reaction} = [1.5\ mol(0\ kJ/mol) + 2\ mol(-241.83\ kJ/mol)]$

$\qquad\qquad - [1\ mol(50.6\ kJ/mol) + 0.5\ mol(11.1\ kJ/mol)] = -539.8\ kJ/mol$

(b) When hydrazine burns in oxygen, the reaction is:

$$N_2H_4\ (l) + O_2\ (g) \rightarrow N_2\ (g) + 2\ H_2O\ (g)$$

$\Delta H_{reaction} = [1\ mol(0\ kJ/mol) + 2\ mol(-241.83\ kJ/mol)]$

$\qquad\qquad - [1\ mol(50.6\ kJ/mol) + 1\ mol(0\ kJ/mol)] = -534.3\ kJ/mol$

The reaction with N_2H_4 is slightly more exothermic because that compound is slightly less stable than the elements from which it forms. (N_2H_4 is used in preference to O_2 because the reaction occurs on contact, while O_2 requires an ignition device.)

6.79   Consult Figure 6-22 to find information about energy sources in various years.
(a) Before 1850, there was only one major energy source, wood.
(b) Before 1980, natural gas production peaked in about 1972, at 23×10^{18} kJ/yr.
(c) To find the percentage due to fossil fuels, first measure the amounts from each source, then take the appropriate fraction. Here are the amounts:

Source	Amount (10^{18} kJ)
Wood	1.8
Hydro	1.8
Nuclear	0.0
Coal	13.9
Gas	7.3
Petroleum	16.4
Total	41.2

$$\text{Fossil fuel \%} = (100\%)\left(\frac{13.9 + 7.3 + 16.4}{41.2}\right) = 91.2\ \%$$

6.81   The process shown is condensation of a gas to form a solid, the reverse of sublimation. Sublimation always is endothermic, as energy must be provided to overcome intermolecular forces of attraction; thus the depicted process is exothermic.

(a) ΔH_{sys} is negative (exothermic process)

(b) ΔE_{surr} is positive (surroundings must absorb the energy released in the process)

(c) $\Delta E_{universe} = 0$ (total energy always is conserved)

6.83   To work a problem involving heat transfers, it is useful to set up a block diagram illustrating the process. In this problem, a rhodium block transfers energy to ice:

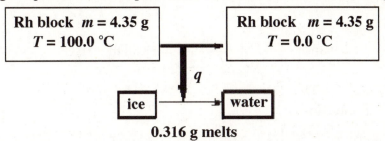

Thus, $q_{ice} = -q_{Rh}$ $q_{Rh} = n_{Rh}\,C\Delta T$, and $q_{ice} = n_{ice}\Delta H_{fus}$

Substituting gives $n_{ice}\Delta H_f = -n_{Rh}\,C\Delta T$

Solve for C, the heat capacity of rhodium:

$$C = \frac{-n_{ice}\Delta H_{fus}}{n_{Rh}\Delta T}$$

Here are the data needed for the calculation:

$$n_{Rh} = \frac{m}{MM} = \frac{4.35\ \text{g}}{102.91\ \text{g/mol}} = 0.04227\ \text{mol} \qquad n_{ice} = \frac{m}{MM} = \frac{0.319\ \text{g}}{18.01\ \text{g/mol}} = 0.01771\ \text{mol}$$

$\Delta T = (0.0\ ^\circ\text{C} - 100.0\ ^\circ\text{C}) = -100.0\ ^\circ\text{C}$

$\Delta H_f = 6.01\ \text{kJ/mol} = 6.01 \times 10^3\ \text{J/mol}$

Substitute and evaluate C:

$$C = \frac{-n_{ice}\Delta H_{fus}}{n_{Rh}\Delta T} = \frac{-(0.01771\ \text{mol})\left(6.01 \times 10^3\ \text{J/mol}\right)}{(0.04227\ \text{mol})(-100.0\ ^\circ\text{C})} = 25.2\ \text{J/mol}\ ^\circ\text{C}$$

6.85   Data in Problem 6.57 indicate that glucose combustion releases energy amounting to 15.7 kJ/g. First determine how much glucose would be required at 100% efficiency, then correct for inefficiency and sugar content:

$$\text{Mass required} = 220 \text{ kJ}\left(\frac{1 \text{ g glucose}}{15.7 \text{ kJ}}\right)\left(\frac{100\%}{30\%}\right)\left(\frac{100 \text{ g cereal}}{35 \text{ g glucose}}\right) = 133 \text{ g cereal}$$

6.87   To work a problem involving heat transfers, it is useful to set up a block diagram illustrating the process. In this problem, a silver spoon absorbs energy from coffee (water):

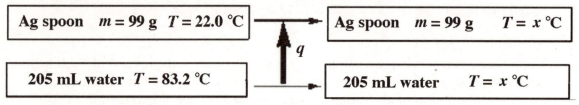

Thus, $q_{\text{water}} = -q_{\text{Ag}}$

$q_{\text{Ag}} = (nC_{\text{Ag}}\Delta T)_{\text{Ag}}$ $\qquad\qquad\qquad$ $q_{\text{water}} = (nC_{H_2O}\Delta T)_{\text{water}}$

For Ag:

$$n = \frac{m}{MM} = 99 \text{ g}\left(\frac{1 \text{ mol}}{107.9 \text{ g}}\right) = 0.918 \text{ mol Ag}$$

$C_{\text{Ag}} = 25.351 \text{ J/mol °C}$ $\qquad\qquad$ $\Delta T = (x - 22.0) \text{ °C}$

For water:

$$n = \frac{m}{MM} = 205 \text{ mL}\left(\frac{1.00 \text{ g}}{1 \text{ mL}}\right)\left(\frac{1 \text{ mol}}{18.02 \text{ g}}\right) = 11.38 \text{ mol } H_2O$$

$C_{H_2O} = 75.291 \text{ J/mol °C}$ $\qquad\qquad$ $\Delta T = (x - 83.2) \text{ °C}$

Substitute and solve for x:

$$(0.918 \text{ mol})\left(\frac{25.351 \text{ J}}{1 \text{ mol K}}\right)(x - 22.0) \text{ K} = -(11.38 \text{ mol})\left(\frac{75.291 \text{ J}}{1 \text{ mol K}}\right)(x - 83.2) \text{ K}$$

$23.27 x - 512 = -856.8 x + 71294$

$880.1 x = 71806$ $\qquad\qquad\qquad\qquad$ $x = 81.6 \text{ °C}$

The final temperature of the coffee is 81.6 ° (not a very efficient way of cooling!).
To determine whether an Al spoon would be more or less effective, compare nC for the two spoons:

$$\text{Al: } = 99 \text{ g}\left(\frac{1 \text{ mol}}{26.982 \text{ g}}\right)\left(\frac{24.35 \text{ J}}{1 \text{ mol °C}}\right) = 89 \text{ J/°C}$$

$$\text{Ag: } = 99 \text{ g}\left(\frac{1 \text{ mol}}{107.9 \text{ g}}\right)\left(\frac{25.351 \text{ J}}{1 \text{ mol °C}}\right) = 23 \text{ J/°C}$$

An Al spoon requires more energy per degree temperature increase, so an Al spoon would be more effective in cooling the coffee (but because the density of Al is smaller than that of Ag, an aluminum spoon *of the same mass* would be quite a bit larger in size).

6.89   The expression that always is correct for q is (e) $q = \Delta E - w$, which is derived from the
first law of thermodynamics. For the others:
(a) $\Delta E \neq q$ when volume changes   (b) $\Delta H \neq q$ when pressure changes
(c) $q_v \neq q$ when volume changes   (d) $q_p \neq q$ when pressure changes

6.91   (a) Work done on the system compresses it, so the new figure should show a smaller
volume (see figure below).
(b) An exothermic reaction increases the temperature of the system, leading to expansion
and a larger volume:

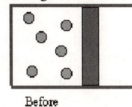 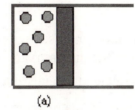 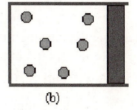

Before   (a)   (b)

6.93   To work a problem involving heat transfers, it is useful to set up a block diagram
illustrating the process. In this problem, a copper block transfers energy to water:

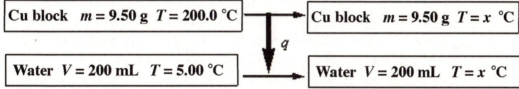

| Cu block $m = 9.50$ g $T = 200.0$ °C | → | Cu block $m = 9.50$ g $T = x$ °C |

q

| Water $V = 200$ mL $T = 5.00$ °C | → | Water $V = 200$ mL $T = x$ °C |

Thus, $q_{water} = -q_{Cu}$ $q_{Cu} = (nC\Delta T)_{Cu}$, and $q_{water} = (nC\Delta T)_{water}$

For Cu, $n = \dfrac{m}{MM} = 9.50$ g $\left(\dfrac{1\ mol}{63.546\ g}\right) = 0.1495$ mol Cu

$C = 24.435$ J/mol °C $\Delta T = (x - 200.0)$ °C

For water, $n = \dfrac{m}{MM} = 200$ mL $\left(\dfrac{1.00\ g}{1\ mL}\right)\left(\dfrac{1\ mol}{18.02\ g}\right) = 11.10$ mol water

$C = 75.291$ J/mol °C $\Delta T = (x - 5.00)$ °C

Substitute and solve for x:

$(0.1495\ mol)\left(\dfrac{24.435\ J}{1\ mol\ ^\circ C}\right)(x - 200.0)\ ^\circ C = -(11.10\ mol)\left(\dfrac{75.291\ J}{1\ mol\ ^\circ C}\right)(x - 5.00)\ ^\circ C$

$3.653\ x - 730.6 = -835.7\ x + 4179$ $839.4\ x = 4910$
$x = 5.85$ °C

6.95   This problem asks how long it will take to heat a sample at a given rate of energy input.
To determine the time, we must first determine how much heat is required to heat the
soup, which can be done using Equation 6-2:

$$q_{water} = (nC\Delta T)_{water}$$

Then determine the time from the power rating of the microwave oven, which is a rate in
joules/second:

Power = energy/time   *so*   time = energy/power

Here are the data for the soup (same as pure water):

$$n = \frac{m}{MM} = 575 \text{ mL} \left(\frac{1.00 \text{ g}}{1 \text{ mL}}\right)\left(\frac{1 \text{ mol}}{18.02 \text{ g}}\right) = 31.91 \text{ mol water}$$

$C = 75.291 \text{ J/mol °C}$ $\Delta T = (85 - 25) = 60. \text{ °C}$

$q = (31.91 \text{ mol})(75.291 \text{ J/mol °C})(60. \text{ °C}) = 1.44 \times 10^5 \text{ J}$

The microwave oven supplies heat at a rate of 750 watts = 750 J/s

time = energy/power = $(1.44 \times 10^5 \text{ J})/(750 \text{ J/s}) = 192 \text{ s}$

Round to two significant figures, because ΔT is only known to two significant figures:

$t = 1.9 \times 10^2 \text{ s}$

6.97   To determine the missing information on this diagram, we must recognize that the starting materials (at the bottom of the diagram) are elements in their standard states. Therefore ΔH_1^o and ΔH_2^o are related to standard heats of formation.

For ΔH_1^o, 3 O_2 are simple spectators; the reaction is $2 N_2 + O_2 \rightarrow 2 N_2O$, for which

$\Delta H_1^o = 2 [\Delta H_f^o (N_2O)] = (2 \text{ mol})(81.6 \text{ kJ/mol}) = 163.2 \text{ kJ}$

We have a value for ΔH_2^o, but we need to identify the products. Consult Appendix D for ΔH_f^o values of nitrogen oxides and find a combination that adds to give 77.5 kJ. Values larger than 77.3 kJ/mol can be excluded, leaving these possibilities:

NO_2: 33.2 kJ/mol N_2O_4: 11.1 kJ/mol N_2O_5: 13.3 kJ/mol

By inspection, we see that 77.5 kJ = (2 mol)(33.2 kJ/mol) + (1 mol)(11.1 kJ/mol). Thus, the products are $2 NO_2 + N_2O_4$, using up all four N atoms and all eight O atoms.

From the diagram, we can see that $\Delta H_2^o + \Delta H_3^o = \Delta H_1^o$, from which

$\Delta H_3^o = \Delta H_1^o - \Delta H_2^o = 163.2 - 77.5 = 85.7 \text{ kJ}$.

6.99   (a) Knowing ΔH^o, use Equation 6-11 to determine ΔE:

$\Delta H^o_{reaction} \cong \Delta E^o_{reaction} + RT\Delta n_{gases}$ $\Delta E^o_{reaction} \cong \Delta H^o_{reaction} - RT\Delta n_{gases}$

In this reaction, $\Delta n_{gases} = 1 \text{ mol}$

$\Delta E^o_{reaction} \cong 39 \text{ kJ} - (8.314 \text{ J mol}^{-1} \text{ K}^{-1})(15 + 273 \text{ K})(1 \text{ mol})(10^{-3} \text{ kJ/J}) = 37 \text{ kJ}$

(b) Use Equation 6-12 to evaluate the standard enthalpy of formation of the anion:

$\Delta H^o_{reaction} = \Sigma \text{ coeff}_p \Delta H^o_f (\text{products}) - \Sigma \text{ coeff}_r \Delta H^o_f (\text{reactants})$

39 kJ = [(1 mol)(-1207.6 kJ/mol) + (1 mol)(-393.5 kJ/mol)

+ (1 mol)(-285.83 kJ/mol)] − [(1 mol)(-543.0 kJ/mol) + (2 mol)(ΔH_f^o of HCO_3^-)]

39 kJ = -1886.93 kJ - [-543.0 kJ + (2 mol)(ΔH_f^o of HCO_3^-)]

(2 mol)(ΔH_f^o of HCO_3^-) = −1383 kJ ΔH_f^o of $HCO_3^- = −692 \text{ kJ/mol}$

7.1   Calculate the molar volume of each metal from its density and molar mass, then convert to atomic volume using Avogadro's number. Then estimate atomic diameter from the atomic volume by assuming that each atom occupies a cubical volume:

(a) $V_{atom} = \dfrac{[MM\,(\text{g/mol})][10^{-3}\ \text{kg/g}]}{[\rho(\text{kg/m}^3)][N_A]}$

(b) $d \cong (V)^{1/3}$

(c) Layer thickness = (# of atoms)(atomic diameter)

Ag: $V_{atom} = \left(\dfrac{107.868\ \text{g}}{1\ \text{mol}}\right)\left(\dfrac{10^{-3}\ \text{kg}}{1\ \text{g}}\right)\left(\dfrac{1\ \text{m}^3}{1.050 \times 10^4\ \text{kg}}\right)\left(\dfrac{1\ \text{mol}}{6.022 \times 10^{23}\ \text{atom}}\right)$

$= 1.706 \times 10^{-29}\ \text{m}^3/\text{atom}$

$d \cong (V)^{1/3} = (1.706 \times 10^{-29}\ \text{m}^3)^{1/3} = 2.574 \times 10^{-10}\ \text{m}$

Thickness = $(6.5 \times 10^6\ \text{atoms})(2.574 \times 10^{-10}\ \text{m}) = 1.7 \times 10^{-3}\ \text{m}$

Pb: $V_{atom} = \left(\dfrac{207.2\ \text{g}}{1\ \text{mol}}\right)\left(\dfrac{10^{-3}\ \text{kg}}{1\ \text{g}}\right)\left(\dfrac{1\ \text{m}^3}{1.134 \times 10^4\ \text{kg}}\right)\left(\dfrac{1\ \text{mol}}{6.022 \times 10^{23}\ \text{atom}}\right)$

$= 3.034 \times 10^{-29}\ \text{m}^3/\text{atom}$

$d \cong (V)^{1/3} = (3.034 \times 10^{-29}\ \text{m}^3)^{1/3} = 3.119 \times 10^{-10}\ \text{m}$

Thickness = $(6.5 \times 10^6\ \text{atoms})(3.119 \times 10^{-10}\ \text{m}) = 2.0 \times 10^{-3}\ \text{m}$

7.3   Gas pressure results from transfer of momentum between atoms and container walls, indicating that atoms have momentum and mass; mass spectrometers sort atomic and molecular ions according to their mass-to-charge ratios; all matter is made up of atoms and has mass.

7.5   At the atomic level, a layer of metal atoms appears as spheres nestled closely together:

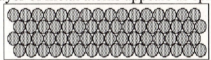

7.7   Use the equation, $v = \dfrac{c}{\lambda}$ to do these calculations:

(a) $v = \left(\dfrac{2.998 \times 10^8\ \text{m/s}}{4.33\ \text{nm}}\right)\left(\dfrac{1\ \text{nm}}{10^{-9}\ \text{m}}\right) = 6.92 \times 10^{16}\ \text{Hz}$

(b) $v = \dfrac{2.998 \times 10^8\ \text{m/s}}{2.35 \times 10^{-10}\ \text{m}} = 1.28 \times 10^{18}\ \text{Hz}$

(c) $v = \left(\dfrac{2.998 \times 10^8\ \text{m/s}}{735\ \text{mm}}\right)\left(\dfrac{10^3\ \text{mm}}{1\ \text{m}}\right) = 4.08 \times 10^8\ \text{Hz}$

(d) $v = \left(\dfrac{2.998 \times 10^8\ \text{m/s}}{4.57\ \mu\text{m}}\right)\left(\dfrac{10^6\ \mu\text{m}}{1\ \text{m}}\right) = 6.56 \times 10^{13}\ \text{Hz}$

7.9   Use the equation, $\lambda = \dfrac{c}{v}$ to do these calculations:

(a) $\lambda = \left(\dfrac{2.998 \times 10^8 \text{ m/s}}{4.77 \text{ GHz}} \right)\left(\dfrac{1 \text{ GHz}}{10^9 \text{ Hz}} \right) = 6.29 \times 10^{-2} \text{ m}$

(b) $\lambda = \left(\dfrac{2.998 \times 10^8 \text{ m/s}}{28.9 \text{ kHz}} \right)\left(\dfrac{1 \text{ kHz}}{10^3 \text{ Hz}} \right)\left(\dfrac{10^2 \text{ cm}}{1 \text{ m}} \right) = 1.04 \times 10^6 \text{ cm}$

(c) $\lambda = \left(\dfrac{2.998 \times 10^8 \text{ m/s}}{60. \text{ Hz}} \right)\left(\dfrac{10^3 \text{ mm}}{1 \text{ m}} \right) = 5.0 \times 10^9 \text{ mm}$

(d) $\lambda = \left(\dfrac{2.998 \times 10^8 \text{ m/s}}{2.88 \text{ MHz}} \right)\left(\dfrac{10^6 \mu m}{1 \text{ m}} \right)\left(\dfrac{1 \text{ MHz}}{10^6 \text{ Hz}} \right) = 1.04 \times 10^8 \ \mu m$

7.11   The energy of one photon of light is described by the equation, $E = hv = \dfrac{hc}{\lambda}$. To obtain energy per mol of photons, multiply by Avogadro's number. Begin by converting the wavelength into meters and then determine the energy:

(a) $\lambda = 490.6 \text{ nm}\left(\dfrac{10^{-9} \text{ m}}{1 \text{ nm}} \right) = 4.906 \times 10^{-7} \text{ m}$

$E_{photon} = \dfrac{(6.626 \times 10^{-34} \text{ J s})(2.998 \times 10^8 \text{ m/s})}{4.906 \times 10^{-7} \text{ m}} = 4.049 \times 10^{-19} \text{ J/photon}$

$E_{molar} = 4.049 \times 10^{-19} \text{ J} \left(\dfrac{6.022 \times 10^{23} \text{ photons}}{1 \text{ mol}} \right)\left(\dfrac{10^{-3} \text{ kJ}}{1 \text{ J}} \right) = 2.438 \times 10^2 \text{ kJ/mol}$

(b) $\lambda = 25.5 \text{ nm}\left(\dfrac{10^{-9} \text{ m}}{1 \text{ nm}} \right) = 2.55 \times 10^{-8} \text{ m}$

$E_{photon} = \dfrac{(6.626 \times 10^{-34} \text{ J s})(2.998 \times 10^8 \text{ m/s})}{2.55 \times 10^{-8} \text{ m}} = 7.79 \times 10^{-18} \text{ J/photon}$

$E_{molar} = 7.79 \times 10^{-18} \text{ J} \left(\dfrac{6.022 \times 10^{23} \text{ photons}}{1 \text{ mol}} \right)\left(\dfrac{10^{-3} \text{ kJ}}{1 \text{ J}} \right) = 4.69 \times 10^3 \text{ kJ/mol}$

(c) $E_{photon} = (6.62608 \times 10^{-34} \text{ J s})(2.5437 \times 10^{10} \text{ s}^{-1}) = 1.6855 \times 10^{-23} \text{ J/photon}$

$E_{molar} = (1.6855 \times 10^{-23} \text{ J})\left(\dfrac{6.022 \times 10^{23} \text{ photons}}{1 \text{ mol}} \right)\left(\dfrac{10^{-3} \text{ kJ}}{1 \text{ J}} \right) = 1.0150 \times 10^{-2} \text{ kJ/mol}$

7.13   First, calculate the energy of one photon; then use $E_{total} = n\,E_{photon}$ to determine n (the number of photons):

$$\lambda = 337.1 \text{ nm} \left(\frac{10^{-9} \text{ m}}{1 \text{ nm}} \right) = 3.371 \times 10^{-7} \text{ m}$$

$$E_{photon} = \frac{(6.626 \times 10^{-34} \text{ J s})(2.998 \times 10^8 \text{ m/s})}{3.371 \times 10^{-7} \text{ m}} = 5.893 \times 10^{-19} \text{ J/photon}$$

$$n = 10.0 \text{ mJ} \left(\frac{10^{-3} \text{ J}}{1 \text{ mJ}} \right) \left(\frac{1 \text{ photon}}{5.893 \times 10^{-19} \text{ J}} \right) = 1.70 \times 10^{16} \text{ photons}$$

7.15   (a) Convert from molar energy to energy per photon by dividing by Avogadro's number, then use $E = \dfrac{hc}{\lambda}$ to find $\lambda = \dfrac{hc}{E}$:

$$E_{photon} = \left(\frac{745 \text{ kJ}}{1 \text{ mol}} \right) \left(\frac{10^3 \text{ J}}{1 \text{ kJ}} \right) \left(\frac{1 \text{ mol}}{6.022 \times 10^{23} \text{ photons}} \right) = 1.237 \times 10^{-18} \text{ J/photon}$$

$$\lambda = \frac{(6.626 \times 10^{-34} \text{ J s})(2.998 \times 10^8 \text{ m/s})}{1.237 \times 10^{-18} \text{ J}} = 1.61 \times 10^{-7} \text{ m}$$

$$v = \frac{c}{\lambda} = \frac{2.998 \times 10^8 \text{ m/s}}{1.61 \times 10^{-7} \text{ m}} = 1.86 \times 10^{15} \text{ Hz}$$

(b) $E_{photon} = 3.55 \times 10^{-19} \text{ J}$

$$\lambda = \frac{(6.626 \times 10^{-34} \text{ J s})(2.998 \times 10^8 \text{ m/s})}{3.55 \times 10^{-19} \text{ J}} = 5.60 \times 10^{-7} \text{ m}$$

$$v = \frac{c}{\lambda} = \frac{2.998 \times 10^8 \text{ m/s}}{5.60 \times 10^{-7} \text{ m}} = 5.35 \times 10^{14} \text{ Hz}$$

7.17   (a) $\lambda = \dfrac{c}{v} = \dfrac{2.998 \times 10^8 \text{ m/s}}{1.30 \times 10^{15} \text{ s}^{-1}} = 2.31 \times 10^{-7} \text{ m}$

(b) First determine the energy of the photon, then use $E_{binding} = E_{photon} - E_{kinetic}$
$E_{photon} = hv = (6.626 \times 10^{-34} \text{ J s})(1.30 \times 10^{15} \text{ s}^{-1}) = 8.61 \times 10^{-19} \text{ J}$
$E_{binding} = 8.61 \times 10^{-19} \text{ J} - 5.2 \times 10^{-19} \text{ J} = 3.4 \times 10^{-19} \text{ J}$

(c) The longest wavelength that will eject electrons corresponds to a photon with energy equal to the binding energy: $\lambda = \dfrac{hc}{E_{binding}}$

$$\lambda = \frac{(6.626 \times 10^{-34} \text{ J s})(2.998 \times 10^8 \text{ m/s})}{3.4 \times 10^{-19} \text{ J}} = 5.8 \times 10^{-7} \text{ m}$$

7.19   Refer to Figure 7-9 for an example of an energy level diagram. The binding energy for cesium (Prob. 7.17) is 3.4×10^{-19} J, and the binding energy for chromium (Prob. 7.18) is 7.21×10^{-19} J. Your energy level diagram should show that chromium has a substantially greater binding energy than cesium. Note, if both metals are exposed to photons of the same wavelength, the photoelectrons of Cs will have substantially greater kinetic energy than those emitted by Cr:

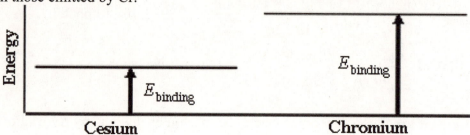

7.21   Use Figure 7-4 to answer these questions:
(a) Radio waves have wavelengths from 1 to 1000 m.
(b) Light with a wavelength of 5.8×10^{-7} m is yellow.
(c) A frequency of 4.5×10^{8} Hz lies in the radar region.

7.23   Figure 7-14 does not show an emission wavelength directly to the ground state from the state that emits 404 nm light, but the combination of emissions at 436 nm and 254 nm connects this state to the ground state. Use $E = \dfrac{hc}{\lambda}$ for each photon:

$$\Delta E (\text{kJ/mol}) = \frac{hcN_A}{\lambda_1} + \frac{hcN_A}{\lambda_2} = hcN_A\left(\frac{1}{\lambda_1} + \frac{1}{\lambda_2}\right)$$

$\Delta E =$

$$6.626 \times 10^{-34} \text{ J s}\left(\frac{2.998 \times 10^{8} \text{ m}}{1 \text{ s}}\right)\left(\frac{6.022 \times 10^{23} \text{ atoms}}{1 \text{ mol}}\right)\left(\frac{1}{436 \text{ nm}} + \frac{1}{254 \text{ nm}}\right)\left(\frac{10^{9} \text{ nm}}{1 \text{ m}}\right)$$

$\Delta E = 7.45 \times 10^5$ J/mol
Convert to kJ/mol:

$$\Delta E = = 7.45 \times 10^5 \text{ J/mol}\left(\frac{10^{-3} \text{ kJ}}{1 \text{ J}}\right) = 745 \text{ kJ/mol}$$

7.25   Energies and wavelengths for transitions in hydrogen atoms can be calculated from the equation for hydrogen atom energy levels:

$$E_n = \frac{-2.18 \times 10^{-18} \text{ J}}{n^2}$$

$$\Delta E_{8-1} = (E_8 - E_1) = \frac{-2.18 \times 10^{-18} \text{ J}}{8^2} - \frac{-2.18 \times 10^{-18} \text{ J}}{1^2} = 2.146 \times 10^{-18} \text{ J}$$

$$\lambda = \frac{hc}{E} = \frac{(6.626 \times 10^{-34} \text{ J s})(2.998 \times 10^{8} \text{ m/s})}{2.146 \times 10^{-18} \text{ J}} = 9.257 \times 10^{-8} \text{ m}$$

Convert to nm:

$$\lambda = 9.257 \times 10^{-8} \text{ m} \left(\frac{10^9 \text{ nm}}{1 \text{ m}} \right) = 92.57 \text{ nm}$$

$$\Delta E_{9-1} = (E_9 - E_1) = \frac{-2.18 \times 10^{-18} \text{ J}}{9^2} - \frac{-2.18 \times 10^{-18} \text{ J}}{1^2} = 2.153 \times 10^{-18} \text{ J}$$

$$\lambda = \frac{hc}{E} = \frac{(6.626 \times 10^{-34} \text{ J s})(2.998 \times 10^8 \text{ m/s})}{2.153 \times 10^{-18} \text{ J}} = 9.227 \times 10^{-8} \text{ m}$$

Convert to nm:

$$\lambda = 9.227 \times 10^{-8} \text{ m} \left(\frac{10^9 \text{ nm}}{1 \text{ m}} \right) = 92.27 \text{ nm}$$

These photons lie in the ultraviolet region of the spectrum (see Figure 7-4).

7.27 As Figure 7-13 shows, absorption transitions must originate in the $n = 1$ level and can end on any higher level. From higher levels, in contrast, transitions can occur to any lower level, not just the $n = 1$ level. Thus there are many more possibilities for emission than for absorption from the ground state.

7.29 Use the mass of the electron and Avogadro's number to calculate the molar mass:

$$m_{\text{molar}} = 9.1093897 \times 10^{-31} \text{ kg} \left(\frac{10^3 \text{ g}}{1 \text{ kg}} \right) \left(\frac{6.0221367 \times 10^{23} \text{ e}^-}{1 \text{ mol}} \right)$$

$$= 5.4857990 \times 10^{-4} \text{ g/mol}$$

7.31 To determine the wavelength of an electron from the kinetic energy, first determine the speed of the electron and then convert to wavelength. The speed of an electron is determined using $u = \sqrt{\dfrac{2 E_{\text{kinetic}}}{m}}$. Remember, for the units to work out, mass must be in kg and the energy in J such that $\dfrac{\text{J}}{\text{kg}} = \dfrac{\text{kg m}^2/\text{s}^2}{\text{kg}} = \text{m}^2/\text{s}^2$. For wavelength calculations, use $\lambda = \dfrac{h}{mu}$:

(a) $u = \sqrt{\dfrac{2(1.15 \times 10^{-19} \text{ J})}{9.109 \times 10^{-31} \text{ kg}}} = 5.02 \times 10^5 \text{ m/s}$

$$\lambda = \frac{6.626 \times 10^{-34} \text{ J s}}{(9.109 \times 10^{-31} \text{ kg})(5.02 \times 10^5 \text{ m/s})} = 1.45 \times 10^{-9} \text{ m or } 1.45 \text{ nm}$$

(b) $E_{electron} = \left(\dfrac{3.55 \text{ kJ}}{1 \text{ mol}}\right)\left(\dfrac{10^3 \text{ J}}{1 \text{ kJ}}\right)\left(\dfrac{1 \text{ mol}}{6.022 \times 10^{23} \text{ photons}}\right) = 5.90 \times 10^{-21}$ J

$u = \sqrt{\dfrac{2(5.90 \times 10^{-21} \text{ J})}{9.109 \times 10^{-31} \text{ kg}}} = 1.14 \times 10^5$ m/s

$\lambda = \dfrac{6.626 \times 10^{-34} \text{ J s}}{(9.109 \times 10^{-31} \text{ kg})(1.14 \times 10^5 \text{ m/s})} = 6.38 \times 10^{-9}$ m or 6.38 nm

(c) $E_{electron} = \left(\dfrac{7.45 \times 10^{-3} \text{ J}}{1 \text{ mol}}\right)\left(\dfrac{1 \text{ mol}}{6.022 \times 10^{23} \text{ electrons}}\right) = 1.24 \times 10^{-26}$ J

$u = \sqrt{\dfrac{2(1.24 \times 10^{-26} \text{ J})}{9.109 \times 10^{-31} \text{ kg}}} = 1.65 \times 10^2$ m/s

$\lambda = \dfrac{6.626 \times 10^{-34} \text{ J s}}{(9.109 \times 10^{-31} \text{ kg})(1.65 \times 10^2 \text{ m/s})} = 4.41 \times 10^{-6}$ m or 4.41 μm

7.33   First determine the speed of the electron from its wavelength and then use the speed to determine the kinetic energy. Kinetic energy and wavelength of an electron are related through $\lambda = \dfrac{h}{mu}$ and $E_{kinetic} = \dfrac{mu^2}{2}$:

(a) $\lambda = 3.75 \text{ nm}\left(\dfrac{10^{-9} \text{m}}{1 \text{ nm}}\right) = 3.75 \times 10^{-9}$ m

$u = \dfrac{6.626 \times 10^{-34} \text{ J s}}{(9.109 \times 10^{-31} \text{ kg})(3.75 \times 10^{-9} \text{ m})} = 1.94 \times 10^5$ m/s

$E_{kinetic} = \dfrac{(9.109 \times 10^{-31} \text{ kg})(1.94 \times 10^5 \text{ m/s})^2}{2} = 1.71 \times 10^{-20}$ J

(b $u = \dfrac{6.626 \times 10^{-34} \text{ J s}}{(9.109 \times 10^{-31} \text{ kg})(4.66 \text{ m})} = 1.56 \times 10^{-4}$ m/s

$E_{kinetic} = \dfrac{(9.109 \times 10^{-31} \text{ kg})(1.56 \times 10^{-4} \text{ m/s})^2}{2} = 1.11 \times 10^{-38}$ J

(c) $\lambda = 8.85 \text{ mm}\left(\dfrac{10^{-3} \text{ m}}{1 \text{ mm}}\right) = 8.85 \times 10^{-3}$ m

$u = \dfrac{6.626 \times 10^{-34} \text{ J s}}{(9.109 \times 10^{-31} \text{ kg})(8.85 \times 10^{-3} \text{ m})} = 8.22 \times 10^{-2}$ m/s

$E_{kinetic} = \dfrac{(9.109 \times 10^{-31} \text{ kg})(8.22 \times 10^{-2} \text{ m/s})^2}{2} = 3.08 \times 10^{-33}$ J

7.35 The designation $6p$ specifies that $n = 6$ and $l = 1$; the other two quantum numbers m_l and m_s, can take any of their possible values. Remember that m_l must be an integer with value between $+l$ and $-l$, while m_s is either $+1/2$ or $-1/2$. There are 6 valid sets:

n	l	m_l	m_s
6	1	+1	+1/2
6	1	+1	−1/2
6	1	0	+1/2
6	1	0	−1/2
6	1	−1	+1/2
6	1	−1	−1/2

7.37 Remember that l is restricted to positive integers less than n, m_l is restricted to integers between $-l$ and $+l$, and $m_s = +1/2$ or $-1/2$. For $n = 3$, l can be 0 with $m_l = 0$, l can be 1 with $m_l = +1$, 0, or -1; and l can be 2 with $m_l = +2$, +1, 0, −1, or −2. For each possibility, $m_s = +1/2$ or $-1/2$.

7.39 Remember that n must be a positive integer, l is restricted to zero and positive integers less than n, m_l is restricted to integers between $-l$ and $+l$, and $m_s = +1/2$ or $-1/2$.
(a) non-existent: m_s must be +1/2 or −1/2; (b) actual; (c) non-existent: l must be less than n; and (d) actual.

7.41 The designation $3d$ specifies that $n = 3$ and $l = 2$; the largest possible value for m_l is $+l$, in this case +2; and "spin up" means $m_s = +1/2$. Thus, the values are 3, 2, 2, +1/2.

7.43 The designation $n = 2$, $l = 1$ refers to a $2p$ orbital. This orbital has no radial nodes, so its electron density rises and then falls. Figure 7-22 shows a contour diagram. For this orbital, an electron density drawing looks very similar to a contour drawing.

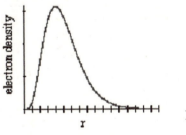

Electron density   Contour

x- and z- axis y-axis

7.45 The p_y orbital runs lengthwise along the y-axis. It has the same view along the x- and z-axes. Along the y-axis, you will only see one lobe of the orbital:

7.47   Orbital (a) has the shape characteristic of a *p* orbital, while orbitals (b) and (c) have shapes characteristic of *d* orbitals. Because orbital (a) appears smaller than the 3*s* orbital, it has $n < 3$ and is a 2*p* orbital; because orbitals (b) and (c) have similar sizes as the 3*s* orbital, they have $n = 3$ and are 3*d* orbitals: (a) 2*p* orbital, $n = 2, l = 1$; (b) and (c) 3*d* orbital, $n = 3, l = 2$.

7.49   Electron density plots accurately show how the wave function varies along one axis, but they fail to show whether or not an orbital has spherical symmetry, and they do not directly show electron densities. Thus, they fail to give a good picture of how an orbital appears in three dimensions.

7.51   A contour drawing of an *s* orbital is a spherical surface. The orbitals should grow larger as *n* increases:

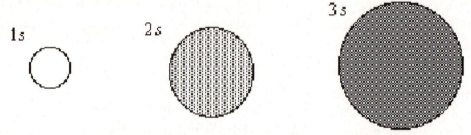

7.53   Each picture should show a molecule absorbing a photon and fragmenting appropriately: N_2:

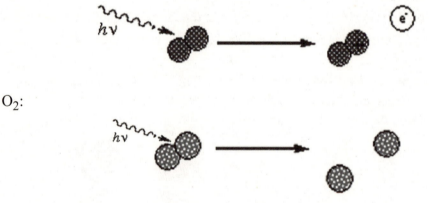

O_2:

7.55   Calculate the fragmentation energy using $E = \dfrac{hcN_A}{\lambda}$:

$$\lambda = 340 \text{ nm} \left(\frac{10^{-9} \text{ m}}{1 \text{ nm}} \right) = 3.40 \times 10^{-7} \text{ m}$$

$$hcN_A = 6.626 \times 10^{-34} \text{ J s} \left(\frac{2.998 \times 10^{8} \text{m}}{1 \text{ s}} \right) \left(\frac{10^{-3} \text{ kJ}}{1 \text{ J}} \right) \left(\frac{6.022 \times 10^{23} \text{ photons}}{1 \text{ mol}} \right)$$

$$hcN_A = 1.196 \times 10^{-4} \text{ kJ m/mol}$$

The minimum energy to fragment an ozone molecule is:

$$\Delta E = \frac{1.196 \times 10^{-4} \text{ kJ m/mol}}{3.40 \times 10^{-7} \text{ m}} = 352 \text{ kJ/mol}$$

The kinetic energy of the fragments is the difference between the energy of the photon causing the fragmentation and the minimum energy required for fragmentation:

$$\lambda = 250 \text{ nm}\left(\frac{10^{-9} \text{ m}}{1 \text{ nm}}\right) = 2.50 \times 10^{-7} \text{ m}$$

$$E_{kinetic} = E_{photon} - \Delta E$$

$$= \frac{(6.626 \times 10^{-34} \text{ J s})(2.998 \times 10^8 \text{ m/s})}{2.50 \times 10^{-7} \text{ m}} - \left(\frac{352 \times 10^3 \text{ J}}{1 \text{ mol}}\right)\left(\frac{1 \text{ mol}}{6.022 \times 10^{23} \text{ photons}}\right)$$

$$E_{kinetic} = (7.95 \times 10^{-19} \text{ J}) - (5.85 \times 10^{-19} \text{ J}) = 2.10 \times 10^{-19} \text{ J}$$

7.57 Use information in the Chemistry and the Environment Box to determine which region of the atmosphere absorbs the given wavelengths:
(a) Below 200 nm, O_2, N_2, and O_3 all absorb effectively (thermosphere).
(b) From 240-310 nm, only O_3 absorbs effectively (stratosphere).

7.59 (a) According to Figure 7-24, a pressure of 10^{-3} atm occurs at ~50 km.
(b) The balloon is on the border of the stratosphere and mesosphere.
(c) The chemistry of this region is the formation of ozone.

7.61 To calculate photon properties, use $\lambda v = c$ and the equations for photons in Table 7-1:

(a) $\lambda = \dfrac{c}{v} = \dfrac{2.998 \times 10^8 \text{ m/s}}{4.5 \times 10^{13} \text{ s}^{-1}} = 6.7 \times 10^{-6} \text{ m}$

(b) $E = hv = (6.626 \times 10^{-34} \text{ J s})(4.5 \times 10^{13} \text{ s}^{-1}) = 3.0 \times 10^{-20} \text{ J}$

(c) $t = \dfrac{\text{distance}}{\text{speed}} = 2.86 \times 10^5 \text{ mi}\left(\dfrac{1.609 \text{ km}}{1 \text{ mi}}\right)\left(\dfrac{10^3 \text{ m}}{1 \text{ km}}\right)\left(\dfrac{1 \text{ s}}{2.998 \times 10^8 \text{ m}}\right) = 1.53 \text{ s}$

7.63 Calculate the wavelength using $\lambda = \dfrac{hcN_A}{E}$:

$$hcN_A = 6.626 \times 10^{-34} \text{ J s}\left(\frac{2.998 \times 10^8 \text{ m}}{1 \text{ s}}\right)\left(\frac{10^{-3} \text{ kJ}}{1 \text{ J}}\right)\left(\frac{6.022 \times 10^{23} \text{ photons}}{1 \text{ mol}}\right)$$

$$hcN_A = 1.196 \times 10^{-4} \text{ kJ m/mol}$$

$$\lambda = \left(\frac{1.196 \times 10^{-4} \text{ kJ m}}{1 \text{ mol}}\right)\left(\frac{1 \text{ mol}}{243 \text{ kJ}}\right) = 4.92 \times 10^{-7} \text{ m or 492 nm}$$

This wavelength is in the visible region of the spectrum, which reaches the surface of the Earth. Thus, Cl_2 in the troposphere will be dissociated by sunlight.

7.65   A 4p electron must have $n = 4$, $l = 1$, but m_l has three possible values and m_s has two possible values. There are six possible sets of values:

n	l	m_l	m_s
4	1	1	1/2
4	1	1	−1/2
4	1	0	1/2
4	1	0	−1/2
4	1	−1	1/2
4	1	−1	−1/2

7.67   When the frequency is doubled and the amplitude is kept the same the peaks are narrower and occur twice as often:

Original

2 x frequency

7.69   The longest wavelength that can eject an electron from the metal surface is the wavelength that just overcomes the binding energy. Calculate the wavelength using

$$\lambda = \frac{hcN_A}{E}:$$

$hcN_A = 1.196 \times 10^{-4}$ kJ m/mol

$$\lambda = \left(\frac{1.196 \times 10^{-4} \text{ kJ m}}{1 \text{ mol}}\right)\left(\frac{1 \text{ mol}}{216.4 \text{ kJ}}\right) = 5.527 \times 10^{-7} \text{ m or } 552.7 \text{ nm}$$

7.71   The photoelectric effect is described by an energy balance equation:

$E_{kinetic} = E_{photon} - E_{binding}$

(a) $E_{binding} = E_{photon} - E_{electron} = 6.00 \times 10^{-19} \text{ J} - 2.70 \times 10^{-19} \text{ J} = 3.30 \times 10^{-19} \text{ J}$

(b) $\lambda = \dfrac{hc}{E} = \dfrac{(6.626 \times 10^{-34} \text{ J s})(2.998 \times 10^{8} \text{ m/s})}{6.00 \times 10^{-19} \text{ J}} = 3.31 \times 10^{-7} \text{ m}$

(c) For electrons, $\lambda = \dfrac{h}{mu}$:

$$u = \sqrt{\frac{2E_{kinetic}}{m}} = \sqrt{\frac{2(2.70 \times 10^{-19} \text{ J})}{9.109 \times 10^{-31} \text{ kg}}} = 7.70 \times 10^{5} \text{ m/s}$$

$$\lambda = \frac{(6.626 \times 10^{-34} \text{ kg m}^2/\text{s})}{(9.109 \times 10^{-31} \text{ kg})(7.70 \times 10^{5} \text{ m/s})} = 9.45 \times 10^{-10} \text{ m}$$

7.73   For electromagnetic radiation, $E = h\nu$ and $\lambda = \dfrac{c}{\nu}$:

$$\lambda = \left(\frac{2.998 \times 10^8 \text{ m/s}}{27.3 \text{ MHz}}\right)\left(\frac{1 \text{ MHz}}{10^6 \text{ Hz}}\right)\left(\frac{1 \text{ Hz}}{1 \text{ s}^{-1}}\right) = 11.0 \text{ m}$$

$$E = (6.626 \times 10^{-34} \text{ J s})(27.3 \text{ MHz})\left(\frac{10^6 \text{ Hz}}{1 \text{ MHz}}\right)\left(\frac{1 \text{ s}^{-1}}{1 \text{ Hz}}\right) = 1.81 \times 10^{-26} \text{ J}$$

7.75   Make use of the figure legends to determine the answers to the questions:
(a) At an altitude of 60 km, $P \sim 2 \times 10^{-4}$ atm; (b) Examples are N_2, O_2, and O_3;
(c) At a pressure of 8 torr (or 0.0105 atm), the altitude is $\sim$35 km;
(d) Stratosphere

7.77   Calculate the frequencies using $\nu = \dfrac{c}{\lambda}$ and the energies using $E = \dfrac{hcN_A}{\lambda}$:

$hcN_A = 1.196 \times 10^{-4}$ kJ m/mol. The calculations for 487 nm are shown:

$$\lambda = 487 \text{ nm}\left(\frac{1 \text{ m}}{10^9 \text{ nm}}\right) = 4.87 \times 10^{-7} \text{ m}$$

$$\nu = \left(\frac{2.998 \times 10^8 \text{ m s}^{-1}}{4.87 \times 10^{-7} \text{ m}}\right) = 6.16 \times 10^{14} \text{ s}^{-1}$$

$$E = \frac{hcN_A}{\lambda} = \frac{1.196 \times 10^{-4} \text{ kJ m mol}^{-1}}{4.87 \times 10^{-7} \text{ m}} = 246 \text{ kJ/mol}$$

λ (nm)	487	514	543	553	578
ν (10^{14} s^{-1})	6.16	5.83	5.52	5.42	5.19
E (kJ/mol)	246	233	220	216	207

7.79   Energies and frequencies for transitions in hydrogen atoms can be calculated from the equation for hydrogen atom energy levels:

$$E_n = \frac{-2.18 \times 10^{-18} \text{ J}}{n^2}$$

$$\Delta E_{5-1} = (E_5 - E_1) = \frac{-2.18 \times 10^{-18} \text{ J}}{5^2} - \frac{-2.18 \times 10^{-18} \text{ J}}{1^2} = 2.09 \times 10^{-18} \text{ J}$$

$$\lambda = \frac{hc}{E} = \frac{(6.626 \times 10^{-34} \text{ Js})(2.998 \times 10^8 \text{ m/s})}{2.09 \times 10^{-18} \text{ J}} = 9.50 \times 10^{-8} \text{ m}$$

$$\lambda = 9.50 \times 10^{-8} \text{ m}\left(\frac{10^9 \text{ nm}}{1 \text{ m}}\right) = 95.0 \text{ nm}$$

These photons fall in the ultraviolet region.

7.81   According to Figure 7-4, visible light has wavelengths in the $4 - 7.2 \times 10^{-7}$ m range. The highest frequency of visible light is $v = \dfrac{2.998 \times 10^8 \text{ m/s}}{4 \times 10^{-7} \text{m}} \sim 7.5 \times 10^{14} \text{ s}^{-1}$. This is smaller than the threshold frequency, so Mg cannot be used in photoelectric devices.

7.83   Volume is related to radius through the equation, $V = \dfrac{4}{3}\pi r^3$:

$$V_{atom} = \frac{4\pi(10^{-10} \text{ m})^3}{3} = 4 \times 10^{-30} \text{ m}^3 \qquad V_{nucleus} = \frac{4\pi(10^{-15} \text{ m})^3}{3} = 4 \times 10^{-45} \text{ m}^3$$

$$\frac{4 \times 10^{-45} \text{ m}^3}{4 \times 10^{-30} \text{ m}^3} = \frac{1}{1 \times 10^{15}} \text{ is the fraction of the volume that the nucleus occupies.}$$

7.85   First calculate the energy of one photon of wavelength 510 nm. Then divide the detection limit by this energy to find the number of photons:

$$\lambda = 510 \text{ nm} \left(\frac{10^{-9} \text{ m}}{1 \text{ nm}} \right) = 5.10 \times 10^{-7} \text{ m}$$

$$E_{photon} = \frac{hc}{\lambda} = \frac{(6.626 \times 10^{-34} \text{ J s})(2.998 \times 10^8 \text{ m/s})}{5.10 \times 10^{-7} \text{ m}} = 3.90 \times 10^{-19} \text{ J/photon}$$

$$\# = \frac{2.35 \times 10^{-18} \text{ J}}{3.90 \times 10^{-19} \text{ J/photon}} = 6 \text{ photons}$$

7.87   The wavelengths associated with particles are calculated from $\lambda = \dfrac{h}{mu}$:

5.000 % of the speed of light is $u = \left(\dfrac{5.000 \%}{100\%} \right)(2.998 \times 10^8 \text{ m/s}) = 1.499 \times 10^7 \text{ m/s}$

electron: $\lambda = \dfrac{h}{mu} = \dfrac{(6.626 \times 10^{-34} \text{ J s})}{(9.109 \times 10^{-31} \text{ kg})(1.499 \times 10^7 \text{ m/s})} = 4.853 \times 10^{-11} \text{ m}$

proton: $\lambda = \dfrac{h}{mu} = \dfrac{(6.626 \times 10^{-34} \text{ J s})}{(1.673 \times 10^{-27} \text{ kg})(1.499 \times 10^7 \text{ m/s})} = 2.642 \times 10^{-14} \text{ m}$

7.89   The energy level diagram shows the energy differences among the various levels. By examining it, we can see that the smallest energy difference is associated with $Photon_1$ and the largest energy difference with $Photon_2$. The energy of a photon is inversely proportional to its wavelength, so $Photon_1$ has the longest wavelength and $Photon_2$ the shortest wavelength. Thus, the assignments are as follows:

      $Photon_1 = 565$ nm (*d to c*) $Photon_2 = 121$ nm (*c to b*)
      $Photon_3 = 152$ nm (*b to a*)

Level b: $\lambda = 152$ nm $\left(\dfrac{10^{-9} \text{ m}}{1 \text{ nm}}\right) = 1.52 \times 10^{-7}$ m

$$E = \frac{hc}{\lambda} = \frac{(6.626 \times 10^{-34} \text{ J s})(2.998 \times 10^8 \text{ m/s})}{1.52 \times 10^{-7} \text{ m}} = 1.31 \times 10^{-18} \text{ J}$$

Level c: $\lambda = 121$ nm $\left(\dfrac{10^{-9} \text{ m}}{1 \text{ nm}}\right) = 1.21 \times 10^{-7}$ m

$$E = 1.31 \times 10^{-18} \text{ J} + \frac{(6.626 \times 10^{-34} \text{ J s})(2.998 \times 10^8 \text{ m/s})}{1.21 \times 10^{-7} \text{ m}} = 2.95 \times 10^{-18} \text{ J}$$

Level d: $\lambda = 565$ nm $\left(\dfrac{10^{-9} \text{ m}}{1 \text{ nm}}\right) = 5.65 \times 10^{-7}$ m

$$E = 2.95 \times 10^{-18} \text{ J} + \frac{(6.626 \times 10^{-34} \text{ J s})(2.998 \times 10^8 \text{ m/s})}{5.65 \times 10^{-7} \text{ m}} = 3.30 \times 10^{-18} \text{ J}$$

7.91   (a) Calculate the energies of the photons using $E = \dfrac{hc}{\lambda}$:

$$E_{488} = \frac{(6.626 \times 10^{-34} \text{ J s})(2.998 \times 10^8 \text{ m/s})}{4.88 \times 10^{-7} \text{ m}} = 4.07 \times 10^{-19} \text{ J}$$

$$E_{514} = \frac{(6.626 \times 10^{-34} \text{ J s})(2.998 \times 10^8 \text{ m/s})}{5.14 \times 10^{-7} \text{ m}} = 3.86 \times 10^{-19} \text{ J}$$

(b) Each transition leaves the ion in the same state, so the state from which 488 nm emission occurs must be slightly higher in energy than that from which 514 nm emission occurs:

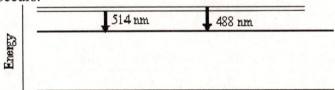

(c) Calculate frequency and wavelength using $v = \dfrac{E}{h}$ and $\lambda = \dfrac{c}{v}$:

$$v = \frac{E}{h} = \frac{2.76 \times 10^{-18} \text{ J}}{6.626 \times 10^{-34} \text{ J s}} = 4.17 \times 10^{15} \text{ s}^{-1}$$

$$\lambda = \frac{c}{v} = \frac{2.998 \times 10^8 \text{ m/s}}{4.17 \times 10^{15} \text{ s}^{-1}} = 7.19 \times 10^{-8} \text{ m or } 71.9 \text{ nm}$$

7.93   (a) Calculate photon energies using $E = \dfrac{hcN_A}{\lambda}$, $hcN_A = 1.196 \times 10^{-4}$ kJ m/mol

$$E_{589.6} = \frac{1.196 \times 10^{-4} \text{ kJ m/mol}}{5.896 \times 10^{-7} \text{ m}} = 202.8 \text{ kJ/mol}$$

$$E_{590.0} = \frac{1.196 \times 10^{-4} \text{ kJ m/mol}}{5.900 \times 10^{-7} \text{ m}} = 202.7 \text{ kJ/mol}$$

(b) The two levels are <u>very close together at about 203 kJ/mol</u> above the ground state:

(c) The energy required to ionize the atom from this state is the difference in energies:

$E = 486 - 202.7 = 283$ kJ/mol. Calculate the wavelength using $\lambda = \dfrac{hcN_A}{E}$:

$$\lambda = \frac{hcN_A}{E} = \left(\frac{1.196 \times 10^{-4} \text{ kJ m}}{1 \text{ mol}}\right)\left(\frac{1 \text{ mol}}{283 \text{ kJ}}\right) = 4.23 \times 10^{-7} \text{ m or 423 nm}$$

7.95 (a) Binding energy is the energy required to eject an electron from a metal.
For metal A, electrons are not ejected until 4.0×10^{14} s^{-1} is reached, therefore the binding energy is the energy corresponding to this frequency:
$E = h\nu = (6.626 \times 10^{-34} \text{ J s})(4.0 \times 10^{14} \text{ s}^{-1}) = 2.7 \times 10^{-19}$ J
Similarly for metal B at 6.5×10^{14} s^{-1}
$E = h\nu = (6.626 \times 10^{-34} \text{ J s})(6.5 \times 10^{14} \text{ s}^{-1}) = 4.3 \times 10^{-19}$ J
Therefore, metal B has the higher binding energy because higher frequency photons are required to generate photoelectrons.
(b) Kinetic energy = photon energy – binding energy

$$\lambda = 125 \text{ nm}\left(\frac{10^{-9} \text{ m}}{1 \text{ nm}}\right) = 1.25 \times 10^{-7} \text{ m}$$

$$E_{\text{photon}} = \frac{hc}{\lambda} = \frac{(6.626 \times 10^{-34} \text{ J s})(2.998 \times 10^{8} \text{ m/s})}{1.25 \times 10^{-7} \text{ m}} = 1.59 \times 10^{-18} \text{ J}$$

Hence, using the photon energy above and the binding energy calculated in part (a), the kinetic energies are:
Metal A: 1.59×10^{-18} J $- 0.27 \times 10^{-18}$ J $= 1.32 \times 10^{-18}$ J
Metal B: 1.59×10^{-18} J $- 0.43 \times 10^{-18}$ J $= 1.16 \times 10^{-18}$ J
(c) The wavelength range over which electrons can be ejected from one metal but not the other corresponds to the frequency range of 4.0×10^{14} s^{-1} to 6.5×10^{14} s^{-1}:

At 4.0×10^{14} s^{-1}, wavelength, $\lambda = \dfrac{c}{\nu} = \dfrac{2.998 \times 10^{8} \text{ m/s}}{4.0 \times 10^{14} \text{ s}^{-1}} = 7.5 \times 10^{-7}$ m or 750 nm

At 6.5×10^{14} s^{-1}, wavelength, $\lambda = \dfrac{c}{\nu} = \dfrac{2.998 \times 10^{8} \text{ m/s}}{6.5 \times 10^{14} \text{ s}^{-1}} = 4.6 \times 10^{-7}$ m or 460 nm

The wavelength range is 460 nm to 750 nm.

8.1   Orbital stability in multi-electron atoms depends on the value of n (stability decreases with increasing value), Z (stability increases with increasing value), l (stability decreases with increasing value), and amount of screening (stability increases as other electrons are removed):

(a) He $1s$ is more stable than He $2s$ because of its lower value of n.

(b) Kr $5s$ is more stable than Kr $5p$ because of its lower value of l.

(c) He$^+$ $2s$ is more stable than He $2s$ because it is less screened by $1s$ electrons.

8.3   A hydrogen atom contains just one electron, so there is no screening effect. In the absence of screening, orbital energy depends only on n and Z, so all $n = 3$ orbitals have identical energy. In a helium atom, on the other hand, an electron in an $n = 3$ orbital is screened from the nucleus by the second electron, and the amount of screening decreases as l increases.

8.5   (a) The ionization energy of the He $2p$ orbital (0.585×10^{-18} J) is not much larger than that of the H $2p$ orbital (0.545×10^{-18} J). The similar values indicate nearly equal effective nuclear charges due to nearly complete screening; in the absence of screening, the He $2p$ orbital would have four times the ionization energy as the H $2p$ orbital.

(b) The ionization energy of the He$^+$ $2p$ orbital (2.18×10^{-18} J) is four times larger than that of the H $2p$ orbital (0.545×10^{-18} J). A four-fold increase when Z doubles indicates a Z^2 dependence.

8.7

8.9   The element below lead in the periodic table would have atomic number 114 (count from Ds, element 110: 2 elements needed to complete the d block, 2 more to reach Column 14 in the p block). The block that is filling would match that of lead, but with one higher n-value: $7p$.

8.11   Element 111 would fall directly below gold in Column 11, in row 7 ($n = 6$ for the d block).

8.13 The column location of an element is the indicator of how many valence electrons it has. Remember that s electrons as well as those in the filling block count as valence electrons: O, fourth column of p block, $4p + 2s = 6$ valence electrons; V, third column of d block, $3d + 2s = 5$ valence electrons; Rb, 1 valence electron; Sn, second column of p block, $2s + 2p = 4$ valence electrons; and Cd, end of d block, 2 valence electrons.

8.15 Use the aufbau principle to determine the orbital occupancies and quantum numbers for the valence electrons in a configuration. When there are equivalent configurations, unpair as many electrons as possible in accordance with Hund's rule.

(a) Be, 4 electrons, [He] $2s^2$

n	l	m_l	m_s
2	0	0	+1/2
2	0	0	−1/2

(b) O, 8 electrons, [He] $2s^2 2p^4$:

n	l	m_l	m_s
2	0	0	+1/2
2	0	0	−1/2
2	1	1	+1/2
2	1	0	+1/2
2	1	−1	+1/2
2	1	1	−1/2

(c) Ne, 10 electrons, [He] $2s^2 2p^6$:

n	l	m_l	m_s
2	0	0	+1/2
2	0	0	−1/2
2	1	1	+1/2
2	1	1	−1/2
2	1	0	+1/2
2	1	0	−1/2
2	1	−1	+1/2
2	1	−1	−1/2

(d) P, 15 electrons, [Ne] $3s^2 3p^3$:

n	l	m_l	m_s
3	0	0	+1/2
3	0	0	−1/2
3	1	1	+1/2
3	1	0	+1/2
3	1	−1	+1/2

8.17 Atoms with unpaired electrons in their configurations are paramagnetic. Of the atoms in Problem 8.15, O and P have unpaired electrons. Here are the orbital occupancy diagrams for the least stable occupied orbitals:

8.19   (a) This is Pauli-forbidden: s orbitals can hold no more than two electrons.
       (b) This is Pauli-forbidden: s orbitals can hold no more than two electrons.
       (c) This is an excited-state configuration: the $1s$ and $2s$ orbitals are not full.
       (d) This is an excited-state configuration: the $2s$ orbital is not full.
       (e) This is the ground-state configuration.
       (f) This configuration uses a non-existent orbital: there is no $1p$ orbital.
       (g) This configuration uses a non-existent orbital: there is no $2d$ orbital.

8.21   The element with more unpaired electrons will have the higher spin. Mo has ground-state configuration [Kr] $5s^1 4d^5$; that of Tc is [Kr] $5s^2 4d^5$:

8.23   Write configurations using the standard filling procedure, but watch for exceptions:
       (a) C, $Z = 6$, [He] $2s^2 2p^2$; (b) Cr, $Z = 24$, exception, [Ar] $4s^1 3d^5$; (c) Sb, $Z = 51$,
       [Kr] $5s^2 4d^{10} 4p^3$; (d) Br, $Z = 35$, [Ar] $4s^2 3d^{10} 4p^5$

8.25   The ground state for N is $1s^2 2s^2 2p^3$. Excited states with no electron having $n > 2$ can be formed by moving electrons out of the $1s$ and/or $2s$ orbital and placing them in the $2p$ orbital. There are seven ways to do this:
       $1s^1 2s^2 2p^4$; $2s^2 2p^5$; $1s^1 2s^1 2p^5$; $1s^2 2s^1 2p^4$; $1s^2 2p^5$; $2s^1 2p^6$; and $1s^1 2p^6$

8.27   Ionization energy decreases with increasing n and increases with increasing Z. Cs, which has the largest n-value, has the smallest ionization energy; K, which has the next larger n-value, has the next smallest ionization energy; because of its larger Z value, Ar has a larger ionization energy than Cl: Ar > Cl > K > Cs

8.29   The value of IE_2 is almost ten times that of IE_1, indicating that the second electron removed is a core electron rather than a valence electron. The elements in Column 1 contain only one valence electron, so this is a Column 1 element. An electron affinity around –50 kJ/mol matches this assignment (see Figure 8-17). (The element is Cs)

8.31   The valence configurations are N, $2s^2 2p^3$; Mg, $3s^2$; and Zn, $4s^2 3d^{10}$:

The added electron adds to an already-occupied orbital. Electron-electron repulsion makes this an unfavorable process.

The added electron adds to the next higher orbital, which is much higher in energy.

The added electron adds to the next higher orbital, which is much higher in energy.

8.33   Br⁻ has 36 electrons. The following are isoelectronic ions with less than 4 units of net charge: As^{3-}, Se^{2-}, Rb^+, Sr^{2+}, and Y^{3+}. For isoelectronic ions, size decreases with increasing Z: $Y^{3+} < Sr^{2+} < Rb^+ < Br^- < Se^{2-} < As^{3-}$

8.35   Stable anions form from elements in Columns 16 (dianions) and 17 (monoanions). Stable cations form from various metals:
Ca, Cu, Cs, and Cr, all metals, may be found in ionic compounds as cations.
Cl may be found in ionic compounds as a –1 anion.
C is not found as an atomic ion.

8.37   To form [K^+I^-] from neutral gaseous atoms we must remove an electron from potassium, add an electron to iodine and then form the coulombic interaction. The sequence of reactions for doing this calculation is (all gas phase):
$K \rightarrow K^+ + e^-$, $\Delta E = IE_1 = 418.8$ kJ/mol $I + e^- \rightarrow I^-$, $\Delta E = EA = -295.3$ kJ/mol
$K^+ + I^- \rightarrow KI$, $\Delta E = E_{electrical}$ (use Equation 7-1):

$$E_{electrical} = \frac{k(q^+)(q^-)}{r} = \frac{(1.389 \times 10^5 \text{ kJ pm/mol})(+1)(-1)}{(133 \text{ pm} + 220 \text{ pm})} = -393 \text{ kJ/mol}$$

The total energy is the sum of the above energies:
$\Delta E_{total} = (418.8$ kJ/mol$) + (-295.3$ kJ/mol$) + (-393$ kJ/mol$) = -270.$ kJ/mol

8.39   (a) The larger the ionic charges, the larger the lattice energy, so $Ba^{3+} O^{3-}$ would have the greatest lattice energy; (b) The first ionization energy always is smallest, and the first electron affinity always is most negative, so Ba^+O^- would have the least energy to form the ions; (c) The compound that actually exists is $Ba^{2+} O^{2-}$. The gain in lattice energy for this compound more than offsets the additional energy required to form the ions, but the energy required to remove the third electron from Ba is prohibitive.

8.41   First identify the chemical equation for the formation of LiF (s):
          Li (s) + 1/2 F_2 (g) → LiF (s)
Now determine the steps needed for this reaction. Li (Column 1) forms a +1 cation and F (Column 17)forms a –1 anion:
(1)   Li (s) → Li (g) $\Delta E = \Delta H_{vap} - 2.5 = 159.3 - 2.5$ kJ/mol = 156.8 kJ/mol
(2)   Li (g) → Li^+ (g) + e^- $\Delta E = IE = 520.2$ kJ/mol
(3)   1/2 F_2 (g) → F (g) $\Delta E = 1/2$ bond energy = 155/2 kJ/mol = 77.5 kJ/mol
(4)   F (g) + e^- → F^- (g) $\Delta E = EA = -328.0$ kJ/mol
(5) F^- (g) + Li^+ (g) → LiF (s) $\Delta E = -$ lattice energy = -1036 kJ/mol
Summing up all of these energies gives the overall energy change for the formation of lithium fluoride:
          (156.8 + 520.2 + 77.5 + -328.0 + -1036) kJ/mol = -610 kJ/mol

8.43   A Born-Haber diagram
        should show the energy of
        vaporization of the metal,
        ionization energy of the
        metal, energy to break the
        molecular bonds, electron
        affinity of anions, and the
        lattice energy that brings
        the ions of the salt together:

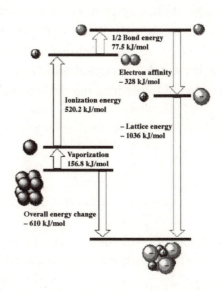

8.45   Use periodic variations in lattice energy to predict the expected value. The lattice energy
        values decrease by nearly the same amount from K^+ to Rb^+ as they decrease from Rb^+ to
        Cs^+. We predict that the decrease from Rb_2O to Cs_2O will be about the same as from
        K_2O to Rb_2O, 75 kJ/mol. The predicted value is $2163 - 75 = 2100$ kJ/mol, rounded to
        two significant figures because we do not expect this prediction to be exact.

8.47   Non-metals are found in the upper right portion of the periodic table, metalloids along a
        diagonal running through the p block, and all other elements are metals. In Column 16,
        Te is classified as a metalloid. The elements above it, O, S, and Se, are non-metals, and
        the element below it, Po, is a metal.

8.49   Non-metals are found in the upper right portion of the periodic table, metalloids along a
        diagonal running through the p block, and all other elements are metals. Thus, C and Cl
        are non-metals, and Ca, Cu, Cs, and Cr are metals.

8.51   The metalloids occupy a diagonal strip across the p block of the periodic table. The
        metals immediately to the left of this strip may be metalloid-like: Al, Ga, Sn, and Bi.

8.53   (a) When orbitals are nearly equal in energy, exceptions to the normal filling order exist.
            Normally, the $4s$ orbital fills immediately after $3p$, starting with element 19.
            Exceptions that indicate nearly equal energies of $3d$ and $4s$ are Cr, $4s^1\ 3d^5$; and Cu,
            $4s^1\ 3d^{10}$.
        (b) The first two elements in Column 6, Cr and Mo, have $s^1\ d^5$ configurations, while the
            second two elements in this column, W and Sg, have $s^2\ d^4$ configurations.
        (c) The $6d$ and $5f$ orbitals fill in the second f-block, the actinides. In this block, four
            elements have both these orbitals partly filled, indicating nearly equal energy: Pa, U,
            Np, and Cm.

8.55   To determine the correct configuration of a cation, determine the configuration of the neutral atom, then remove valence electrons, removing s electrons before d electrons:

Mn (25 electrons): $1s^2\,2s^2\,2p^6\,3s^2\,3p^6\,4s^2\,3d^5$

Mn^{2+} (remove 2 $4s$ electrons from Mn): $1s^2\,2s^2\,2p^6\,3s^2\,3p^6\,3d^5$

The least stable occupied orbitals are the five $3d$ orbitals, among which electrons are distributed to produce maximum spin:

n	l	m_l	m_s
3	2	2	+1/2
3	2	1	+1/2
3	2	0	+1/2
3	2	-1	+1/2
3	2	-2	+1/2

8.57   Total electron spin depends on the number of unpaired electrons. In partially filled shells, each unpaired electron contributes 1/2 to the total spin:

P: [Ne] $3s^2\,3p^3$; each $3p$ orbital is half filled, so net spin = 3(1/2) = 3/2

Br^-: [Kr]; all orbitals are filled, so net spin = 0

Cu^+: [Ar] $3d^{10}$; all orbitals are filled, so net spin = 0

8.59   Ionization energy decreases with increasing value of n and, for the same value of n, increases with increasing Z (increasing electrical attraction). Na is the only species in this set with an $n = 3$ valence electron, so it has the smallest IE. Na^+ has the largest Z, so it has the largest IE. O has its last two electrons paired, so electron-electron repulsion reduces its IE below that of N despite its higher Z value: Na < O < N < Ne < Na^+

8.61   Size increases with increasing n value and, for the same value of n, decreases with increasing Z (increasing electrical attraction). Cl, Cl^-, and K^+ all have $n = 3$ for the valence electrons, while Br^- has $n = 4$, so Br^- is the largest of this set. Among the others, the extra electron in Cl^- makes it larger than Cl, and K^+ has higher Z than Cl:
Br^- > Cl^- > Cl > K^+

8.63   Use the periodic table to determine the correct configurations:
$Z = 9$ is F: [He] $2s^2\,2p^5$; $Z = 20$ is Ca: [Ar] $4s^2$; and $Z = 33$ is As: [Ar] $4s^2\,3d^{10}\,4p^3$

8.65   Use the values in Table 8-4 to calculate the averages. The average difference between alkali fluorides and alkali chlorides is 125 kJ/mol, and the average difference between alkali bromides and alkali iodides is 42.6 kJ/mol. Both sets of values show the same periodic trend, decreasing with the size of the alkali cation. The decreases occur because a larger cation cannot get as close to the anion as can a smaller cation.

8.67   Remember that in the transition metal cations, the ns orbital is less stable than $(n-1)d$:
Cu^+ is $4s^0\,3d^{10}$ Mn^{2+} is $4s^0\,3d^5$ Au^{3+} is $6s^0\,5d^8$

8.69   (a) In a one-electron atom or ion, orbital energy depends only on n and Z, both of which
            are the same for the hydrogen atom $2s$ and $2p$ orbitals.
        (b) In multi-electron atoms, orbitals with the same n value but different l values are
            screened to different extents. The orbital with the lower l value is less screened, hence
            more stable.

8.71   This problem provides an exercise in graph reading:
        (a) The greatest drop within a row is from element 80 (Hg) to 81 (Tl)
        (b) Cs, element 55, has the lowest value shown in the figure
        (c) The value changes least from $Z = 56$ to 71, $Z = 39$ to 46, and $Z = 20$ to 29
        (d) Elements with values between 925 and 1050 kJ/mol are 15 (P), 16 (S), 33 (As),
            34 (Se), 53 (I), 80 (Hg), 85 (At), and 86 (Rn)

8.73   Unpaired electrons occur only among the valence orbitals. Construct the configuration of
        neutral S, then remove one electron to form S^+ and add one electron to form S^-:
        S: [Ne] $3s^2 3p^4$ S^+: [Ne] $3s^2 3p^3$ S^-: [Ne] $3s^2 3p^5$

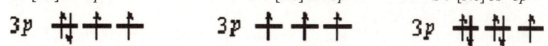

        S^+, with three unpaired electrons, has more unpaired than S or S^-.

8.75   The ground state places electrons in the most stable orbitals in a way that is consistent
        with the Pauli principle and maximizes electron spin. Thus, (c) is the ground-state
        configuration. Views (a) and (b) are excited states, but (d) is non-existent because it has
        two electrons with identical descriptions ($3s$, spin + 1/2).

8.77   Use periodic trends and the periodic table to identify the elements involved:
        (a) Mg has the second smallest radius among alkaline earths, and S has an anion that is
            isoelectronic with Ar, the Row 3 noble gas: MgS;
        (b) K, beginning of Row 4, has a cation isoelectronic with Ar, at the end of Row 3; and
            the Row 2 element with the highest electron affinity is F: KF;
        (c) Be is the alkaline earth element with the highest second ionization energy, and it
            combines in 1:2 ratio with elements from Row 17: $BeCl_2$

8.79   Identify the orbitals using shapes and relative sizes, given that c has $n = 3$: a and b are
        both spherical, with b smaller than a; a is the same size as c, so a is $3s$ and b is $2s$; c has
        the unique shape of the $3d_{z^2}$ orbital:
        (a) b is most stable (smallest n value), and a is more stable than c
        (b) $n = 3, l = 0, m_l = 0, m_s = +1/2; n = 3, l = 0, m_l = 0, m_s = -1/2$
        (c) Any element from magnesium to calcium has $3s^2$ but $3d^0$: $Z = 12$ to $Z = 20$
        (d) The $3d$ transition metal cations have partially occupied $3d$ orbitals, for example, Fe^{3+}
        (e) There are eight other $n = 3$ orbitals: $3s$, three $3p$, and four additional $3d$
        (f) Whenever an electron is removed from a doubly-occupied orbital, electron-electron
            repulsion decreases for the remaining electron, so the orbital becomes smaller.

8.81   In an excited state configuration, one electron is shifted to a less stable orbital. Start by constructing the ground state configuration, then move an electron from the most stable occupied to the least stable unoccupied orbital:

Be: [He] $2s^2$; the next orbital is $2p$:   [He] $2s^1 2p^1$

O^{2-}: [He] $2s^2 2p^6$; the next orbital is $3s$:   [He] $2s^2 2p^5 3s^1$

Br^-: [Ar] $4s^2 3d^{10} 4p^6$; the next orbital is $5s$: [Ar] $4s^2 3d^{10} 4p^5 5s^1$

Ca^{2+}: [Ar]; the next orbital is $4s$:   [Ne] $3s^2 3p^5 4s^1$

Sb^{3+}: [Kr] $4d^{10} 5s^2$; the next orbital is $5p$: [Kr] $4d^{10} 5s^1 5p^1$

8.83   Examine the configurations of the atoms and ions to determine the reason for the ionization energy differences:

Li: $1s^2 2s^1$ Be: $1s^2 2s^2$

Li^+: $1s^2$ Be^+: $1s^2 2s^1$

The first ionization energies involve the $2s$ orbital for both atoms, so Be, with larger Z, has a larger *IE*. The second electron removed from Li is a core electron, so the *IE* is much greater than the second *IE* for Be.

8.85   Consult Figures 7-20, 7-21, and 8-6 for examples of this kind of drawing.

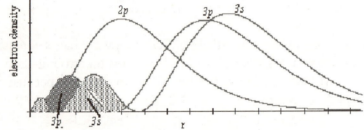

The shaded portions of the $3s$ and $3p$ indicate the regions that are ineffectively screened by the $2p$ orbital. $3s$ and $3p$ electrons have greater probability of being in the larger areas mostly outside the $2p$ orbital. Thus, the $2p$ orbital effectively screens electrons in both the $3s$ and $3p$ orbitals.

8.87   Francium is an alkali metal (Column 1) whose properties should closely resemble those of cesium: low *IE*, highly reactive, forms a cation with +1 charge, soft metal, melts at a low temperature.

8.89   Removing electrons from any atom or ion reduces screening and electron-electron repulsion and therefore stabilizes the orbitals. The Li^{2+} cation has a more stable $2s$ orbital.

8.91   Consult any of the graphs showing how electron density varies with distance from the nucleus. Notice that the density value decreases gradually. At what distance from the nucleus has the electron density decreased sufficiently that we consider it to be zero? That question has no unambiguous answer.

9.1   Determine a configuration from the position of an element in the periodic table. The
       electrons with the highest principal quantum number will be involved in bond formation:
       (a) O: $1s^2\ 2s^22p^4$; its six $n = 2$ electrons will be involved in bond formation.
       (b) P: $1s^2\ 2s^22p^63s^23p^3$; its five $n = 3$ electrons will be involved in bond formation.
       (c) B: $1s^2\ 2s^22p^1$; its three $n = 2$ electrons will be involved in bond formation.
       (d) Br: $1s^2\ 2s^22p^63s^23p^63d^{10}\ 4s^2\ 4p^5$; its seven $n = 4$ electrons will be involved in bond
           formation.

9.3   One valence $5p$ orbital of the iodine atom is directed along the axis between the two
       atoms and can overlap with the $1s$ orbital of the hydrogen atom to form a bond:

       5p orbital is shaded

9.5   The Group number tells us how many valence electrons any element has:
       (a) Al, Group 13, $13 - 10 = 3$ valence electrons
       (b) As, Group 15, $15 - 10 = 5$ valence electrons
       (c) F, Group 17, $17 - 10 = 7$ valence electrons
       (d) Sn, Group 14, $14 - 10 = 4$ valence electrons

9.7   Bonds form by sharing electrons in valence orbitals. The configuration of Li is [He] $2s^1$.
       The electron in the $2s$ orbital of a lithium atom and the electron in the $1s$ orbital of a
       hydrogen atom are shared between the two atoms, forming a bond between the nuclei:

9.9   Electronegativities describe the tendency of each element to attract bonding electrons
       from another. In any pair, the element with higher electronegativity attracts bonding
       electrons more strongly. Use Figure 9-7 to obtain the electronegativity of each element:
       (a) N (3.0) will attract electrons more than C (2.5)
       (b) S (2.5) will attract electrons more than H (2.1)
       (c) I (2.5) will attract electrons more than Zn (1.6)
       (d) S (2.5) will attract electrons more than As (2.0)

9.11   Electronegativity differences determine the direction of bond polarities. The more
       electronegative atom has the negative charge:
       (a) δ^+ Si $-$ O δ^-; (b) δ^- N $-$ C δ^+; (c) δ^+ Cl $-$ F δ^-, and (d) δ^- Br $-$ C δ^+

9.13   Bond polarity increases with the difference in electronegativity of the bond-forming elements. In these compounds, H is the less electronegative element, so bond polarity increases with the electronegativity of the other element: The electronegativity order is $P < S < N < O$, so the order of bond polarity is $PH_3 < H_2S < NH_3 < H_2O$

9.15   Determine numbers of valence electrons from the position of each element in the periodic table, adding one for each negative charge and subtracting one for each positive charge. (a) H_3PO_4: each H has one valence electron, P (Group 15) has five, and each O (Group 16) has six, for a total of 32 valence electrons; (b) $(C_6H_5)_3C^+$: each H has one valence electron and each C (Group 14) has four; subtract one for the positive charge, giving a total of 90 valence electrons; (c) $(NH_2)_2CO$: each H has one valence electron, each N (Group 15) has five, C (Group 14) has four, and O has six, for a total of 24 valence electrons; and (d) SO_4^{2-}: S and O (Group 16) have six valence electrons; add two for the two negative charges, giving a total of 32 valence electrons

9.17   Build a molecular framework from information contained in the chemical formula. Each chemical bond requires two valence electrons:

(a) $(CH_3)_3CBr$: There are 13 bonds, which require 26 valence electrons:

(b) $(CH_3CH_2CH_2)_2NH$: There are 21 bonds, which require 42 valence electrons:

(c) $HClO_3$: remember that H bonds to O in oxoacids. There are four bonds, which require eight valence electrons:

(d) $OP(OCH_3)_3$: the parentheses identify three OCH_3 groups bonded to P. There are 16 bonds, which require 32 valence electrons:

9.19   Use the procedure in your textbook for determining the Lewis structures.
1.   Each species has 8 valence electrons.
2.   Each species has its H atoms bonded to N, and each N-H bond requires two electrons.
3.   There are no outer atoms other than H.
4.   Place remaining electrons on the N atom:

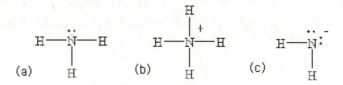

Optimize electron configurations of the inner atoms: N has an octet in each structure, so all three are correct.

9.21   Use the procedure in your textbook for determining the Lewis structures.
1.   Count the valence electrons.
(a) PBr_3: P (Group 15) has 5 electrons, Br (Group 17) has 7, for a total of 26
(b) SiF_4: Si (Group 14) has 4 electrons, F (Group 17) has 7, for a total of 32
(c) BF_4^-: B (Group 13) has 3 electrons, F (Group 17) has 7, plus one for the anion's charge, for a total of 32
2.   Assemble the bonding framework. In each of these molecules, the unique atom is inner. Each bond uses two valence electrons:
(a) PBr_3: three P – Br bonds use six electrons
(b) SiF_4: four Si – F bonds use eight electrons
(c) BF_4^-: four B – F bonds use eight electrons
3.   Add 3 nonbonding electron pairs to all non-hydrogen outer atoms.
(a) PBr_3: three pairs on each Br atom use 18 electrons; there are two left
(b) SiF_4: three pairs on each F atom use 24 electrons; there are none left
(c) BF_4^-: three pairs on each F atom use 24 electrons; there are none left

4.   Add any remaining electrons to the inner atom:

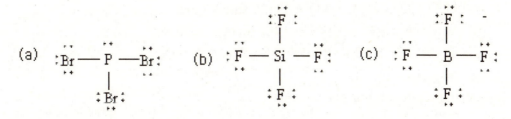

5.   Optimize electron configurations of the inner atoms: P and Si have FC = 0, so structures (a) and (b) are correct. B has FC = -1, but there is no way to reduce this charge, so structure (c) is correct.

9.23   Use the procedure in your textbook for determining the Lewis structures.
1. Count the valence electrons.
  (a) H_3CNH_2: C (Group 14) has 4 electrons, N (Group 15) has 5, and each H has 1, for a total of 14
  (b) CF_2Cl_2; C (Group 14) has 4 electrons, F and Cl (Group 17) have 7, for a total of 32
  (c) OF_2: O (Group 16) has 6 electrons, F (Group 17) has 7, for a total of 20
2. Assemble the bonding framework. Each bond uses two valence electrons.
  (a) The formula of H_3CNH_2 indicates its framework:

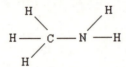

Six bonds use 12 valence electrons, leaving two to be placed
  (b) CF_2Cl_2: C is inner, with four bonds to the outer halogens that use eight electrons
  (c) OF_2: two O – F bonds use four electrons
3. Add 3 nonbonding electron pairs to all non-hydrogen outer atoms:
  (a) H_3CNH_2: there are no outer atoms other than H; there are two electrons left
  (b) CF_2Cl_2: three pairs on each halogen atom use 24 electrons; there are none left
  (c) OF_2: three pairs on each F use 12 electrons; there are four left
4. Add any remaining electrons to an inner atom:

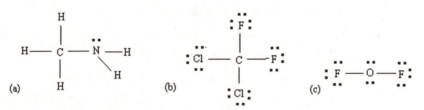

5. Optimize electron configurations of the inner atoms: All inner atoms are from the second row and have octets, so these are the correct structures.

9.25   Use the procedure in your textbook for determining the Lewis structures.
1. Count valence electrons:
  (a) $(CH_3)_2CO$ has $3(4) + 6(1) + 6 = 24$ valence electrons
  (b) CH_3CN has $2(4) + 3(1) + 5 = 16$ valence electrons
  (c) CH_2CHCH_3 has $3(4) + 6(1) = 18$ valence electrons
  (d) CH_3CHNH has $2(4) + 5(1) + 5 = 18$ valence electrons
2. Use the chemical formula to determine the framework:

3. Add three electron pairs to each outer atom except H. Only (a) and (b) have such atoms:

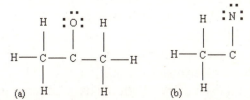

(a)   (b)

At the end of this step, all the valence electrons have been placed for (a) and (b), but there are two electrons left to place on (c) and (d)

4. Place remaining electrons on inner atoms, starting with the most electronegative:

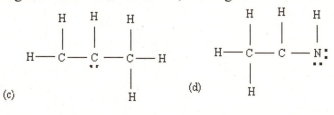

(c)   (d)

5. Optimize electron configurations of the inner atoms: Each structure has an inner atom with less than an octet of electrons. Shift lone pairs to make multiple bonds until all inner atoms have octets:

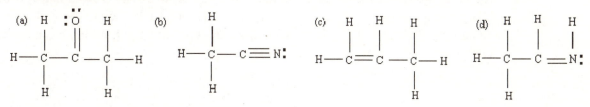

6. There are no equivalent structures.

9.27   Use the procedure in your textbook for determining Lewis structures.
1. Count valence electrons:
    (a) IF_5: 6(7) = 42 valence electrons
    (b) SO_3: 4(6) = 24 valence electrons
    (c) $OPCl_3$: 6 + 5 + 3(7) = 32 valence electrons
    (d) XeF_2: 8 + 2(7) = 22 valence electrons
2. Use the chemical formula to determine the framework. Each bond uses two valence electrons.
    (a) IF_5: I is inner and forms five bonds to F atoms, using 10 valence electrons
    (b) SO_3: S is inner and forms three bonds to O atoms, using six valence electrons
    (c) $OPCl_3$: P is inner and forms four bonds to the outer atoms, using eight valence electrons
    (d) XeF_2: Xe is inner and forms two bonds to F atoms, using four valence electrons

3. Add three electron pairs to each outer atom except H.
   (a) Three pairs on each of five F atoms uses 30 more electrons, leaving
       $42 - 10 - 30 = 2$ electrons
   (b) Three pairs on each of three O atoms uses 18 more electrons, leaving
       $24 - 6 - 18 = 0$ electrons
   (c) Three pairs on each of four outer atoms uses 24 more electrons, leaving
       $32 - 8 - 24 = 0$ electrons
   (d) Three pairs on each of two F atoms uses 12 more electrons, leaving
       $22 - 4 - 12 = 6$ electrons

4. Add remaining electrons to inner atoms. Each structure has only one inner atom, so
   all the remaining electrons must go on that atom regardless of octet considerations:

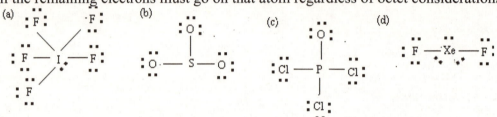

5. Optimize electron configurations of the inner atoms: Each inner atom is beyond row
   2, so optimize based on formal charge. In structures (a) and (d), the formal charge of
   the inner atom is zero, so these are the correct structures. The S atom has a formal
   charge of $6 - 3 = 3$, so make three double bonds, one from each O atom. The P atom
   has a formal charge of $5 - 4 = 1$, so make a double bond to O, which has a formal
   charge of -1:

6. There are no equivalent structures.

9.29 Use the procedure in your textbook for determining the Lewis structures.
   1. Count valence electrons, adding one for each negative charge:
      (a) NO_3^-: $5 + 3(6) + 1 = 24$ valence electrons
      (b) HSO_4^-: $5(6) + 1 + 1 = 32$ valence electrons
      (c) CO_3^{2-}: $4 + 3(6) + 2 = 24$ valence electrons
      (d) ClO_2^-: $7 + 2(6) + 1 = 20$ valence electrons
   2. Use the chemical formula to determine the framework. Each bond uses two valence
      electrons:
      (a) NO_3^-: N is inner and forms bonds to three O atoms, using 6 valence electrons
      (b) HSO_4^-: S is inner and forms bonds to four O atoms, one of which bonds to H,
          using ten valence electrons
      (c) CO_3^{2-}: C is inner and forms bonds to three O atoms, using 6 valence electrons
      (d) ClO_2^-: Cl is inner and forms bonds to 2 Cl atoms, using four valence electrons

3. Add three electron pairs to each outer atom except H.
   (a) Three pairs on each of 3 O atoms uses 18 electrons, leaving
       $24 - 6 - 18 = 0$ electrons
   (b) Three pairs on each of 3 O atoms (the fourth is inner, bonded to H and S) uses
       18 electrons, leaving $32 - 10 - 18 = 4$ electrons
   (c) Three pairs on each of 3 O atoms uses 18 electrons, leaving
       $24 - 6 - 18 = 0$ electrons
   (d) Three pairs on each of 2 O atoms uses 12 electrons, leaving
       $20 - 4 - 12 = 4$ electrons
4. Add remaining electrons to inner atoms to complete octets.
   (a) No electrons to add
   (b) Add the four electrons to the inner O, the more electronegative atom
   (c) No electrons to add
   (d) Add the four electrons to Cl:

5. Optimize electron configurations of the inner atoms. The inner atoms in (a) and (c)
   are second row, so complete their octets. The inner atoms in (b) and (d) are third row,
   so reduce their formal charges to zero by making two double bonds to each:

6. The outer oxygen atoms all are equivalent, so all but (d) have three equivalent
   structures.

9.31   Molecular shapes are determined by the steric numbers (SN) of inner atoms, which can be found from the Lewis structures. The Lewis structures of these molecules, determined in Problems 9.21 and 9.23, show that each inner atom has SN = 4, so each has tetrahedral molecular group geometry:

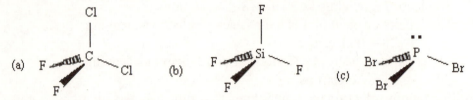

(a) CF_2Cl_2 has tetrahedral shape, with F atoms at two apices and Cl atoms at the other two; (b) SiF_4 has tetrahedral shape; (c) PBr_3, with a lone pair of electrons on the inner atom, has trigonal pyramidal shape

9.33   Construct the bonding framework from the formula, knowing that the more electronegative atoms (Cl) will be in outer positions. In 1,2-dichloroethane, each carbon atom is bonded to the other, and there is one Cl atom bonded to each carbon atom. There are $2(7) + 2(4) + 4(1) = 26$ valence electrons. The bonding framework contains 7 bonds, and there are three lone pairs on each Cl atom, accounting for all the valence electrons. Thus, the bonding framework is the correct Lewis structure of 1,2-dichloroethane. Each carbon atom has SN = 4 and tetrahedral geometry. Your ball-and-stick model should reflect this:

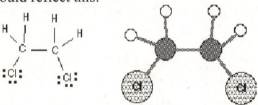

9.35   To draw structural isomers of an alkane, start with the "straight chain" compound, then rearrange the bonding arrangement to make other isomers. It is convenient to work with only the carbon skeleton. You can add H atoms to complete the Lewis structures:

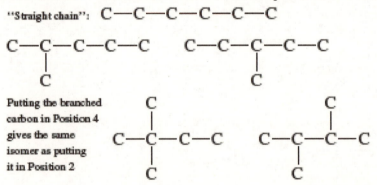

9.37   Determine the Lewis structure following the standard procedures.
1) There are $2(4) + 7(1) + 5 = 20$ valence electrons.
2) The chemical formula indicates that the bonding framework includes two $-CH_3$ groups attached to the nitrogen atoms. The bonding framework contains 9 bonds = 18 electrons, leaving 2 valence electrons:
3) All the outer atoms are hydrogen, so no electrons can be placed on outer atoms.
4) Place the remaining pair o the nitrogen atom. The resulting structure has an octet around N, so it is the correct Lewis structure.
Each inner has SN = 4, so the electron group geometry about each inner atom is tetrahedral. The shape about the N atom is trigonal pyramidal, like ammonia:

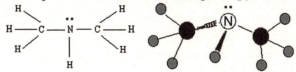

9.39   Determine the Lewis structure following the standard procedures:
1) There are $4(4) + 4 + 12(1) = 32$ valence electrons.
2) The chemical formula indicates that the bonding framework includes four $-CH_3$ groups attached to the silicon atom. The bonding framework has 16 bonds = 32 electrons. All electrons are placed, so this structure is the correct Lewis structure. Each inner atom has SN = 4, and the geometry about each inner atom is tetrahedral:

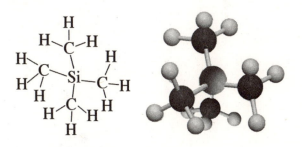

9.41   Determine molecular shapes from electron pair geometry, taking account of lone pairs:
(a) Two lone pairs and three ligands gives SN = 5 and trigonal bipyramidal electron group geometry. Two equatorial positions are occupied by the lone pairs, so this is T-shaped.
(b) A steric number of five means trigonal bipyramidal electron group geometry. One equatorial position is occupied by the lone pair, so this is seesaw shaped.
(c) Steric number of three means trigonal planar electron group geometry, and the molecular shape is the same when there are no lone pairs.
(d) A steric number of six means octahedral electron group geometry. One lone pair gives the shape of a square pyramid.

9.43   Determine the Lewis structures following the standard procedures.
GeF_4:
1) There are $4 + 4(7) = 32$ valence electrons.
2) Four electron pairs are used in forming the bonding framework, leaving $32 - 4(2) = 24$ electrons.
3) Place 6 electrons around each outer F atom, leaving $24 - 4(6) = 0$ electrons.
4) There are no remaining electrons.
5) The resulting structure has $FC_{Ge} = 4 - 4 = 0$, so the structure is correct.

SeF_4:

1) There are $6 + 4(7) = 34$ valence electrons.
2) Four electron pairs are used in forming the bonding framework, leaving $34 - 4(2) = 26$ electrons.
3) Place three pairs of electrons around each outer F atom, leaving $26 - 4(6) = 2$ remaining electrons.
4) Place the remaining two electrons on the inner Se atom.
5) The resulting structure has $FC_{Se} = 6 - 4 - 2 = 0$, so the structure is correct.

XeF_4:

1) There are $8 + 4(7) = 36$ valence electrons.
2) Four electron pairs are used in forming the bonding framework, leaving $36 - 4(2) = 28$ electrons.
3) Place three pairs of electrons around each outer F atom, leaving $28 - 4(6) = 4$ remaining electrons.
4) Place the remaining four electrons on the inner Xe atom.
5) The resulting structure has $FC_{Xe} = 8 - 4 - 4 = 0$, so the structure is correct.

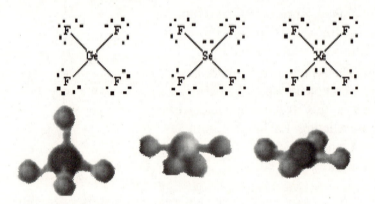

In GeF_4, $SN_{Ge} = 4$, so this molecule is tetrahedral. In SeF_4, $SN_{Se} = 5$ with a lone pair, so this molecule has seesaw geometry. In XeF_4, $SN_{xe} = 6$ with two lone pairs, so this molecule has square planar geometry.

9.45   Determine the Lewis structures following the standard procedures:

SO_2:

1) There are $3(6) = 18$ valence electrons.
2) Two electron pairs are used in forming the bonding framework, leaving $18 - 2(2) = 14$ electrons.
3) Place 6 electrons around each outer O atom, leaving $14 - 2(6) = 2$ electrons.
4) Place the remaining pair on the S atom.
5) The resulting structure has $FC_S = 6 - 4 = 2$. Make two double bonds to complete the Lewis structure.

SbF_5:

  1) There are $5 + 5(7) = 40$ valence electrons.
  2) Five electron pairs are used in forming the bonding framework, leaving $40 - 5(2) = 30$ electrons.
  3) Place 6 electrons around each outer F atom, leaving $30 - 5(6) = 0$ electrons.
  4) There are no electrons left to place on the inner atom.
  5) The resulting structure has $FC_{Sb} = 5 - 5 = 0$, so the structure is correct.

ClF_4^+:

  1) There are $7 + 4(7) - 1 = 34$ valence electrons.
  2) Four electron pairs are used in forming the bonding framework, leaving $34 - 4(2) = 26$ electrons.
  3) Place 6 electrons around each outer F atom, leaving $26 - 4(6) = 2$ electrons.
  4) Place the remaining electron pair on the inner Cl atom.
  5) The resulting structure has $FC_{Cl} = 7 - 4 - 2 = +1$, the same as the overall charge, so the structure is correct.

ICl_4^-:

  1) There are $5(7) + 1 = 36$ valence electrons.
  2) Four electron pairs are used in forming the bonding framework, leaving $36 - 4(2) = 28$ electrons.
  3) Place 6 electrons around each outer F atom, leaving $28 - 4(6) = 4$ electrons.
  4) Place the remaining electron pairs on the inner I atom.
  5) The resulting structure has $FC_I = 7 - 4 - 4 = -1$, the same as the overall charge, so the structure is correct.

(a) SN = 3, giving trigonal planar electron group geometry. One lone pair gives a bent shape, with ideal angle of 120°; (b) SN = 5, giving trigonal bipyramidal shape and ideal angles of 90° and 120°; (c) SN = 5, giving trigonal bipyramidal electron group geometry. One lone pair gives a seesaw shape, with ideal angles of 90° and 120°; (d) SN = 6, giving octahedral electron group geometry. Two lone pairs results in a square planar shape, with ideal bond angles of 90°.

9.47 Determine Lewis structures using the standard procedures. (Here we do not number the steps.) Only asymmetric molecules have dipole moments. Use the Lewis structures to determine steric numbers and ascertain whether or not the molecules are asymmetric:

SiF_4: There are $4 + 4(7) = 32$ valence electrons. 4 pairs are used in the bonding framework, and 3 pairs are placed on each outer F atom. This leaves $FC_{Si} = 4 - 4 = 0$, so this is the correct Lewis structure.

H_2S: There are $6 + 2 = 8$ valence electrons. 2 pairs are used in the bonding framework, the remaining 4 electrons are placed on the S atom.

XeF_2: There are $8 + 2(7) = 22$ electrons. 2 pairs are used in the bonding framework, three pairs are placed on each outer F atom, the remaining 6 electrons are placed on the inner Xe atom. The resulting structure has $FC_{Xe} = 8 - 8 = 0$, so this is the correct Lewis structure.

$GaCl_3$: There are $3 + 3(7) = 24$ valence electrons. 3 pairs are used in the bonding framework. Add 3 pairs to each outer atom (using the remaining electrons). Gallium is in the fourth row, Group 13. Its formal charge is $3 - 3 = 0$, so this is the correct structure.

NF_3: there are $5 + 3(7) = 26$ valence electrons. 3 pairs are used in the bonding framework. Add 3 pairs to each outer atom. The remaining pair goes on the inner N atom, giving N an octet, so this is the correct structure:

SiF_4 is tetrahedral and symmetric; it has no dipole moment; H_2S, like H_2O, is bent and has a dipole moment; XeF_2 has $SN = 5$, with its 3 lone pairs in equatorial positions, so the molecule is linear without a dipole moment; $GaCl_3$ has $SN = 3$, so the molecule is trigonal planar and has no dipole moment; NF_3, like NH_3, is pyramidal and has a dipole moment.

9.49   The Lewis structure of CO_2 shows no lone pairs on the C atom, giving $SN = 2$ and linear shape. The two $C = O$ bonds point opposite each other, so bond polarities cancel. The Lewis structure of SO_2 (Problem 9.43) shows a lone pair on the S atom, resulting in a bent molecule whose polar bonds do not cancel each other.

9.51   Determine Lewis structures in the usual fashion:

(a) PF_5 has no lone pairs on its inner atom, and all its ligands are identical, so it is trigonal bipyramidal with ideal bond angles of 90° and 120°.

(b) CH_3I has $SN = 4$ and tetrahedral electron group, with no lone pairs, but the very large I atom repels the H atoms, making the $I - C - H$ angles larger than 109.5° and the $H - C - H$ angles smaller than 109.5°.

(c) BrF_5 has octahedral electron
group geometry and a lone pair
on Br, making the F – Br – F
angles to the F atom at the apex
smaller than 90°:

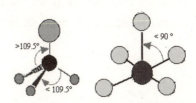

9.53   The determinants of bond length, in order of importance, are principal quantum number
of the valence orbitals, bond order, and bond polarity. Among these bonds, H–N is
shortest because it involves an $n = 1$ orbital and Cl–N is longest because it involves an
$n = 3$ orbital. Of the three bonds involving $n = 2$ orbitals, the N–N single bond is longest
and the polar C≡N triple bond is shortest: H–N < C≡N < N≡N < N–N < Cl–N

9.55   Bond strengths, like bond lengths, depend on principal quantum number, bond order, and
bond polarity, but the correlation is not as strong as for bond lengths, so tabulated values
must be consulted. Here are the values from Table 6-2 of your textbook, listed in
increasing order:

Bond	C–C	<	H–N	<	C=C	<	C=O	<	N≡N
Energy (kJ/mol)	345		390		615		750		945

Strong bonding and bond polarity of the $n = 1$ orbital makes H–N > C–C; multiple
bonding makes C=C > H–N; bond polarity makes C=O > C=C; and multiple bonding
makes N≡N > C=O.

9.57   Determine Lewis structures following the standard procedures. Draw ball-and-stick
models that illustrate the molecular geometries determined by the steric numbers of the
inner atoms.
(a) Cl_2O has $2(7) + 6 = 20$ valence electrons. Its Lewis structure has two bonding pairs,
three lone pairs on each chlorine atom, and two lone pairs on the inner oxygen atom.
$SN_O = 4$, leading to tetrahedral electron pair geometry and bent molecular shape:

(b) C_6H_6 has $6(4) + 6(1) = 30$ valence electrons. After constructing the bonding
framework, six electrons remain to be placed. Place these on alternate carbon atoms;
then complete octets around the other carbon atoms by shifting lone pairs to make
double bonds. There are two equivalent resonance structures:

Provisional structure   resonance structures

$SN_C = 3$, leading to trigonal planar geometry about each carbon atom and a flat, hexagonal molecular shape.

(c) C_2H_4O has $2(4) + 4(1) + 6 = 18$ electrons. When the bonding framework is complete, four electrons remain. These are placed on the oxygen atom. The three bond angles of the triangular ring must sum to 180°, even though $SN = 4$ for each atom.

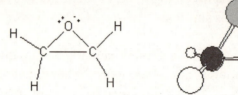

9.59   Determine the Lewis structures using standard procedures. Each of the molecules in this problem has four halogen atoms bonded to an inner atom. Both CI_4 and $SiCl_4$ have 32 electrons and SeF_4 has 34 valence electrons. The 32-electron species have the same Lewis structure and the same geometry.

CI_4 and $SiCl_4$:
1) There are $6 + 4(7) = 32$ valence electrons.
2) Four electron pairs are used in forming the bonding framework, leaving $32 - 4(2) = 24$ electrons.
3) Place three pairs of electrons around each outer atom, leaving $24 - 4(6) = 0$ electrons.
4) There are no remaining electrons.
5) The resulting structure has $FC_{Si} = 4 - 4 = 0$ and C has an octet, so the structures are correct.

SeF_4:
1) There are $6 + 4(7) = 34$ valence electrons.
2) Four electron pairs are used in forming the bonding framework, leaving $34 - 4(2) = 26$ electrons.
3) Place 6 electrons around each outer F atom, leaving $26 - 4(6) = 2$ electrons.
4) Place the remaining two electrons on the inner Se atom.
5) The resulting structure has $FC_{Se} = 6 - 4 - 2 = 0$, so the structure is correct.

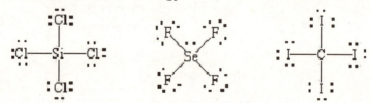

The 32-electron species have $SN = 4$ and are tetrahedral; SeF_4 is a seesaw.

9.61   To find exceptions to normal trends, consult the appropriate listings in tabulated values. Table 9-2 lists bond lengths by *n*-value. For *n* = 2 it lists four X–X bonds, C–C, N–N, O–O, and F–F; and three X–Y bonds, C–N, C–O, and C–F. Consult Table 6-2 to find the exceptions:
(a) C–N is shorter but weaker than C–C; (b) C–N is longer but stronger than N–N.

9.63   Determine the number of different isomers by constructing the different possible geometric possibilities. Two X atoms can be placed at opposite ends of one Cartesian axis; all structures with this arrangement are equivalent, because the molecule can be rotated about an axis to superimpose these positions. The three X atoms can all be placed at right angles to one another, giving a different isomer. Thus, there are two isomers:

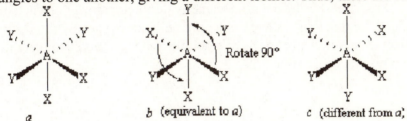

9.65   The empirical chemical formula of the silicon-oxygen network of zircon, an orthosilicate, is $SiO_4{}^{4-}$. Each silicon atom has SN = 4 and tetrahedral geometry, and each O atom is outer, so the molecular geometry of the $SiO_4{}^{4-}$ anions is tetrahedral. Orthosilicates are ionic, containing discrete (not connected) $SiO_4{}^{4-}$ anions and metal cations.

9.67   Determine Lewis structures following the standard procedures.
(a) Bromate, $BrO_3{}^-$, contains $7 + 3(6) + 1 = 26$ valence electrons. Three pairs are required for the bonding framework, nine pairs go on the oxygen atoms, and the remaining pair is placed on the bromine atom. The resulting structure has $FC_{Br} = +2$, shift two electron pairs to make two double bonds and reduce FC_{Br} to 0:

$FC_{Br} = 7 - 5 = +2$
Shift 2 electron pairs
to make 2 double bonds

There are two additional resonance structures

(b) Nitrite, $NO_2{}^-$, contains $5 + 2(6) + 1 = 18$ valence electrons. Two pairs are required for the bonding framework, six pairs go on the O atoms, and the remaining pair is placed on N. Shift one electron pair to make a double bond and complete the octet on nitrogen. There are two equivalent structures. $FC_N = 5 - 5 = 0$:

2 equivalent structures

137

(c) Phosphate, PO_4^{3-}, contains $5 + 4(6) + 3 = 32$ valence electrons. Four pairs are required for the bonding framework and 12 pairs go on the oxygen atoms, giving a provisional structure with $FC_P = 5 - 4 = +1$. Shift one electron pair to make a double bond and reduce the formal charge to zero. There are four resonance structures.

There are two additional equivalent structures

(d) Hydrogen carbonate, HCO_3^-, contains $1 + 4 + 3(6) + 1 = 24$ valence electrons. Four pairs are required for the bonding framework, three pairs go on each outer oxygen atom, and the remaining two pairs go on the inner oxygen atom. Shift one electron pair from an outer O atom to form a double bond to C and complete its octet. There are two resonance structures:

9.69   Consult Table 9-3 for the correspondence between molecular shapes and lone pairs. (a) This is the seesaw shape, derived from the trigonal bipyramid when there is one lone pair. The ideal bond angles are $90°$ and $120°$, and an example is SF_4. (b) This is the square planar shape, derived from the octahedron when there are two lone pairs, which take opposing axial positions. The ideal and actual bond angles are $90°$, and an example is XeF_4. (c) This is the tetrahedron, the shape associated with $SN = 4$ and no lone pairs. The ideal bond angles are $109.5°$, and an example is CH_4.

9.71   The determinants of bond length, in order of relative importance, are principal quantum number of the valence orbitals, bond order, and bond polarity. The bonds found in these molecules are H–O, H–C, C=O, and C≡N. Of these, H–O and H–C are shorter than the others because they form from an $n = 1$ orbital; H–O is shorter than H–C because it is a more polar bond. C=O has bond order 2 and is longer than C≡N because bond order is a stronger determinant of length than bond polarity.

9.73   Follow the standard procedure to obtain the Lewis structure of a molecule, then use the Lewis structure to determine the geometry around inner atoms.

(a) This molecule has 4 N (5 e⁻) $+ 2$ C (4 e⁻) $+ 2$ O (6 e⁻) $+ 4$ H (1 e⁻) $= 44$ valence electrons, 22 of which are used for the 11 bonds in the framework. Add 6 electrons around each outer O atom, then place the remaining 5 pairs on inner N atoms, which are more electronegative than C atoms, giving a provisional structure in which the three second-row atoms marked with *3* have only three pairs of electrons. Complete their octets by shifting electron pairs to form double bonds in the final Lewis structure.

Provisional structure   Final structure

(b) The end N atoms have SN = 4 and tetrahedral electron pair geometry. Because of their lone pairs, these N atoms have trigonal pyramidal geometries. The C atoms and the center N atoms have SN = 3 and trigonal planar electron pair geometry. The geometry about the C atoms is the same as their electron pair geometry, whereas the lone pairs on the center N atoms result in bent shapes about these atoms.

9.75   VSEPR theory predicts tetrahedral bond angles, 109.5°, for both these molecules. A bond angle of 104.5° is a bit smaller than this, because of the larger repulsion of lone pairs. An angle of 92.2° indicates that the bonding can be well described using *p* orbitals rather than spacing the electron pairs as far apart as possible. Space-filling models of the two molecules shows that the smaller oxygen atom cannot accommodate two H atoms at right angles, but the larger S atom can.

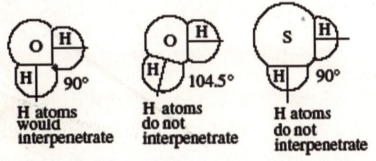

9.77   The electronegativity of hydrogen (2.1) is the smallest among the elements involved in these chemical bonds, so the ranking of bond polarity is from smallest to largest electronegativity of the bonding partner. Remember that electronegativity decreases down a column and increases across a row: Si-H < C-H < N-H < O-H < F-H

9.79   All positions around a tetrahedral center are equivalent, so there is only one isomer of CH_2Br_2. The two structures show different views of the same compound.

9.81   The six-membered ring of benzyne requires bond angles of 120°. The two carbon atoms involved in the triple bond have SN = 2, for which the optimum VSEPR angle is 180°. Thus, these atoms react readily, either by adding another bonded atom to generate SN = 3 or by breaking the ring so the bond angle can become 180°.

9.83   Determine the Lewis structures using standard procedures. Both molecules have
2(5) + 6 = 16 valence electrons. Two pairs are required for the bonding framework, and
each outer atom receives three pairs:

:N——N——O: :N——O——N:

Because these molecules contain only row two atoms, we need to complete the octets.
This can be accomplished by shifting two electron pairs to make double bonds:

N-N-O   N-O-N

N——N——O   N——O——N

Molecules have dipole moments only if they are unsymmetric. A linear N–O–N structure
would not have a dipole moment, because the N–O dipole would exactly cancel the O–N
dipole. Thus N_2O must have the structure N–N–O if it is to have a dipole moment.

9.85   A molecule has a dipole moment if it is non-symmetric. In isomer a, the two polar C–Cl
bonds point in opposite directions and cancel, thus a is nonpolar. Both b and c are polar,
because their dipoles do not cancel; b has the larger dipole moment, because its dipoles
point more nearly in the same direction.

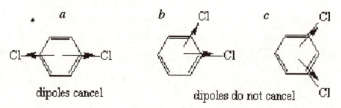

a dipoles cancel b c dipoles do not cancel

9.87   The Lewis structures of molecules with formula XF_3 show octets around the inner atom
and $FC_X = 0$, making them stable. Compounds with formula XF_5 also have $FC_X = 0$ but
have five electron pairs associated with the inner atom. This is possible for phosphorus, a
third row element that has d orbitals available for bonding. It is not possible for nitrogen,
a second row element that lacks valence d orbitals:

X = N or P

9.89   The shape of a molecule indicates the steric number of its inner atom and the number of
atoms bonded to the inner atom. Consult Table 9-3 of your textbook for details:
Tellurium, in column 16 of the periodic table, has six valence electrons, and each fluorine
contributes seven valence electrons. Species with an even number of fluorine atoms are
neutral, but those with an odd number must have one unit of charge (+1 or –1) in order to
contain an even number of electrons.
(a) bent indicates TeF_2

(b) T-shape indicates SN = 5 on Te, three F atoms, and two lone pairs on Te, for a total of $4 + 3(8) = 28$ valence electrons. TeF_3 would have $6 + 3(7) = 27$ valence electrons, so the formula of this species must be TeF_3^-

(c) square pyramid indicates SN = 6 on Te, five F atoms, and one lone pair on Te, for a total of $2 + 5(8) = 42$ valence electrons. TeF_5 would have $6 + 5(7) = 41$ valence electrons, so the formula of this species must be TeF_5^-

(d) trigonal bipyramid indicates SN = 5 on Te and five F atoms, for a total of $5(8) = 40$ valence electrons. TeF_5 would have $6 + 5(7) = 41$ valence electrons, so the formula of this species must be TeF_5^+

(e) octahedron indicates SN = 6 and six F atoms, formula TeF_6

(f) seesaw indicates SN = 5 on Te, four F atoms, and one lone pair on Te, formula TeF_4

9.91   (a) Square planar XY_2Z_2 molecules may have like atoms adjacent to or opposite each other:

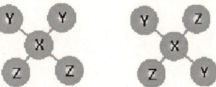

(b) When like atoms are opposite each other, their bond polarities oppose each other and cancel, so this isomer has no dipole moment. When like atoms occupy adjacent positions, the bond polarities do not entirely cancel, generating a net dipole moment:

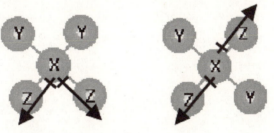

9.93   A compound XY_7 is possible if X has d orbitals available for bonding, but it is sterically crowded because of the large number of Y atoms around the central X atom. For this compound to exist, Y must be as small as possible and X must be as large as possible. Thus, Y is the halogen with the smallest size, fluorine; and X is iodine, the halogen with the largest size (apart from astatine, which has no stable isotopes). The compound is IF_7.

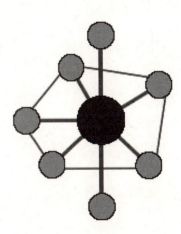

10.1   A diatomic halogen molecule forms its bond by overlap of the valence p orbitals that point along the bond axis. In Br_2, the valence orbitals have $n = 4$, so the bond forms by overlap of two $4p$ orbitals.

10.3   Hydrogen always uses its $1s$ orbital to form bonds. A Group 1 element has only one valence electron, in an ns orbital. Lithium is a second-row element, with $n = 2$ for its valence electron. Thus, the bond in LiH forms by overlap of the hydrogen $1s$ orbital and the lithium $2s$ orbital:

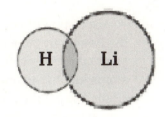

10.5   Bond angles near 90° signal interactions of valence p orbitals from the inner atom. An outer halogen atom always uses one of its valence p orbitals to form a bond. Antimony has $n = 5$ valence orbitals, and fluorine has $n = 2$ valence orbitals, so each of the three bonds in SbF_3 can be described as resulting from overlap between a $5p$ orbital from Sb and a $2p$ orbital from F. There are three identical bonds that point at near-right angles to one another:

10.7   The steric number of an inner atom uniquely determines its hybridization:
(a) SN = 2 + 2 = 4, requiring sp^3 hybridization; (b) SN = 3 + 1 = 4, requiring sp^3 hybridization; (c) SN = 3 + 0 = 3, requiring sp^2 hybridization; and (d) SN = 5 + 1 = 6, requiring sp^3d^2 hybridization.

10.9   The name of a hybrid orbital set uses the letters of the atomic orbitals, with superscripts indicating the number of each type, used to make the set: (a) sp^3; (b) sp; and (c) sp^3d^2.

10.11   The steric number of an inner atom uniquely determines its hybridization. Use the Lewis structure of the molecule to identify steric numbers:

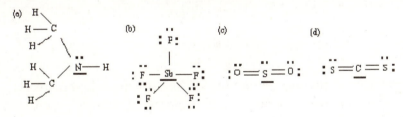

(a) SN = 4, sp^3 hybrids; (b) SN = 5, sp^3d hybrids; (c) SN = 3, sp^2 hybrids; and (d) SN = 2, sp hybrids.

10.13   The steric number of an inner atom uniquely determines its hybridization. Use the Lewis structure of the molecule to find steric numbers: (a) SN = 4, sp^3 hybrids; (b) SN = 5, sp^3d hybrids; (c) SN = 6, sp^3d^2 hybrids; and (d) SN = 3, sp^2 hybrids.

10.15   A description of bonding begins with the Lewis structure of the molecule, from which steric numbers of inner atoms identify hybridization. Outer atoms use atomic orbitals for bond formation. Here is the Lewis structure of chloroform:

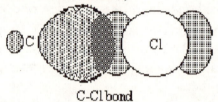

The carbon atom has SN = 4 and uses sp^3 hybrids to form three C – Cl bonds and one C – H bond. Each chlorine uses a $3p$ orbital for bond formation and has lone pairs in the $3s$ and two other $3p$ orbitals. The H atom uses its $1s$ orbital to form the C – H bond. The geometry about the carbon atom is tetrahedral. The bonds can be depicted as follows (the white region around the Cl atom represents its core electrons):

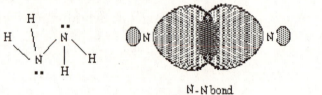

C-H bond   C-Cl bond

10.17   A description of bonding begins with the Lewis structure of the molecule, from which steric numbers of inner atoms translate into hybridization. Outer atoms use atomic orbitals for bond formation. The N atoms in hydrazine have SN = 4, use sp^3 hybrids and have tetrahedral geometry: 4 sp^3 (N) – $1s$ (H) σ bonds, 1 sp^3 (N) – sp^3 (N) σ bond, 2 lone pairs in sp^3 hybrids. Here is the Lewis structure (showing the approximate geometry) and sketches of the bonds:

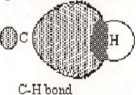 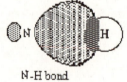

N-N bond   N-H bond

10.19   To determine the bonding pattern from a line structure, first convert the line structure into a molecular structure by adding C and H atoms. Molecular structures contain enough information to deduce the steric numbers of C, N, and O inner atoms without determining the complete Lewis structure. Acetone has three inner C atoms. Two have only single bonds (one C–C and three C–H); these have SN = 4. The atom bonded to O has one π bond and SN = 3. In the entire molecule, there are 6 sp^3 (C) – $1s$ (H) σ bonds, 2 sp^3 (C) – sp^2 (C) σ bonds, 1 sp^2 (C) – $2p$ (O) σ bond, 1 π bond between C and O, 2 lone pairs in $2s$ and $2p$ orbitals on O. Here are sketches of the bonding orbitals:

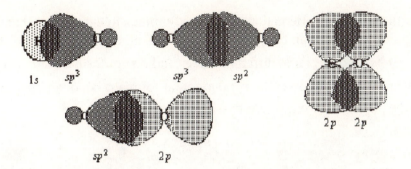

10.21 The line structure of neocembrene in the Chemistry and Life Box shows that it contains only inner carbon atoms and outer hydrogen atoms. Some of the inner atoms have all single bonds, SN = 4, sp^3 hybridization, and tetrahedral geometry; others have one double bond, SN = 3, sp^2 hybridization, and trigonal planar geometry. The line structure at right shows the SN values for the 20 carbon atoms:

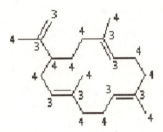

10.23 We determine the steric number of each carbon atom from the line structures of the compounds. These are shown on the line structures below:

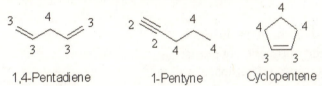

1,4-Pentadiene   1-Pentyne   Cyclopentene

C atoms designated "4" have only single bonds, SN = 4, sp^3 hybridization, and tetrahedral geometry. Those with double bonds, designated "3," have SN = 3, sp^2 hybridization, and trigonal planar geometry. Two C atoms, designated "2," have two double bonds, sp hybridization, and linear geometry. The σ bonds are described by overlap of hybrid orbitals, with each C–H bond described as a hybrid overlapping with a hydrogen $1s$ orbital. The π bonds are described by side-by-side overlap of $2p$ orbitals. In 1–pentyne, the two C atoms designated "2" have three bonds: a σ bond formed from sp hybrids and two π bonds, at right angles to each other, as in acetylene. Here are sketches of the different types of bonds:

σ bonds   π bond

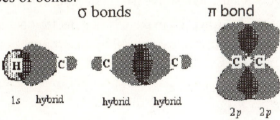

10.25   Examine the shape and orientation of a pair of orbitals to determine what kind of bond, if
any, it will form: (a) $2p_z$ and $2p_z$ point toward each other along the bond axis and form a
σ bond; (b) $2p_y$ and $2p_x$ lie in different planes and do not form a bond; (c) sp^3 and $2p_z$
point toward each other along the bond axis and form a σ bond; and (d) $2p_y$ and $2p_y$ are
in the same plane and overlap side-by-side to form a π bond.

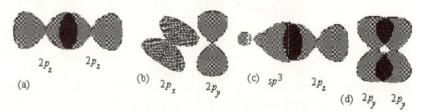

10.27   The orbital energy level diagram for first-row diatomic species has just two levels, each
of which can hold two electrons. The relative bond strengths can be deduced from the
bond orders, which are calculated using Equation 10-1 of your textbook:

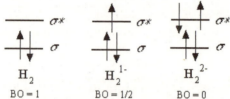

From weakest to strongest bond, the order is $H_2^{2-} < H_2^- < H_2$.

10.29   The stability of a diatomic molecule is
determined by the number of bonding and
antibonding electrons. To compare
species, determine how many valence
electrons each species possesses, then
place the electrons in the available
orbitals, following the Pauli and aufbau
principles. Here are the configurations for
nine and ten valence electrons:

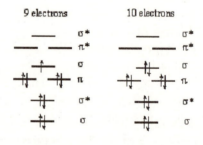

(a) CO has 10 valence electrons. To make CO^+, an electron is removed from a bonding
orbital, which destabilizes the ion, so CO has a stronger bond.

(b) N_2 has 10 valence electrons. To make N_2^+, an electron is removed from a bonding
orbital, which destabilizes the ion, so N_2 has a stronger bond.

(c) CN^- has 10 valence electrons. To make CN, an electron is removed from a bonding
orbital, which destabilizes the species, so CN^- has a stronger bond.

10.31 When two orbitals from different atoms interact, they form two MOs, one showing additive overlap and the other showing subtractive overlap. Additive overlap generates a bonding MO, while subtractive overlap generates an antibonding MO. Axial overlap gives sigma orbitals, and off-axis overlap gives π orbitals:

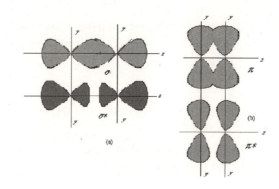

10.33 Determine how many valence electrons the ion possesses, then place the electrons in the available orbitals, following the Pauli and aufbau principles. Stability is determined by the number of bonding and antibonding electrons.

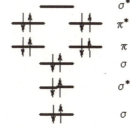

ClO⁻ has 14 valence electrons, six from O, seven from Cl, plus one due to its charge:

There are eight bonding electrons and six antibonding electrons, for a net of two bonding electrons and a single bond for this ionic species.

10.35 Determine the Lewis structure using the usual procedures. The molecule has $2(6) + 4 = 16$ valence electrons:

$$\ddot{\underset{\cdot\cdot}{S}} = C = \ddot{\underset{\cdot\cdot}{S}}$$

The bonding pattern for CS_2 is like that for CO_2, except that the S atoms have $n = 3$ valence orbitals rather than $n = 2$ orbitals. The inner atom can be described using sp hybrids which overlap with $3p$ orbitals from the S atoms to form two σ bonds. There are two delocalized π systems at right angles to each other, each made up of a $2p$ orbital on the C atoms overlapping side-by-side with a $3p$ orbital on each S atom. As in CO_2, 8 electrons occupy the delocalized π orbitals, 4 in bonding and 4 in non-bonding orbitals.

10.37 Determine the Lewis structures using the usual procedures. The NCN^{2-} anion has $2(5) + 4 + 2 = 16$ valence electrons, making it isoelectronic with CO_2:

$$\ddot{\underset{\cdot\cdot}{N}} = C = \ddot{\underset{\cdot\cdot}{N}}$$

The Lewis structure and MO description are the same as that for CO_2. There is no lone pair on the inner C atom, giving SN = 2, sp hybridization, and a linear geometry. There are two sets of three-center delocalized π orbitals:

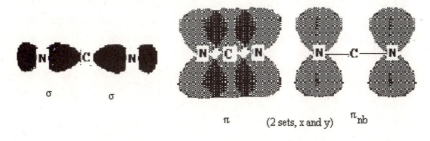

10.39 Determine the Lewis structure using the usual procedures. The molecule has
$3(5) + 1 = 16$ valence electrons. Begin by drawing the bonding framework and placing
lone pairs on the outer atoms:

This uses all the valence electrons. Form double and triple bonds to complete the octet on
the inner N atom:

The N atom bonded to H has SN = 3 and 4 in the two structures, so it has bent geometry
with an angle between 109° and 120° (the experimental value is 112° 39'). The best
hybridization is sp^2, leaving one p orbital free to form a π bond. The other inner N atom
has SN = 2, sp hybridization, and bond angles of 180°. There are two π networks, one
delocalized over all three N atoms and the other localized between the outer N atom and
its adjacent inner N atom.

10.41 Conjugation stabilizes a molecule whenever there are more than two adjacent atoms
involved in π bond formation, thereby leading to delocalized π orbitals. Of the molecules
whose structures appear in the Chemistry and Life Box of your textbook, only xanthin
has a delocalized π system, to which all 26 atoms that are part of the double-bond system
contribute.

10.43 We can determine hybridizations directly
from line structures: Each carbon atom
with no double bonds, and each inner
oxygen atom, has SN = 4, and sp^3 hybrids
can be used to describe the bonding. A C
atom with one double bond has SN = 3
and sp^2 hybridization:

(a) Hybridization is as shown on the line drawing ($3 = sp^3$, $2 = sp^2$).
(b and c) The Lewis structure shows 2 π bonds on adjacent atoms, so as in butadiene
    there are four electrons in the delocalized π system.
(d) The four atoms with double bonds and the four additional atoms bonded directly to
    them all lie in the same plane.

10.45 Determine the Lewis structure using the usual procedures: The molecule has
$3(4) + 2(6) = 24$ valence electrons. Begin by drawing the bonding framework and adding
in lone pairs of electrons to the outer atoms. Use the lone pairs to form double bonds to
complete the octets on the inner carbons:

Each carbon atom has SN = 2, so the molecule is linear and the σ bond network can be described using *sp* hybrids. There are two *sp-sp* bonds between carbon atoms and two *sp-2p* bonds between carbon and oxygen atoms:

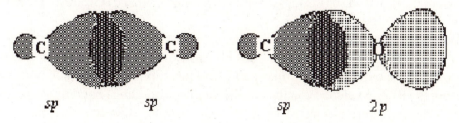

sp   *sp*   *sp*   *2p*

If the molecular axis is the *z*-axis, a p_x and a p_y orbital on each atom overlap side-by-side, giving two sets of delocalized π orbitals, each extending over all five atoms:

10.47   A description of bonding in a molecule starts with its Lewis structure, from which steric number, geometry, hybridization, and extension of π bonds can be deduced:
ClO_4^- has 7 + 4(6)+ 1 = 32 valence electrons.

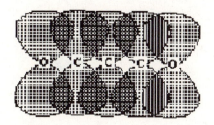

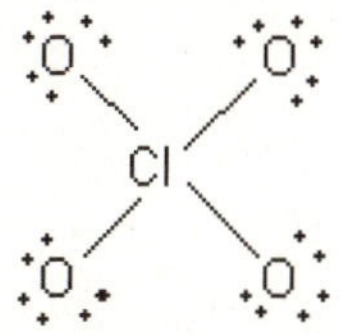

The provisional Lewis structure has $FC_{Cl} = 7 - 4 = +3$, so three electron pairs are shifted to make double bonds:

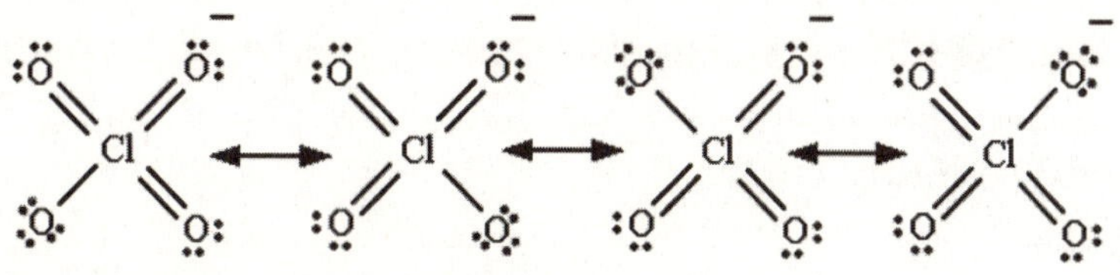

$SN_{Cl} = 4$, indicating that the molecule has tetrahedral geometry and the σ bond framework can be described using sp^3 hybrids on Cl overlapping with $2p$ orbitals from O. The resonance structures signal that the π system is extended over all five atoms between the *p* orbitals on the O atoms and the *d* orbitals of the Cl atom, and there are three bonding π orbitals occupied by six electrons.

10.49 Doped semiconductors are *p*-type if the dopant has fewer valence electrons than the bulk element and *n*-type if the dopant has more valence electrons than the bulk element: (a and c) GaP and CdSe are stoichiometric compounds, hence are undoped semiconductors. (b) InSb (Groups 13 and 15) doped with Te (Group 16) is *n*-type.

10.51 The bonding of any metal can be described by constructing bands from the valence orbitals and then distributing the valence electrons into these bands. Iron and potassium are both row 4 elements with similar band structures, but whereas each potassium atom contributes just one valence electron to the valence (bonding) band of the solid, each iron atom contributes eight electrons. Thus, iron has much greater bonding, making it harder and giving it a higher melting point than potassium.

10.53 The appearance of a band gap diagram depends on the type of doping in the semiconductor. GaAs doped with Zn is a *p*-type semiconductor, meaning that it has vacancies in its valence band:

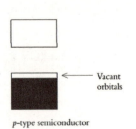

Vacant orbitals

p-type semiconductor

10.55 Describe bonding and geometry starting with a Lewis structure and a count of bonds and non-bonding pairs around inner atoms. The provisional structure of each compound has a positive formal charge on the row 3 sulfur atom, which is reduced to zero by making double bonds to each oxygen atom:

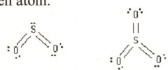

The sulfur atom in each compound has SN = 3, so sp^2 hybrids overlapping with oxygen $2p$ orbitals describe the σ bonds: two in SO_2 and three in SO_3. SO_2 is bent, like O_3, and SO_3 is trigonal planar, like NO_3^-. The Lewis structures indicate the presence of two π bonds in SO_2 and three π bonds in SO_3. All the π orbitals extend over the entire molecule. Because sulfur is a third-row element, its $3d$ orbitals contribute to the extended π bonding orbitals, which form from side-by-side overlap of oxygen $2p$ orbitals with sulfur $3p$ and $3d$ orbitals.

10.57 The band gap of a semiconductor is related to the frequency of light that can promote an electron through the equation $\Delta E_{gap} = E_{photon} = h\nu$

Divide by Avogadro's number to convert the band-gap energy in kilojoules per mole to the energy of one photon:

$$E_{photon} = \frac{(34.7 \text{ kJ/mol})(10^3 \text{ J/kJ})}{6.022 \times 10^{23} \text{ photons/mol}} = 5.76 \times 10^{-20} \text{ J}$$

$$v = \frac{E}{h} = \frac{5.76 \times 10^{-20} \text{ J}}{6.626 \times 10^{-34} \text{ J s}} = 8.70 \times 10^{13}/\text{s}$$

This frequency is in the infrared region of the spectrum.

10.59   To describe bonding, use the Lewis structure to determine steric numbers and hybridization. Then build orbital pictures accordingly. The π system in acrolein is identical to that in methyl methacrylate, which is described in Example 10-12 in your textbook. The outer oxygen atom uses $2p$ atomic orbitals and the three double-bonded carbons are sp^2 hybridized. These assignments lead to the following inventory of σ bonds: 4 sp^2-$1s$ C–H σ bonds, 2 sp^2-sp^2 C–C σ bonds, and 1 sp^2-$2p$ C–O σ bond. This leaves one p orbital on each C atom, plus one from the oxygen atom, to form a delocalized π network extending over four atoms. That network has two bonding π orbitals, each with a pair of electrons. To complete the valence electron inventory, there is one $2s$ and one $2p$ lone pair on the outer O atom.

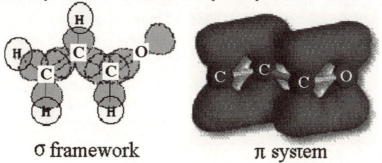

σ framework   π system

10.61   The formula of cinnamic acid reveals that it contains C, H, and O atoms. From the line structure below, we see that there is one outer O atom, whose bonding can be described using $2p$ orbitals. All other C and O atoms are inner. Those with only single bonds have SN = 4, sp^3 hybridization, and tetrahedral geometry. Those with double bonds have

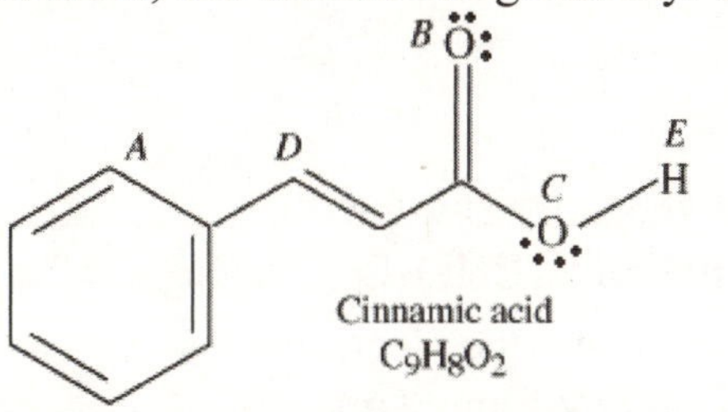

Cinnamic acid
$C_9H_8O_2$

SN = 3, sp^2 hybridization, and trigonal planar geometry:
(a) There are lone pairs only on the two O atoms; (b and c) A: sp^2 and atomic p, 120°; B: $2p$, no angle (outer); C: sp^3, 109.5°; D: sp^2 and atomic p, 120°; E: atomic $1s$, no angle (outer) (d) Cinnamic acid has five π bonds.

10.63 A molecule is paramagnetic if it has unpaired electrons. For diatomic molecules, use MO diagrams to place the valence electrons and determine whether or not they are all paired:

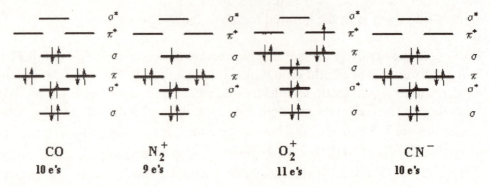

CO
10 e's

N₂⁺
9 e's

O₂⁺
11 e's

CN⁻
10 e's

a) diamagnetic; (b) paramagnetic; (c) paramagnetic; and (d) diamagnetic.

10.65 To determine which orbitals atoms use in a molecule, we must determine steric numbers from either the line structure or the Lewis structure. Any inner C, N, or O atom with no double bonds has SN = 4; with one double bond, an inner C or N atom has SN = 3. Remember that outer atoms (including H) always use atomic orbitals:

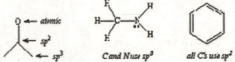

10.67 Band gap diagrams have valence bands and conduction bands, which can be filled, empty, or partially filled. A metallic conductor has no gap between the bands. An *n*-type semiconductor has a full valence band and a partially filled conduction band:

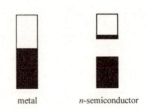

metal   *n*-semiconductor

10.69 A description of bonding and geometry starts with determination of the Lewis structure. $(CH_3)_3C^+$ has 4(4) + 9(1) – 1 = 24 electrons. All the electrons are required to complete the bonding framework, so there are no lone pairs to shift and the provisional structure is correct even though the central carbon atom has only six electrons associated with it (as this suggests, carbocations are highly reactive species):

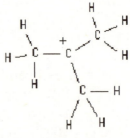

The central carbon has SN = 3, trigonal planar geometry, and its bonding orbitals can be described using sp^2 hybrid orbitals. The methyl carbons have SN = 4, tetrahedral geometry, and their bonding orbitals can be described using sp^3 hybrid orbitals. The vacant *p* orbital on the central C atom makes this cation highly reactive.

10.71   As Figure 10-50 shows, an electrical potential applied to a material distorts the bands. As Figure 10-52 shows, a semiconductor has a relatively small band gap between its filled valence band and empty conduction band. The distortion of the bands allows electrons to "hop" from valence to conduction band without changing energy, leading to electron mobility and conduction:

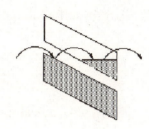

10.73   We obtain the charges on the ions from the chemical formulas: KO_2 contains one K^+ per O_2 unit, so superoxide has one negative charge; K_2O_2 contains two K^+ per O_2 unit, so peroxide has two negative charges. These ions are second-row diatomic species whose electron configurations can be written using the orbital energy level diagrams in Figure 10-35 of your textbook ($Z > 7$). Superoxide has one more electron than O_2, and peroxide has two more:

O_2: $(\sigma_s)^2 (\sigma_s^*)^2 (\pi_x)^2 (\pi_y)^2 (\sigma_p)^2 (\pi_x^*)^1 (\pi_y^*)^1$ Bond order = $1/2(8-4) = 2$

O_2^-: $(\sigma_s)^2 (\sigma_s^*)^2 (\pi_x)^2 (\pi_y)^2 (\sigma_p)^2 (\pi_x^*)^2 (\pi_y^*)^1$ Bond order = $1/2(8-5) = 1.5$

O_2^{2-}: $(\sigma_s)^2 (\sigma_s^*)^2 (\pi_x)^2 (\pi_y)^2 (\sigma_p)^2 (\pi_x^*)^2 (\pi_y^*)^2$ Bond order = $1/2(8-6) = 1$

From the bond orders, we see that O_2 has the strongest, shortest bond and O_2^{2-} the longest, weakest bond. From the number of unpaired electrons, we see that O_2 and O_2^- are both magnetic, with O_2 showing the largest magnetism.

10.75   A description of bonding in a molecule starts with its Lewis structure, from which steric number, geometry, hybridization, and extension of π bonds can be deduced. ClO_3^- has $7 + 3(6) + 1 = 26$ valence electrons. Draw the bonding framework, add 3 lone pairs to each outer oxygen atom and place the two remaining electrons on the inner Cl atom:

The provisional Lewis structure has $FC_{Cl} = 7 - 5 = +2$, so two electron pairs are shifted to make double bonds:

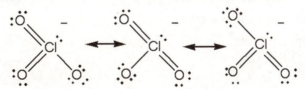

$SN_{Cl} = 4$, indicating that the molecule has tetrahedral geometry and the σ bond framework can be described using sp^3 hybrids on Cl overlapping with $2p$ orbitals from O. The resonance structures signal that the π system is extended over all four atoms, and there are two bonding π orbitals occupied by four electrons.

ClO_2^- has $7 + 2(6) + 1 = 20$ valence electrons. The provisional Lewis structure has two single bonds and two lone pairs on Cl, giving $FC_{Cl} = 7 - 2 - 2(2) = 1$, so one electron pair is shifted to make a double bond:

$SN_{Cl} = 4$, indicating that the electron pair geometry is tetrahedral, the molecule is bent, and the σ bond framework can be described using sp^3 hybrids on Cl overlapping with $2p$ orbitals from O. The resonance structures signal that the π system is extended over all three atoms, with one bonding π orbital occupied by two electrons.

10.77 Bond order measures the amount of electron sharing, based on Lewis structures or orbital configurations. Because bond order is based on models, it cannot be directly measured by experiments. The more electrons shared between atoms, however, the stronger and shorter the bond generally will be, so both bond energy and bond length are correlated with bond order. These bond properties can be measured experimentally, so we obtain confirmation of the bond orders predicted by bonding models by seeing whether experimental bond strengths and lengths match predicted bond orders. Higher bond order should be accompanied by increased bond strength and decreased bond length.

10.79 A description of bonding always starts with the Lewis structure for the molecule. Here is the Lewis structure for acrylonitrile. The steric numbers of the inner atoms are shown:

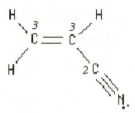

(a) The carbon atoms with SN = 3 have σ bonding that can be described using sp^2 hybrid orbitals, and the σ bonding of the carbon atom with SN = 2 can be described using sp hybrids.

(b) The entire molecule lies in the same plane. The σ bond network falls within this plane, and there is a localized π bond in this plane. Perpendicular to the molecular plane is an extended π network:

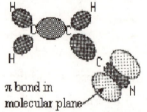

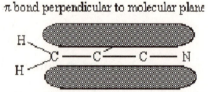

(c) There is an extended π network that includes the three C atoms and the N atom. It has four π electrons in two bonding orbitals that span the four atoms and have their electron density perpendicular to the molecular plane.

10.81 The bonding in second-row diatomic molecules can be described using the orbital energy level diagrams shown in Figure 10-35 of your textbook. CN has 9 valence electrons and the orbital energy level diagram for $Z \leq 7$:

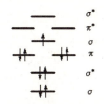

There are 7 bonding electrons and 2 antibonding electrons, for a net of 5, or bond order 2.5. The least stable occupied orbital is the σ_p:

10.83 A description of bonding and geometry starts with determination of the Lewis structure. ClO_2 has $7 + 2(6) = 19$ valence electrons. The bond framework requires two pairs, six additional electrons are placed on each O atom, leaving three lone electrons on the Cl atom. $FC_{Cl} = 7 - 2 - 3 = +2$, so shift two pairs to make double bonds:

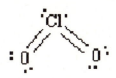

The Cl atom has SN = 4 (remember that a lone electron requires an orbital, just as an electron pair does), so its geometry is tetrahedral, the σ bonds can be described using sp^3 hybrids from Cl, and the molecule has a bond angle near $109°$. There is an extended π system formed from d orbitals on Cl overlapping side-by-side with $2p$ orbitals from the two O atoms. This molecule is considered unusual because it has an odd number of electrons.

10.85 The MO diagrams for these two diatomic molecules, which appear at right, show that they are very different in stabilities. Whereas dilithium resembles dihydrogen, except that it uses $2s$ valence orbitals rather than $1s$ valence orbitals, diberyllium is like dihelium. Thus, we expect that diatomic molecules of any Group 2 element will not exist, and in fact diberyllium has never been prepared.

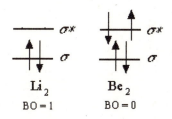

10.87 The bonding in H–X molecules can be represented by overlap between the $1s$ orbital of the H atom and the $n\,p$ orbital of the X atom. As n increases, the p orbital becomes larger but more diffuse because the electron is spread out over a larger volume. Consequently, electron density decreases and the amount of overlap with the $1s$ orbital decreases:

  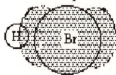 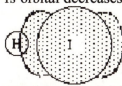

In these sketches, the shading represents the electron density within the orbital. Lighter shading represents lower electron density, hence a weaker bond.

10.89 The differences in behavior between $n = 2$ elements and their $n = 3$ counterparts from the same column of the periodic table can be explained by the difference in side-by-side π overlap for $n = 2$ and $n = 3$ orbitals. Multiple bonds involving π overlap are strong for N and O, but π overlap is very slight for P and S. Consequently, P–P and S–S diatomic molecules are minimally stabilized by π bonding, and these elements are more stable in chains with single bonds.

10.91 The standard procedure gives a Lewis structure from which hybridization, geometry, and π bonding are identified. Ketene has 2 C (4 e⁻) + 1 O (6 e⁻) + 2 H (1 e⁻) =16 valence electrons, 8 of which are used for the 4 bonds in the framework. Add 6 electrons around the outer O atom, then place the remaining pairs on one C atom. Complete the octet around the other C atom by shifting electron pairs to form 2 double bonds in the final Lewis structure:

The steric numbers of the two C atoms differ, as indicated on the Lewis structure. All atoms lie in a common plane, with a σ bond framework described as shown in the sketch. The C atom with SN = 3 has trigonal planar geometry and 120° angles. The C atom with SN = 2 has linear geometry and 180° angles. There is a two-center π bond in the plane of the molecule and an extended π bond perpendicular to the molecular plane:

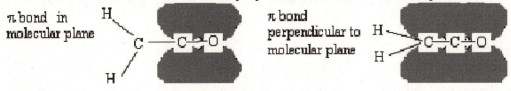

11.1   The condensation temperature of a substance reflects the strength of its interatomic attractions. Xe has the highest condensation temperature, indicating that it has the greatest interatomic attractions. Ar, with the lowest condensation temperature, has the least interatomic attractions. At 140 K, Xe is in a condensed phase because the kinetic energy of the

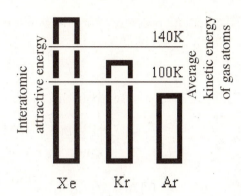

atoms is not enough to overcome the interatomic attractions. Kr and Ar, however, are gases because the kinetic energy of their atoms is sufficient to overcome their interatomic attractions. At 100K, only argon atoms have kinetic energies sufficient to overcome its interatomic attractions, so it is the only gas. Xe and Kr are both in condensed phases.

11.3   For any given substance, intermolecular attractions are constant. Their relative importance depends on two features: How close together molecules are, on average; and how much kinetic energy of motion molecules have, on average.
For any given substance, molecular volume is constant, so its relative importance depends on how close together molecules are.
 (a) Molecules are farther apart, making intermolecular attractions less significant, and molecular size is less significant.
 (b) Molecules are closer together, so intermolecular attractions and molecular sizes are more significant.
 (c) Increasing the temperature at constant pressure leads to a volume increase, so molecules are farther apart, making intermolecular attractions and molecular size less significant.

11.5   Pictures of atomic arrangements should show a high degree of order for a solid, close packing but some disorder for a liquid, and large distances between atoms for a gas:

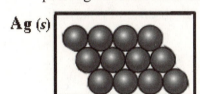 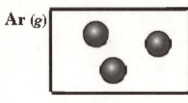 

11.7   To determine deviation from ideal behavior, calculate the pressure using the ideal gas equation and compare the calculated result with the experimental value:

$$P_{ideal} = \frac{nRT}{V} = \frac{(1.00 \text{ mol})(8.206 \times 10^{-2} \frac{\text{L atm}}{\text{mol K}})(40.0 + 273.15 \text{ K})}{1.20 \text{ L}} = 21.4 \text{ atm}$$

$$\% \text{ deviation} = (100\%)\left(\frac{P_{actual} - P_{ideal}}{P_{ideal}}\right) = (100\%)\left(\frac{19.7 \text{ atm} - 21.4 \text{ atm}}{21.4 \text{ atm}}\right) = -7.9\%$$

11.9   To calculate pressure using the van der Waals equation, we must know n, V, T, a, and b:

$$P = \left(\frac{nRT}{V - nb} \right) - \left(\frac{n^2 a}{V^2} \right)$$

Table 11-1 in your textbook lists values for Cl_2: $a = 6.493$ L^2 atm/mol^2, $b = 0.0562$ L/mol

$$n_{Cl_2} = \frac{m}{MM} = 1.25 \text{ kg} \left(\frac{10^3 \text{g}}{1 \text{ kg}} \right) \left(\frac{1 \text{ mol}}{70.906 \text{ g}} \right) = 17.63 \text{ mol}$$

$$(V - nb) = 15.0 \text{ L} - 17.63 \text{ mol} \left(\frac{0.0562 \text{ L}}{1 \text{ mol}} \right) = 14.01 \text{ L}$$

$$P = \frac{(17.63 \text{ mol})(8.206 \times 10^{-2} \frac{\text{L atm}}{\text{mol K}})(295 \text{ K})}{14.01 \text{ L}} - \frac{(17.63 \text{ mol})^2 (6.493 \frac{L^2 \text{atm}}{\text{mol}^2})}{(15.0 \text{ L})^2}$$

$P = 30.46 \text{ atm} - 8.97 \text{ atm} = 21.5 \text{ atm}$

$$P_{ideal} = \frac{nRT}{V} = \frac{(17.63 \text{ mol})(8.206 \times 10^{-2} \frac{\text{L atm}}{\text{mol K}})(295 \text{ K})}{15.0 \text{ L}} = 28.5 \text{ atm}$$

$$\% \text{ deviation} = (100\%) \left(\frac{P_{actual} - P_{ideal}}{P_{ideal}} \right) = (100\%) \left(\frac{21.5 \text{ atm} - 28.5 \text{ atm}}{28.5 \text{ atm}} \right) = -25\%$$

11.11   Ease of liquefaction depends on the magnitude of intermolecular attractions: The larger the attractions, the easier the substance is to liquefy. All these molecules are symmetrical (tetrahedral), so the ranking depends entirely on dispersion forces. Ease of liquefaction will increase with molecular size: CH_4 (hardest to liquefy) < CF_4 < CCl_4 (easiest to liquefy).

11.13   Polarizability increases with the size of the highest occupied orbitals. Draw the molecules with larger orbitals (CCl_4) to show greater distortion than for the molecules with smaller orbitals (CH_4). These molecules are close to spherical in the absence of polarization:

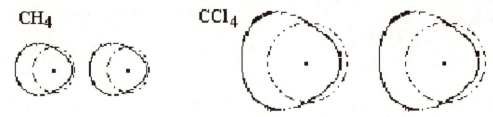

11.15   Boiling point depends on the magnitude of intermolecular attractions: The larger the attractions, the higher the boiling point. The size of intermolecular attractions depends on the amount of dispersion forces (size of molecule and size of orbitals), the presence of polar bonds, and hydrogen bonding. Among these substances, ethanol forms hydrogen bonds, so it has the highest boiling point. Propane, being smaller than n-pentane, has the lowest boiling point:
Propane (lowest bp) < n-pentane < ethanol (highest bp).

11.17 Hydrogen bonding capability requires the presence of an electronegative atom with lone pairs (O, F, or N) and an H-X bond to a highly electronegative atom (X = O, N, or F): (b and d) hydrogen bonding; (a and c) no hydrogen bonding.

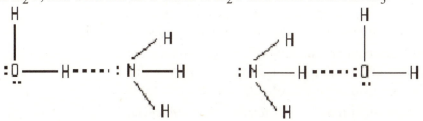

11.19 Hydrogen bonding capability requires the presence of an electronegative atom with lone pairs (O, F, or N) and an H–X bond to a highly electronegative atom (X = O, N, or F): (a) Ammonia has one N atom with a lone pair that can form a hydrogen bond to any H atom of another ammonia molecule:

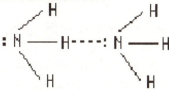

(b) Two types of hydrogen bonds are possible, between an N atom of NH_3 and an H atom of H_2O, and between an O atom of H_2O and an H atom of NH_3.

11.21 The viscosities of a set of similar substances increases with increasing chain length, because of increased dispersion forces (stronger intermolecular attractions make it harder for molecules to slide by one another) and increased "tangling" (longer chains are more entangled). Thus, the order of increasing viscosity is:

$$\text{pentane } (C_5) < \text{gasoline } (C_8) < \text{fuel oil } (C_{12})$$

11.23 An oxide surface has the ability to form hydrogen bonds to water, so the meniscus of water in aluminum tubing will be concave, just like inside glass tubing. The nonmetallic nature of the oxide layer means that there will not be strong liquid-surface forces for mercury inside aluminum tubing, so the meniscus will be convex, just like inside glass tubing.

11.25 Paper towels contain cellulose, which forms good hydrogen-bonding interactions with water, so aqueous solutions wet paper towels well and are absorbed effectively. There are no strong intermolecular interactions between an oil and the fibers of paper towels, so salad oil is not absorbed effectively.

11.27 The vapor pressure of a substance at any given temperature is determined by the strength of intermolecular forces: the stronger the forces, the lower the vapor pressure. (a) benzene has a higher vapor pressure, because chlorobenzene has a polar C-Cl bond that gives it a dipole moment and generates dipolar intermolecular forces; (b) hexane has a higher vapor pressure, because there are hydrogen bonding interactions for 1-hexanol that increase its intermolecular forces; (c) heptane has a higher vapor pressure, because octane, being the larger molecule, has stronger dispersion forces.

11.29 The position of an element in the periodic table, the chemical formula, and a knowledge of polyatomic ions all help in identifying types of solids: Sn, a metal, is a metallic solid; S_8 has a specific molecular formula, so it is a molecular solid; Se is not a metal, so it is a network solid; SiO_2 is described in your textbook as a network solid; and Na_2SO_4 contains Na^+ cations and SO_4^{2-} polyatomic anions, so it is an ionic solid.

11.31 Consult your textbook for information about the different types of solids.
(a) The bonding in metals comes from extended networks of delocalized electrons, while the bonding in network solids includes many individual covalent bonds.
(b) Metals conduct electricity, are malleable and ductile, and are shiny. Network solids are non-conductors or semiconductors, are brittle, and often have dull appearances.

11.33 The position of an element in the periodic table, the chemical formula, and knowledge of polyatomic ions all help in identifying types of solids:
(a) Br_2 is a discrete neutral molecule. It forms a molecular solid.
(b) KBr contains cations and anions. It forms an ionic solid.
(c) Ba is an alkaline earth metal. It forms a metallic solid.
(d) SiO_2 is described in your textbook as a network solid.
(e) CO_2 is a covalent molecule. It forms a molecular solid.

11.35 The density of a solid depends on the nature of the elements it contains and on how compact the bonding network is. A structure with more open bonding pattern has a lower density than one with a more compact bonding pattern. Graphite uses sp^2 hybrids and has an extended π bonding network, making it planar with a relatively open structure between successive planes. Diamond uses sp^3 hybrids and has a compact network of bonds:

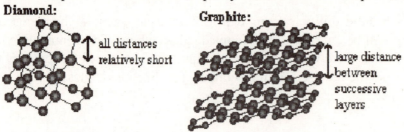

11.37   To determine a chemical formula from a unit cell, count the number of atoms of each element contained within the unit cell. Atoms on a face count 1/2 (each is part of two unit cells), those on an edge count 1/4 (each is part of four unit cells), and those at a corner count 1/8 (each is part of eight unit cells):   There is one Ti atom within the cell
Ca: 1/8 (8 corner atoms) = 1   O: 1/2 (6 face atoms) = 3
The chemical formula is $CaTiO_3$.

11.39   The problem states that carborundum is diamond-like. Diamond is composed entirely of tetrahedral carbon atoms (SN = 4), and the empirical formula of carborundum, SiC, indicates that alternate C atoms should be replaced with Si atoms. Each C atom bonds to four Si atoms and each Si atom bonds to 4 C atoms:

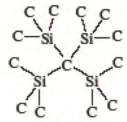

11.41   A unit cell has to be a shape that can be used over and over to build the entire pattern. The upper screen shows a unit cell containing two complete fish, and the lower screen shows a unit cell that contains no complete fish, but notice that it contains 1/2(2 edge) + 1/4(4 corner) = 2 fish overall, just like the other unit cell:

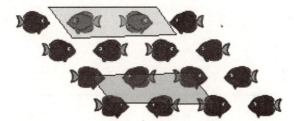

11.43   First determine the number of sulfide ions in the unit cell. A face-centered ionic cube has six face ions and eight corner ions, giving 1/2(6) + 1/8(8) = 4 sulfide ions. The sulfide anion has a charge of –2 and the lithium cation has a charge of +1, so we need twice as many lithium cations as sulfide anions in the unit cell. There are eight lithium cations in the interior of the unit cell.

11.45   The heats of phase changes indicate the strengths of intermolecular forces; the larger the intermolecular forces, the larger the heat of the phase change.
(a) Methane has a lower heat of vaporization than ethane because, being a smaller molecule, it has smaller dispersion forces.
(b) Ethanol has a significantly higher heat of vaporization than diethyl ether because of strong hydrogen bonding between ethanol molecules.
(c) Argon (18 electrons) has a higher heat of fusion than methane (10 electrons) because it has a higher polarizability due to its larger number of electrons.

11.47 A phase diagram is map of the stable
phases at different P and T, and
phase boundary lines meet at the
triple point and cross the horizontal
line corresponding to $P = 1$ atm at
the normal freezing and boiling
points:

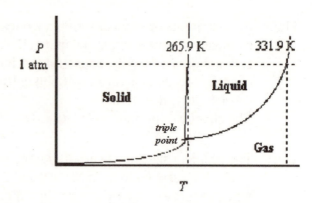

11.49 Phase diagrams provide "road maps" allowing us to determine what phase changes occur
as temperature and pressure vary in particular ways:
(a) At $T = 400$ K, $P = 1.00$ atm, Br_2 is a gas. As it cools under a pressure of 1.00 atm, it
liquefies at 331.9 K and solidifies at 265.9 K.
(b) At $T = 265.8$ K, $P = 1.00 \times 10^{-3}$ atm, Br_2 is a gas. Compressing it at this temperature
causes it to liquefy at about $P = 6 \times 10^{-2}$ atm and solidify at about 0.5 atm.
(c) $P = 2.00 \times 10^{-2}$ atm is below the triple point of Br_2. Heating solid Br_2 at this pressure
from 250 to 400 K causes sublimation to the vapor at about 265 K.

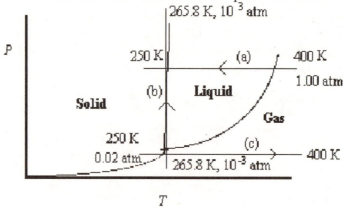

11.51 Summarize the data and the process in a flow chart:

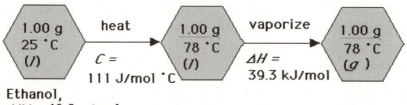

Ethanol,
$MM = 46.0$ g/mol

The heat required is $nC\Delta T$ for the heating and $n\Delta H$ for the vaporization.

The amount of ethanol is $n = \dfrac{1.00 \text{ g}}{46.0 \text{ g/mol}} = 2.17 \times 10^{-2}$ mol

Substitute the values and do the calculation:

$q = (2.17 \times 10^{-2} \text{ mol})[(111 \text{ J/mol °C})(78 - 25 \text{ °C}) + (39.3 \text{ kJ/mol})(10^3 \text{ J/kJ})] = 980.$ J

11.53 All substances are subject to dispersion forces. In addition, look for polarity and hydrogen-bonding ability:

(a) NH_3 is polar and forms hydrogen bonds in addition to its dispersion interactions;

(b) $CHCl_3$ is a polar molecule in addition to having dispersion interactions;

(c) CCl_4 is symmetrical (tetrahedral), so it has only dispersion interactions; and

(d) CO_2 is symmetrical (linear) so it has only dispersion interactions.

11.55 Face-centered cubic crystals are close-packed, each atom having 12 nearest neighbors. In body-centered cubic crystals, each atom has 8 nearest neighbors. Thus, the face-centered cubic structure contains more atoms in a given volume and is denser.

11.57 All substances are subject to dispersion forces. In addition, look for polarity and hydrogen-bonding ability:

(a) CH_3OH can hydrogen bond, so it boils at a higher temperature than CH_3OCH_3;

(b) SiO_2 is a network solid with covalent bonds that must break for it to boil, so it has a higher boiling point than SO_2;

(c) HF forms strong hydrogen bonds, so it boils at a higher temperature than HCl; and

(d) I_2 has larger orbitals, giving it higher polarizability, larger dispersion forces, and a higher boiling point than Br_2.

11.59 Consult the phase diagram for N_2 in Figure 11-40 of your textbook to determine the temperatures and pressures at which phase changes occur.:

(a) A constant temperature process is a vertical line on the phase diagram. $T = 70$ K is in between the triple point (0.124 atm, 63 K) and normal boiling point (1 atm, 77 K) of nitrogen. At 70 K, 1 atm, nitrogen is liquid. When the pressure falls below about 0.5 atm, the liquid boils, and at 70 K, 0.1 atm, the substance is gaseous.

(b) A constant temperature process is a vertical line on the phase diagram. $T = 298$ K is within the gaseous region at all pressures, so compression from 1.00 atm to 50.0 atm leaves the substance gaseous.

(c) A constant pressure process is a horizontal line on the phase diagram, and 1 atm represents "normal" pressure, shown on the figure as a dotted line. When the temperature of the gas reaches 77 K, the gas liquefies, and when the temperature reaches 63 K, the liquid solidifies. At 50 K, N_2 is solid.

11.61 When two substances with otherwise similar structures have different boiling points, look for differences in polarity and/or hydrogen-bonding behavior. The Lewis structures of the two isomers, shown with arrows representing bond polarities, indicate why the isomers have different boiling points:

In the *trans* isomer, the two polar bonds oppose each other and cancel, making this a non-polar compound. In the *cis* isomer, the two polar bonds add, making this a polar compound. Thus, *cis* 1,2-dichloroethylene has the higher boiling point.

11.63   Deviations from ideal gas behavior can be predicted based on the strengths of intermolecular interactions. The larger the intermolecular interactions, the greater will be the deviations from ideality. Both pairs of substances are non-polar, so dispersion interactions dominate their intermolecular forces. The substance with more electrons will have larger dispersion forces and deviate more from ideality: Cl_2.

11.65   The pressure scale of Figure 11-41 is in katm, so a pressure of 1 atmosphere essentially lies along the x axis. There are four phase transitions as silica is heated. At room temperature, the stable phase is α quartz, which converts to β quartz at a temperature of around 600 °C. At around 850 °C, the solid becomes tridymite, and just below 1500 °C, this converts to cristobalite. At around 1700 °C, cristobalite melts to give liquid silica.

11.67   Refer to Figures 11-27 and 11-31 in your textbook for views of body-centered cubic and hexagonal close-packed crystals. The more closely-packed form will be stable at higher pressures, so the body-centered cubic structure is the low temperature, low pressure form.

11.69   Solids are molecular, metallic, network, or ionic: (a) ZrO_2 (mp = 2677 °C) is a network solid, as indicated by its high melting point; (b) Zr is a transition metal, so the element is a metallic solid; and (c) $ZrSiO_4$ contains the silicate anion, so this solid is ionic.

11.71   Phase diagrams provide "road maps" allowing us to determine what phase changes occur as temperature and pressure vary in particular ways. (a) For rhombic sulfur to melt, it must be heated above 153 °C at a pressure that is greater than 1420 atm. (b) For rhombic sulfur to change to monoclinic sulfur, it must be heated to a temperature that depends on the pressure, ranging from 95.3 °C at 5.1×10^{-6} atm to 153 °C at 1420 atm. (c) Rhombic sulfur only sublimes when the pressure is less than 5.1×10^{-6} atm.

11.73   (a) There is a rhenium atom at each corner of the unit cell cube but nowhere else, so the rhenium lattice is simple cubic; (b) Each corner of a cube is shared by eight cubes, and there are eight corners, so there are 8(1/8) = 1 Re atom per unit cell. Each edge of a cube is shared by four cubes, and there are 12 edges, so there are 12(1/4) = 3 O atoms per unit cell. The chemical formula is ReO_3.

11.75   The melting point of a metal is determined by the strength of interatomic forces. The smaller those forces, the lower the melting point of the metal. The low melting points of Hg, Cs, and Ga indicate that these metals have realatively small interatomic forces.

11.77   (a) When sulfur is heated at any constant pressure between 3.2×10^{-5} atm and 1420 atm, there is a temperature at which each phase is stable. Below 95.3 °C, the rhombic phase is stable. At a pressure-dependent temperature between 95.3 °C and 153 °C, the monoclinic phase becomes stable. At a pressure-dependent temperature between 115.2 °C and 153 °C, monoclinic sulfur melts and the liquid becomes stable. At a still higher temperature that depends strongly on the pressure, the vapor phase is stable.

(b) At any temperature between 115.2 °C and 153 °C, there is a pressure at which each phase is stable. Below 3.2×10^{-5} atm, the vapor is stable. Above this pressure, the liquid is stable up to a pressure that depends on the exact temperature, but at higher pressure, the monoclinic solid phase is stable. At extremely high pressure, the rhombic phase becomes stable.

11.79   The magnitudes of dispersion forces depend on how extended the valence electrons are. In a molecule, valence electrons are spread over a larger volume because they are shared between atoms. Sketches of the electron clouds of H_2 and He illustrate this feature:

The more extended valence electron cloud of H_2 leads to stronger dispersion forces than for He, giving H_2 the higher boiling point.

11.81   See Figure 11-28 in your textbook for an example of the face-centered cube. Here is a rendering of the unit cell. The ion in the center of the front face of the unit cell, shown in black, can be used to determine how many nearest neighbors each ion has. The thick lines connect to four ions in the same face, one in the center of the unit cell and one in front of the front face (ion not shown). Thus, each ion has six nearest neighbors of opposite charge:

11.83   To answer these questions, we need to make reasonable extrapolations of the phase diagram for aluminum silicate. (a) The phase diagram extends to 900 °C without showing a liquid phase, so we can say that the melting point is higher than 900 °C; how much higher is not known. (b) The only stable solid phase of aluminum silicate at high temperature is sillimanite, so at all pressures from 0-10 kbar, the liquid will solidify to this form of the solid. (c) The kyanite-sillimanite phase boundary is moving toward higher temperature as the pressure increases, so there may be a sufficiently high pressure where this is the phase that solidifies from liquid aluminum silicate. (d) The stable phase at highest pressure is the densest phase; this is kyanite. The stable phase at lowest pressure is the least dense; this is andalusite.

11.85   This problem asks us to construct a graph that summarizes how the temperature changes as a liquid sample is cooled to below its freezing point. As energy is removed from the sample, the temperature will drop until the liquid begins to freeze. Then the temperature will remain fixed until all the water has frozen. Finally, the temperature of the ice will decrease until the final temperature is reached:

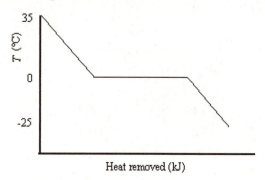

11.87   (a)

(b) The fraction of paired molecules decreases as temperature rises, because energy added to the system breaks some of the hydrogen bonds.

12.1 The solvent is the component that determines the phase of the solution; usually it is the component present in the largest amount. Any other components are solutes. (a) water (normally a liquid) is the solvent and carbon dioxide (normally a gas) is the primary solute; (b) water (normally a liquid) is the solvent and ethanol (normally a liquid) is the primary solute; (c) molecular nitrogen (normally a gas) is the solvent, molecular oxygen (normally a gas) and water (normally a liquid) are the primary solutes.

12.3 Mole fractions are determined by calculating the number of moles of each component of the solution and then using the equation for X:

$$X_A = \frac{n_A}{n_{total}}$$

For CH_3OH: $n = \dfrac{14.5\ g}{32.04\ g/mol} = 0.4526$ mol

For H_2O: $n = \dfrac{101\ g}{18.01\ g/mol} = 5.608$ mol

$n_{total} = 5.608 + 0.4526 = 6.061$ mol

$X_{methanol} = \dfrac{0.4526\ mol}{6.061\ mol} = 0.0747$ $X_{water} = 1 - 0.0747 = 0.925$

12.5 To determine the concentrations that are asked for, we need to know the mass and number of moles of each component. Convert volume into mass using the density equation:

$m = \rho V = (0.818\ g/mL)(85.0\ mL) = 69.53$ g

A table is a useful way to summarize the information that we need:

Substance	Formula	MM	m	n
Acetone	C_3H_6O	58.08 g/mol	69.53 g	1.197 mol
Maleic acid	$C_4H_4O_4$	116.07 g/mol	1.521 g	1.31×10^{-2} mol

Use this information to calculate each concentration:

Molality: $c_m = \dfrac{n_{solute}}{m_{solvent}\,(in\ kg)} = \dfrac{1.31 \times 10^{-2}\ mol}{(69.53\ g)\left(10^{-3}\ kg/g\right)} = 1.88 \times 10^{-1}$ mol/kg

Mole fraction: $X = \dfrac{n_{solute}}{n_{total}} = \dfrac{1.31 \times 10^{-2}\ mol}{(1.197 + 0.0131)\ mol} = 1.08 \times 10^{-2}$

Mass %: $\% = (100\%)\dfrac{m_{solute}}{m_{total}} = (100\%)\dfrac{1.521\ g}{(1.521 + 69.53)\ g} = 2.14\ \%$

12.7   To determine the various concentration values for a solution whose mass percentage is known, take exactly 100 g of the solution as a convenient amount to work with. Determine the masses of each substance in this amount of solution, convert to moles, then apply the appropriate equations:

For H_2O_2: $n = \dfrac{30.\text{ g}}{34.01 \text{ g/mol}} = 0.88 \text{ mol}$

For H_2O: $n = \dfrac{70.\text{ g}}{18.01 \text{ g/mol}} = 3.89 \text{ mol}$

$n_{total} = 0.88 + 3.89 = 4.77 \text{ mol}$

$X_{\text{hydrogen peroxide}} = \dfrac{0.88 \text{ mol}}{4.77 \text{ mol}} = 0.18$ $X_{water} = 1 - 0.18 = 0.82$

To determine the molarity, we need the volume of this amount of solution:

$\rho = \dfrac{m}{V}$ so $V = \dfrac{m}{\rho} = \dfrac{100 \text{ g}}{1.11 \text{ g/mL}} = 90.1 \text{ mL}$

$M = \dfrac{n}{V} = \dfrac{0.88 \text{ mol}}{(90.1 \text{ mL})(10^{-3} \text{ L/mL})} = 9.8 \text{ M}$

To determine the molality, we need the mass of solvent in kg:

$c_m = \dfrac{n_{solute}}{m_{solvent}} = \dfrac{0.88 \text{ mol}}{(70 \text{ g})(10^{-3} \text{ kg/g})} = 13 \text{ mol/kg}$

12.9   To determine other concentration measures for a solution whose molarity is known, work with exactly 1 L of solution and use the density to determine the mass of the solution and of its solvent.

$\rho = \dfrac{m}{V}$ $m = \rho V = (0.9811 \text{ g/mL})(1000 \text{ mL/L})(1 \text{ L}) = 981.1 \text{ g}$

Find the mass of ammonia in 1 L of solution from the molarity and molar mass:

$m_{ammonia} = n\,MM = (2.30 \text{ mol/L})(1 \text{ L})(17.03 \text{ g/mol}) = 39.17 \text{ g}$

$m_{water} = 981.1 \text{ g} - 39.17 \text{ g} = 941.9 \text{ g}$

$n_{water} = \dfrac{941.9 \text{ g}}{18.01 \text{ g/mol}} = 52.30 \text{ mol}$

$X_{ammonia} = \dfrac{2.30 \text{ mol}}{(2.30 + 52.30) \text{ mol}} = 0.042$ $X_{water} = 1 - 0.042 = 0.958$

Mass fraction ammonia $= \dfrac{\text{mass}}{\text{total mass}} = \dfrac{39.17 \text{ g}}{981.1 \text{ g}} = 0.0399$

$c_m = \dfrac{n_{solute}}{m_{solvent}} = \dfrac{2.30 \text{ mol}}{(941.9 \text{ g})(10^{-3} \text{ kg/g})} = 2.44 \text{ mol/kg}$

12.11 Work with moles to determine the amounts of solute and solvent needed when mole fraction is the target.

First determine the number of moles of methanol:

$$n_{methanol} = \frac{25.0 \text{ g}}{32.04 \text{ g/mol}} = 0.780 \text{ mol}$$

(a) Substitute into the mole fraction expression to determine moles of water:

$$X_A = \frac{n_{methanol}}{n_{methanol} + n_{water}} = \frac{0.780 \text{ mol}}{0.780 \text{ mol} + n_{water}} = 0.105$$

$$0.780 \text{ mol} = 0.105 \, (0.780 \text{ mol} + n_{water})$$

Divide both sides by 0.105: $7.43 \text{ mol} = 0.780 \text{ mol} + n_{water}$

$$n_{water} = 7.43 \text{ mol} - 0.780 \text{ mol} = 6.65 \text{ mol}$$

$$m_{water} = (6.65 \text{ mol})(18.01 \text{ g/mol}) = 120. \text{ g}$$

(b) Use the results of part (a) and the equation for molality to calculate c_m:

$$c_m = \frac{n_{solute}}{m_{solvent}} = \frac{0.780 \text{ mol}}{(120. \text{ g})(10^{-3} \text{ kg/g})} = 6.50 \text{ mol/kg}$$

12.13 The immiscibility of liquids can be explained using the concept that "like dissolves like." A molecule of salad oil consists almost entirely of hydrocarbon segments, which have dispersion-type intermolecular forces but low polarity and no hydrogen-bonding capability. A solution of acetic acid in water has high polarity and a large hydrogen-bonding capability. Thus, these two liquids are not like each other and do not mix. In terms of balance of forces, hydrogen bonds would have to be broken for the two liquids to mix, making mixing energetically unfavorable.

12.15 The types of solids that dissolve in a solvent can be predicted using the concept that "like dissolves like." Liquid ammonia is similar to liquid water: It is a polar molecule with a strong hydrogen-bonding capability. Thus, liquid ammonia should dissolve the same sorts of substances as water, ionic salts and polar organic molecules such as alcohols.

12.17 Miscible pairs are similar in the types of intermolecular forces that they experience. H_2O has strong hydrogen bonding, and so does CH_3OH, so this pair is miscible.

C_8H_{18} and CCl_4 both have only dispersion forces, so this pair is miscible.

All the other possible pairs match unlike substances, so none of the other pairs is miscible.

12.19 To match a solute to its most appropriate solvent, use the "like dissolves like" principle:

Solute	I_2	NaCl	Au	paraffin
Solvent	CCl_4	water	Hg	n-octane

CCl_4 is most like I_2; NaCl is a salt, so it only dissolves in water. Au, a metal, requires another metal, Hg; and paraffin, a hydrocarbon, is best matched by n-octane, another hydrocarbon.

12.21   Copper atoms and nickel atoms are nearly the same size, so this alloy will be substitutional. Your picture should show nine Cu atoms for every one Ni atom. Copper forms face-centered cubes, which are most easily shown as hexagonal arrays:

12.23   Methane differs from water in two major respects: it is nonpolar, and it does not form hydrogen bonds. Consequently, methane is a much poorer solvent than water. The lack of hydrogen bonding capability makes it very unlikely that methane could support life, because the complex interactions needed for life processes depend heavily on hydrogen-bonding interactions.

12.25   Your picture should look similar to the one in Figure 12-5 in your textbook, with K^+ cations in place of Na^+ cations:

12.27   In order for solute molecules to escape from a solution, they must overcome the intermolecular forces of attraction holding them in solution. To accomplish this requires energy, which is supplied by the solvent. Thus, the total energy of the remaining solution decreases, and the temperature falls.

12.29   Table 12-2 indicates that the solubility of O_2 in water decreases as the temperature rises from 0 °C to 25 °C. Use Henry's law to calculate the concentration at each temperature. Note that although equilibrium constants are dimensionless, each concentration must be in standard units (M for solutes, atm for gases):

$$\frac{[O_{2(aq)}]_{eq}}{(p_{O_2})_{eq}} = K_H \qquad\qquad [O_{2(aq)}]_{eq} = (p_{O_2})_{eq} K_H$$

The atmosphere is 21 % O_2, so the partial pressure of O_2 is 0.21 atm:

At 0 °C, $[O_{2(aq)}]_{eq} = (0.21 \text{ atm})(2.5 \times 10^{-3}) = 5.2 \times 10^{-4}$ M

At 25 °C, $[O_{2(aq)}]_{eq} = (0.21 \text{ atm})(1.3 \times 10^{-3}) = 2.7 \times 10^{-4}$ M

The percentage change is: $(100 \text{ %}) \dfrac{5.2 \times 10^{-4} \text{M} - 2.7 \times 10^{-4} \text{M}}{5.2 \times 10^{-4} \text{M}} = 48\%$

12.31   Use the information about equilibrium pressure to determine the Henry's law constant. Then use the Henry's law constant to determine the equilibrium pressure for any other concentration. Note that although equilibrium constants are dimensionless, each concentration must be in standard units (M for solutes, atm for gases):

$$[\text{gas } (aq)]_{eq} = K_H(p_{\text{gas}})_{eq} \qquad so \qquad K_H = \frac{[\text{gas } (aq)]_{eq}}{(p_{\text{gas}})_{eq}}$$

$$K_H = \frac{0.500 \text{ M}}{(6.8 \text{ torr})(1 \text{ atm}/760 \text{ torr})} = 55.88$$

$$(p_{\text{gas}})_{eq} = \frac{[\text{gas } (aq)]_{eq}}{K_H} = \frac{2.5 \text{ M}}{55.88}(760 \text{ torr/atm}) = 34 \text{ torr}$$

12.33   The vapor pressure of a solvent above a solution is described by Raoult's law, Equation 12-3 in your textbook:

$$vp_{\text{solution}} = X_{\text{solvent}} \, vp_{\text{pure solvent}}$$

In addition to the vapor pressure of pure solvent, which is given in the problem, we need the mole fraction of the solvent. To calculate this, we must know the number of moles of solute and solvent:

For urea: $n = \dfrac{7.50 \text{ g}}{60.05 \text{ g/mol}} = 0.125 \text{ mol}$

For water, we have $m = (15.0 \text{ mL})(1.00 \text{ g/mL}) = 15.0 \text{ g}$

$$n = \frac{15.0 \text{ g}}{18.01 \text{ g/mol}} = 0.833 \text{ mol}$$

$n_{\text{total}} = 0.125 + 0.833 = 0.958 \text{ mol}$ $X_{\text{water}} = \dfrac{0.833 \text{ mol}}{0.958 \text{ mol}} = 0.869$

$vp_{\text{solution}} = X_{\text{solvent}} \, vp_{\text{pure solvent}} = (0.869)(33.00 \text{ torr}) = 28.7 \text{ torr}$

12.35   Freezing points are calculated using Equation 12-4 from your textbook: $\Delta T_f = K_f c_m$. For water, K_f is 1.858 °C kg/mol. First, the molality of the solution must be calculated:

$$c_m = \frac{\text{mol solute}}{\text{kg solvent}}$$

Consider 100 g of a solution that is 12% by mass ethanol. It contains 12 g ethanol and 88 g of solvent (water):

n solute $= 12 \text{ g}\left(\dfrac{1 \text{ mol}}{46 \text{ g}}\right) = 0.26 \text{ mol}$ m solvent $= 88 \text{ g}\left(\dfrac{10^{-3} \text{ kg}}{1 \text{ g}}\right) = 8.8 \times 10^{-2} \text{ kg}$

$$c_m = \frac{0.26 \text{ mol}}{8.8 \times 10^{-2} \text{ kg}} = 2.95 \text{ mol/kg}$$

$$\Delta T_f = K_f c_m = \left(\frac{1.858 \text{ °C kg}}{1 \text{ mol}}\right)\left(\frac{2.95 \text{ mol}}{1 \text{ kg}}\right) = 5.5 \text{ °C}$$

This is the depression in freezing point, so the new freezing point is $0.0 - 5.5 = -5.5$ °C.

12.37 Boiling points can be calculated for solutions of *non-volatile* solutes, using the boiling point elevation constant, $\Delta T_b = K_b\, c_m$. However, wine contains a *volatile* solute, ethanol, which contributes to the vapor pressure above the solution and lowers the boiling point. Thus, we do not have enough information to calculate the boiling point of wine.

12.39 The freezing point depression of an aqueous solution depends on the *total* molality of solute particles. For an ionic salt, we must take into account the cations and anions, remembering that the net effect will be diminished by ion pairing. The solution with the smallest total molality will have the highest freezing point.
The order of freezing points of these solutions is as follows:
0.25 M NH_3 (highest) > 0.30 M NaCl > 0.25 M $MgCl_2$ > 0.75 M NH_3 (lowest)

12.41 Molar masses can be calculated from osmotic pressures using Equation 12-6 in your textbook.

First convert torr into atm: $P = (64.8\ \text{torr})\left(\dfrac{1\ \text{atm}}{760\ \text{torr}}\right) = 8.53 \times 10^{-2}$ atm.

(a) $MM = \dfrac{mRT}{\Pi V} = \dfrac{(1.00\ \text{g})(8.206 \times 10^{-2}\ \frac{\text{L atm}}{\text{mol K}})(25 + 273\ \text{K})}{(8.53 \times 10^{-2}\text{atm})(1.00\ \text{L})} = 287$ g/mol

(b) The nonpolar tail contains 11 C atoms and 2 H atoms for every C except the first, which has 3 H atoms, giving 23 H atoms altogether. It contributes:
11(12.01 g/mol) + 23(1.008 g/mol) = 155 g/mol, leaving for the polar head group,
$MM = 287 - 155 = 132$ g/mol.

12.43 The boiling point of a solution allows us to calculate the solution's molality, but to calculate the osmotic pressure, we need the solution's molarity. To convert from molality to molarity requires knowledge of the density of the solution.

$$\Delta T_b = K_b\, c_m \qquad\qquad c_m = \frac{\Delta T_b}{K_b}$$

$$c_m = \frac{(101.45\ ^\circ\text{C} - 100.00\ ^\circ\text{C})}{0.512\ ^\circ\text{C kg/mol}} = 2.83\ \text{mol/kg}$$

To convert from molality, consider a solution containing 1 kg of solvent. It contains 2.83 mol of sucrose ($MM = 342$ g/mol), with mass $m = (2.83\ \text{mol})(342\ \text{g/mol}) = 969$ g.
The total mass of this solution is 1969 g, from which we can calculate its volume:

$$V = \frac{m}{\rho} = \frac{1969\ \text{g}}{(1.036\ \text{g/mL})(10^3\ \text{mL/L})} = 1.90\ \text{L}$$

Now we can calculate the molarity:

$$M = \frac{n}{V} = \frac{2.83\ \text{mol}}{1.90\ \text{L}} = 1.49\ \text{M}$$

Knowing the molarity allows us to determine the osmotic pressure of this solution:
$\Pi = MRT = (1.49\ \text{mol/L})(0.08206\ \text{L atm/mol K})(35 + 273\ \text{K}) = 37.6$ atm

12.45 Both fog and smoke are colloidal suspensions in a gas, the Earth's atmosphere. Both are aerosols. The main difference between them is that the particles in fog are a liquid (water), whereas those in smoke are mostly solids (carbon-based compounds).

12.47 Surfactants must have a highly polar (or ionic) end and a nonpolar end. Of these three substances, propanoic acid has too short a nonpolar end and would be the worst. Lauryl alcohol and sodium lauryl sulfate have similar nonpolar portions, but the ionic portion of sodium lauryl sulfate makes it the best surfactant of these three substances:

12.49 Surfactants must have a highly polar (or ionic) end and a nonpolar end. The polar end of the surfactant molecule associates with a polar substance, while the nonpolar end of a surfactant molecule associates with a nonpolar substance. The figure shows the nonpolar ends of sodium stearate immersed in the solvent, indicating that the solvent molecules are nonpolar.

12.51 To determine the concentrations that are asked for, we need to know the mass and number of moles of each component and the volume of the solution. Work with exactly 100 g of solution. Round the final results to two significant figures to match the percentages given in the problem.
A table is a useful way to summarize the information that we need:

Substance	Formula	MM	m	n
Phosphoric acid	H_3PO_4	97.99 g/mol	85 g	0.867 mol
Water	H_2O	18.02 g/mol	15 g	0.832 mol

Convert mass into volume using the density equation:

$$V = \frac{m}{\rho} = \frac{100 \text{ g}}{1.685 \text{ g/mL}} = 59.35 \text{ mL}$$

Use this information to calculate each concentration:

Molality: $c_m = \dfrac{n_{solute}}{m_{solvent}\text{(in kg)}} = \dfrac{0.867 \text{ mol}}{(15 \text{ g})(10^{-3} \text{ kg/g})} = 58 \text{ mol/kg}$

Mole fraction: $X = \dfrac{n_{solute}}{n_{total}} = \dfrac{0.867 \text{ mol}}{(0.867 + 0.832) \text{ mol}} = 0.51$

Molarity: $M = \dfrac{n_{solute}}{V_{solution}\text{(in L)}} = \dfrac{0.867 \text{ mol}}{(59.35 \text{ mL})(10^{-3} \text{ L/mL})} = 15 \text{ M}$

12.53 Hydrophilicity is conveyed by the presence of ionic, polar, or hydrogen-bonding groups. DDT contains none of these, so it is highly hydrophobic. Thus, it is stored in fatty tissues (which are nonpolar) rather than being excreted from the body. Consequently, the effects of DDT are cumulative, especially for organisms high up on the "food chain," such as predator birds.

12.55 This problem deals with the freezing point depression of water using ethyl glycol. Use the following equation to solve this problem: $\Delta T_f = K_f c_m$

$$20\ °C = \left(\frac{1.858\ °C\ kg}{1\ mol}\right) c_m$$

$$c_m = 20\ °C\left(\frac{1\ mol}{1.858\ °C\ kg}\right) = 11\ mol/kg$$

12.57 The solubilities of organic substances in water depend on the presence of polar groups and groups that are capable of forming hydrogen bonds with water. Alcohols contain OH groups that enhance solubility. Among these compounds, benzene (C_6H_6) is nonpolar and has the lowest solubility, while erythritol has four OH groups, making it highly soluble:

benzene (lowest solubility) < pentanol < erythritol (highest solubility)

12.59 Differential solubilities can be explained using the concept that "like dissolves like." Here, iodine is a nonpolar solid that dissolves in nonpolar liquids. Water is highly polar and CCl_4 is nonpolar, so these two liquids are immiscible and I_2 preferentially dissolves in CCl_4.

12.61 The notion of "like dissolves like" can be extended to the wetting behavior of one substance for another. Solder is a metal that will adhere well to another metal only if unlike substances have been removed from the surface. Nonpolar substances such as oils and greases as well as network solids such as metal oxides must be removed by flux.

12.63 The molal boiling point constant for a liquid is related to temperature and concentration through the boiling point elevation equation, Equation 12-5 in your textbook:

$$\Delta T_b = K_b c_m \qquad\qquad K_b = \frac{\Delta T_b}{c_m}$$

(a) Use the data provided in the problem to evaluate the molality and the temperature change:

$\Delta T_b = 47.46\ °C - 46.30\ °C = 1.16\ °C$

Mass of $CS_2 = V\rho = (200.0\ mL)(1.261\ g/mL)10^{-3}\ kg/g) = 0.2522\ kg$

$$c_m = \frac{0.125\ mol}{0.2522\ kg} = 0.496\ mol/kg$$

$$K_b = \frac{\Delta T_b}{c_m} = \frac{1.16\ °C}{0.496\ mol/kg} = 2.34\ °C\ kg/mol$$

(b) Use the result of part (a) to determine the molality of the solution of the unknown compound. From the molality, determine the number of moles, then use the mass to find the molar mass of the unknown compound.

$$\Delta T_b = 47.08\ ^\circ C - 46.30\ ^\circ C = 0.78\ ^\circ C$$

$$c_m = \frac{\Delta T_b}{K_b} = \frac{0.78\ ^\circ C}{2.34\ ^\circ C\ kg/mol} = 0.333\ mol/kg$$

$$c_m = \frac{mol\ solute}{kg\ solvent} \qquad n_{solute} = (kg\ solvent)(c_m)$$

$$n_{solute} = (25.0\ mL)(1.261\ g/mL)(10^{-3}\ kg/g)(0.333\ mol/kg) = 1.05 \times 10^{-2}\ mol$$

Finally, calculate MM using $n = \dfrac{m}{MM}$:

$$MM = \frac{m}{n} = \frac{2.7\ g}{1.05 \times 10^{-2}\ mol} = 2.6 \times 10^2\ g/mol$$

12.65 Brine has much higher osmotic pressure than fresh water, so when a fish is placed in brine, water leaches from the cells, dehydrating them. In this dehydrated condition, bacteria are not able to survive and multiply, so the fish does not spoil.

12.67 When pure water freezes out of a solution, the remaining solution becomes more concentrated, because solvent has been removed but solutes have not. Thus, the remaining solution has a higher molality and a lower freezing point than the original solution. Calculate the molality of the remaining solution from its freezing point:

$$\Delta T_f = 0\ ^\circ C - (-10\ ^\circ C) = 10\ ^\circ C \qquad c_m = \frac{\Delta T_f}{K_f} = 10\ ^\circ C\left(\frac{1\ mol}{1.858\ ^\circ C\ kg}\right) = 5.4\ mol/kg$$

12.69 The solubility properties of water and gasoline can be predicted from the concept of "like dissolves like." Water is highly polar and hydrogen-bonding, while gasoline is a mixture of nonpolar hydrocarbons. Thus, water will not dissolve gasoline. Gasoline, being less dense, floats on water, so if water is used on a gasoline fire, the burning gasoline will be transported to other places by the flowing water. Rather than extinguishing a gasoline fire, water actually causes it to spread!

12.71 Calculate osmotic pressure using Equation 12-6 in your textbook: $\Pi = MRT$. Here, M is the total molarity of the solution. Consider 100 g of brackish water; it contains 0.5 g NaCl and has a volume of about 1.0×10^2 mL = 0.10 L.

$$n_{NaCl} = \frac{m}{MM} = 0.5\ g\left(\frac{1\ mol}{58.44\ g}\right) = 8.6 \times 10^{-3}\ mol,\ and.\ n_{total} = 2\ n_{NaCl} = 1.7 \times 10^{-2}\ mol$$

$$\Pi = \frac{(1.7 \times 10^{-2}\ mol)(8.206 \times 10^{-2}\ \frac{L\ atm}{mol\ K})(298\ K)}{0.10\ L} = 4\ atm$$

This result has only one significant figure, because the concentration (0.5 %) is stated to only one significant figure.

12.73   This is a problem involving the use of Henry's law, Equation 12-2 in your textbook:

$$[\text{gas } (aq)]_{eq} = K_H(p_{\text{gas}})_{eq}$$

According to Table 12-2, K_H for CO_2 is 3.4×10^{-2} at 25 °C and 7.8×10^{-2} at 0 °C.

Also according to the table, solution concentrations are in mol/L, gas pressures in atm.
(a) Begin by calculating the concentration of CO_2 in the solution:

$$[\text{gas } (aq)]_{eq} = K_H(p_{\text{gas}})_{eq} = (3.4 \times 10^{-2})(1.10) = 3.74 \times 10^{-2} \text{ mol/L.}$$

Use this concentration to calculate the amount of CO_2 in 225 mL of solution:

$$n = MV = (3.74 \times 10^{-2} \text{ mol/L})(225 \text{ mL})(10^{-3} \text{ L/mL}) = 8.42 \times 10^{-3} \text{ mol}$$

Do a mole-mass conversion to determine the mass:

$$m = n\,MM = (8.42 \times 10^{-3} \text{ mol})(44.01 \text{ g/mol}) = 0.37 \text{ g}$$

(b) Use the concentration determined in part (a) to find the pressure above the solution at 0 °C:

$$[\text{gas } (aq)]_{eq} = K_H(p_{\text{gas}})_{eq} \qquad\qquad (p_{\text{gas}})_{eq} = \frac{[\text{gas } (aq)]_{eq}}{K_H}$$

$$(p_{\text{gas}})_{eq} = \frac{3.74 \times 10^{-2} \text{ mol/L}}{7.8 \times 10^{-2} \text{ mol/L atm}} = 0.48 \text{ atm}$$

13.1   Convert line drawings to structural formulas following the standard rules, adding a C atom at each vertex and line end and adding –H bonds until each C atom has 4 bonds. Identify functional groups from memory or from Table 13-1 in your textbook:

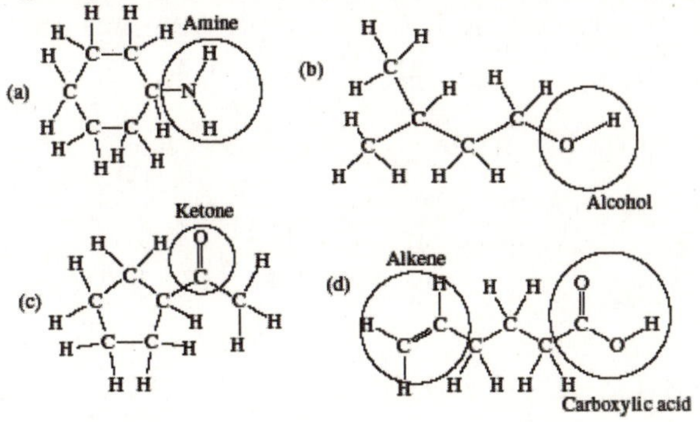

13.3   Identify functional groups from memory and convert structures to line drawings following the standard rules:

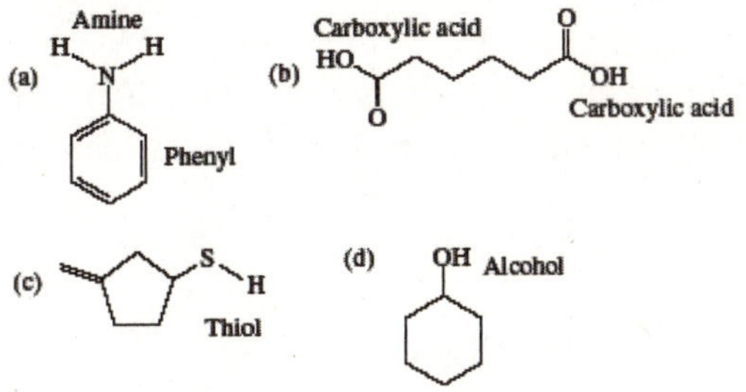

13.5   Convert line and ball-and-stick drawings to structural formulas following the standard rules, adding a C atom at each vertex and line end and adding –H bonds until each C atom has 4 bonds. Identify linkage groups from memory or Table 13-1 in your textbook:

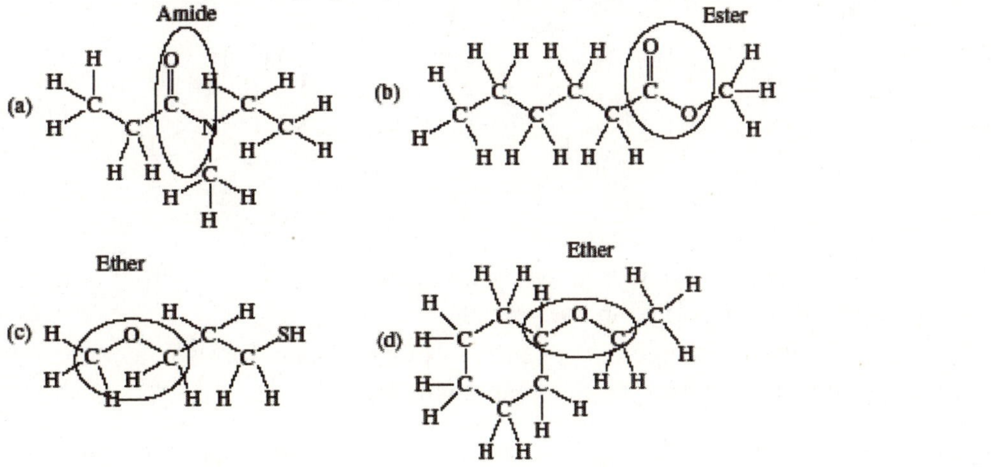

13.7   To draw examples of compound types, first identify the chemical formulas of its functional groups. Remember that carbon always forms four bonds. There are many possible correct answers to this problem. We provide just one example for each.

(a) an amine contains N bonded to C. The simplest amines contain $-NH_2$ units:

(b) An ester contains $C-CO_2-C$. There must be five additional C atoms, arranged in any fashion:

(c) The functional group in an aldehyde is $O=C-H$, $MM = 29$ g/mol. To have a molar mass greater than 80, an aldehyde must contain at least four additional C atoms:

(d) The ether linkage is $-O-$, and phenyl is the benzene ring:

13.9   In a condensation reaction, two monomers react to form a larger unit, eliminating a small molecule such as H_2O in the process.

(a) The amide linkage in the center of this molecule forms in a condensation reaction between a carboxylic acid and an amine:

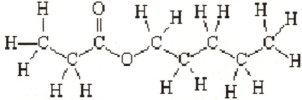

(b) The ester linkage in this molecule forms in a condensation reaction between a carboxylic acid and an alcohol:

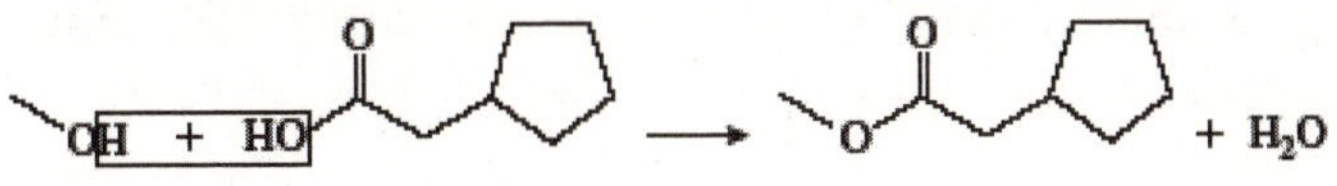

(c) The ether linkage in this molecule forms in a condensation reaction between two alcohols:

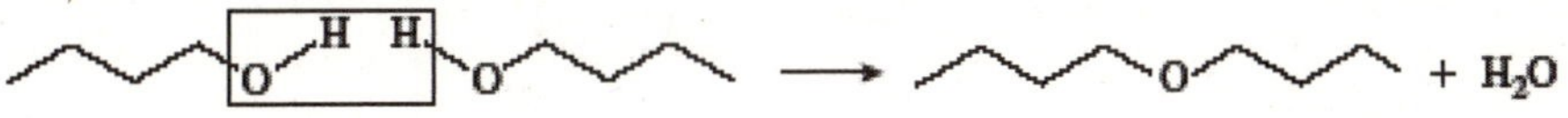

13.11 All amino acids can condense with each other in two ways. In addition, the extra carboxylic acid of glutamic acid could undergo condensation, so there are three possible products:

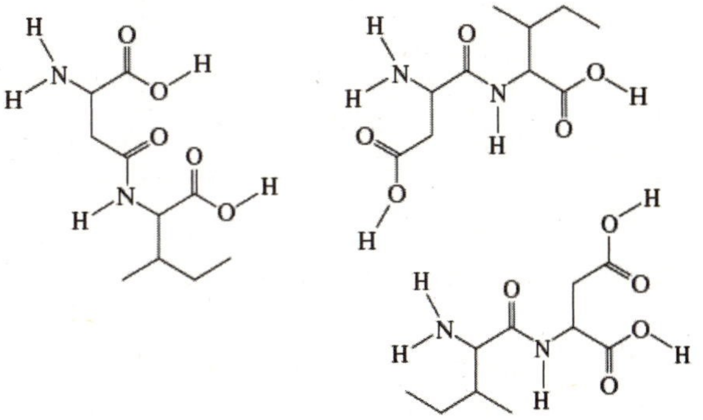

13.13 Polymers made from substituted ethylenes form by breakage of the C=C double bond to form two new single C–C bonds to other monomers. In a copolymer, each monomer occurs randomly:

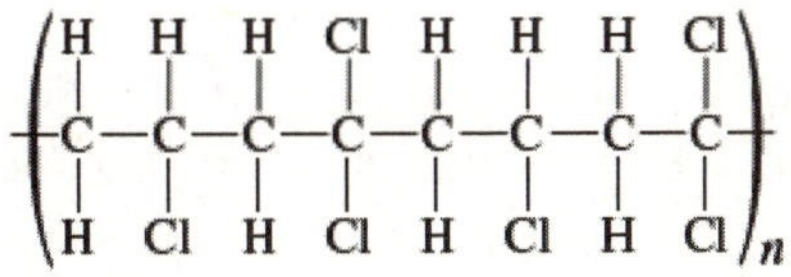

13.15 To deconstruct an alkane polymer into its monomers, break C–C single bonds and form double bonds:

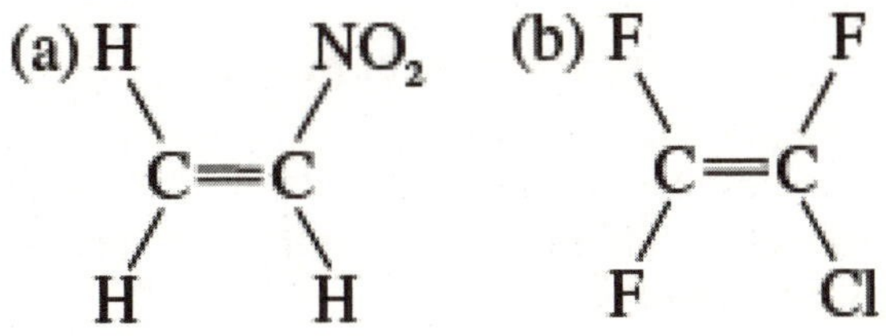

13.17   Polybutadiene forms from butadiene by a combination of double bond breakage and migration. The resulting polymer differs from polyethylene in having one C=C double bond in each repeat unit, whereas polyethylene is entirely CH_2 units connected by C–C single bonds (each arrow represents shifting one electron):

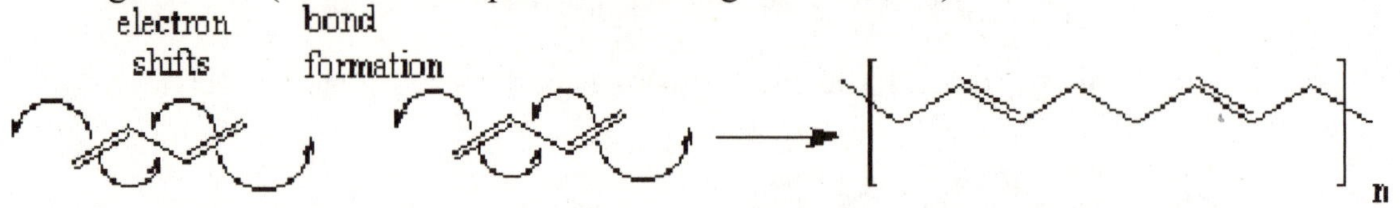

13.19   To determine the monomers from which a condensation polymer has formed, decompose the polymer into its monomeric parts and add components of a small molecule, for example –H and –OH. Here, the structure shows a single component whose repeat unit is $(CH_2)_{10}$, and the linkage is an amide, which breaks down into an amine and a carboxylic acid:

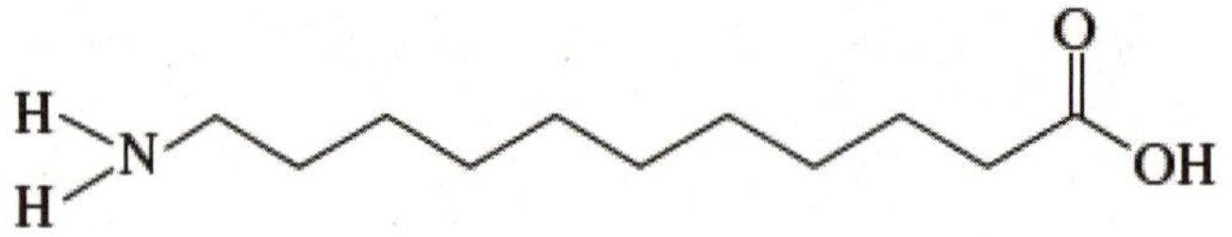

13.21   Construct polyethylene oxide by breaking a C–O bond and linking the fragments:

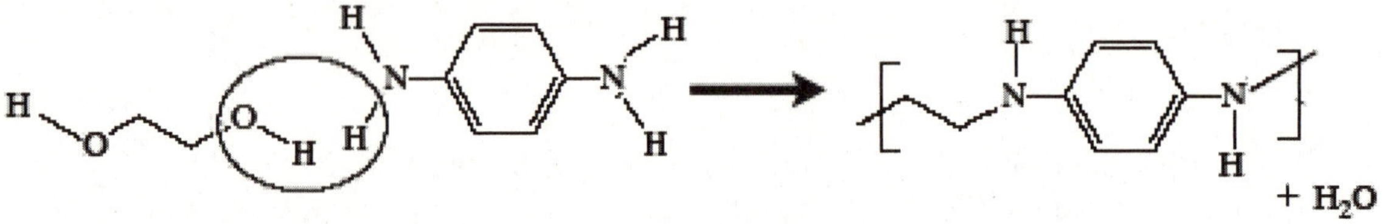

13.23   Cross-linking leads to increased rigidity, because chemical bonds between polymer chains restrict the ability of polymer chains to slide past one another. Thus, relatively rigid tires have more extensive cross-linking than flexible surgeon's gloves.

13.25   The categories of polymers and their characteristic properties are as follows: plastics, which exist as blocks or sheets; fibers, which can be drawn into long threads; and elastomers, which can be stretched without breaking.
(a) Balloons must stretch, so they are made of elastomers.
(b) Rope is made of fibers.
(c) Camera cases are rigid, so they are made of plastics.

13.27   Dioctylphthalate is an example of a liquid plasticizer, which reduces the amount of cross-linking in a polymer as well as adding a fluid component; both these changes result in improved flexibility of the polymer.

13.29   To draw the structures of sugars that are related to glucose, start with the glucose structure and make modifications as needed. For α-talose, move the OH group on carbons 2 and 4 from down positions to up positions.

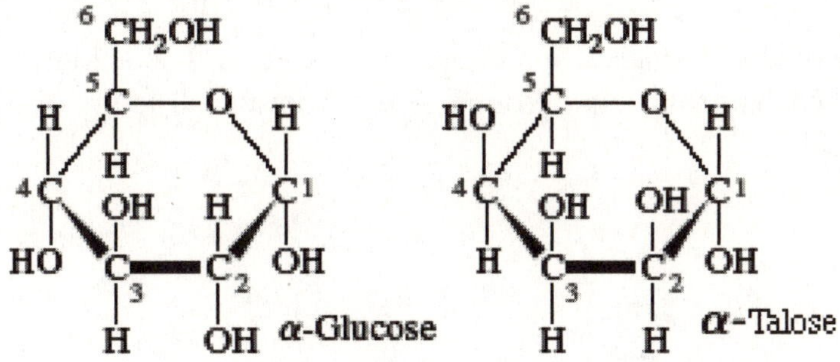

α-Glucose   α-Talose

13.31   To switch between α and β isomers move the OH group on carbon 1 from a down position to an up position (or vice versa when going from β to α). To draw β-talose, start with the structure of α-talose from problem 13.29 and move the OH on carbon 1 from a down position to an up position.

β-Talose

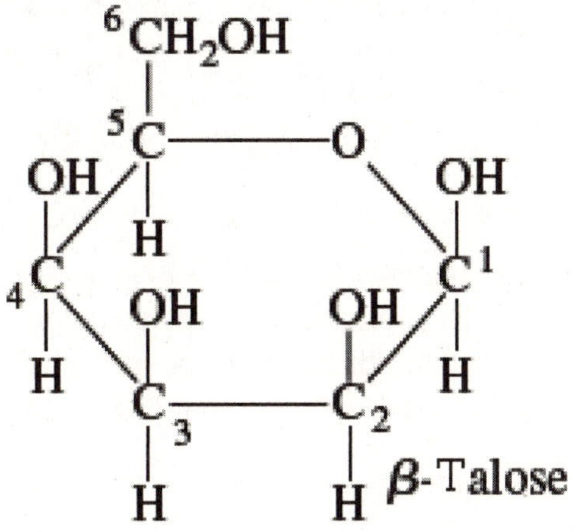

13.33   As described in your textbook, polymers of α-glucose coil upon themselves. Glycogen is this type of polymer. Polymers of β-glucose, of which cellulose is an example, form planar sheets. It is easier to see the distinction between the two polymers by looking at them from a side-on view (as shown). In the drawings, the dark solid lines indicate the direction of the continuing chain:

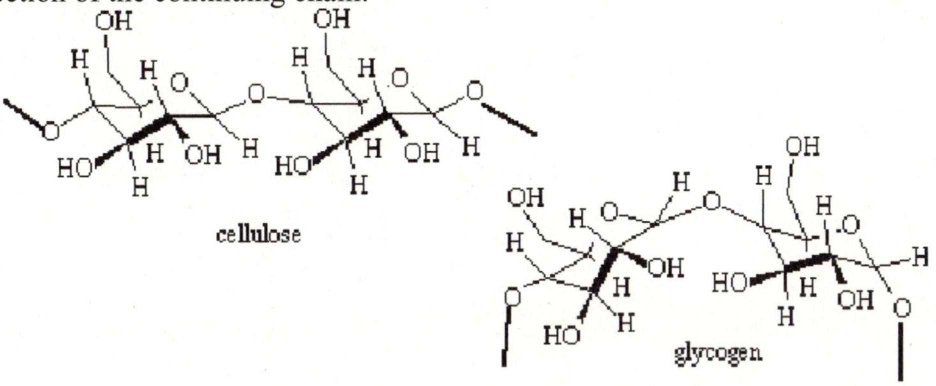

cellulose

glycogen

13.35 The complementary strands of DNA form from hydrogen-bond linkages between specific nucleic acids. A pairs with T and G pairs with C, so the complementary sequence of A-A-T-G-C-A-C-T-G is T-T-A-C-G-T-G-A-C.

13.37 The structure of DNA consists of a phosphate-sugar-nucleic acid trio bonded to others through its phosphate and sugar groups. Apart from the structure of the nucleic acid, each unit is identical:

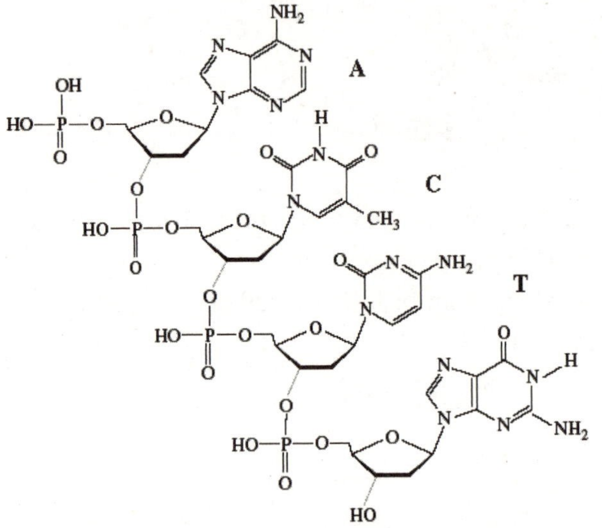

13.39 The strands of DNA fit together as a result of the hydrogen-bonding interactions illustrated in Figure 13-31. The complementary sequence for A-T-C is T-A-G. The backbones of complementary strands run in opposite directions. For clarity in showing the hydrogen-bonding interactions, we represent the backbone with a solid line rather than showing its details:

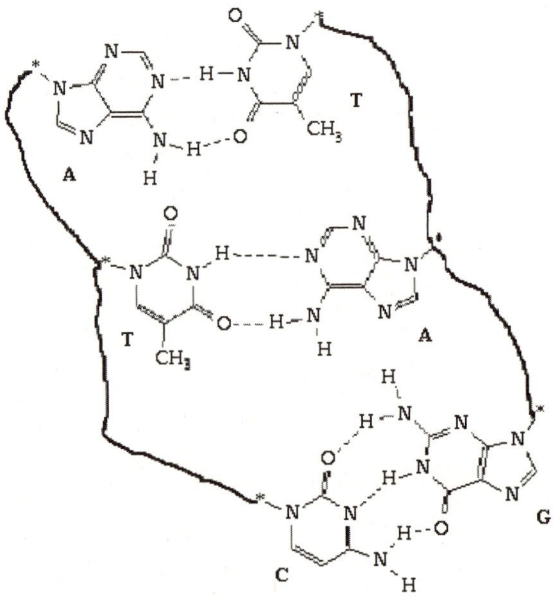

13.41   Consult Figure 13-35 for the structures of the different amino acids, all of which have the same amino acid backbone.

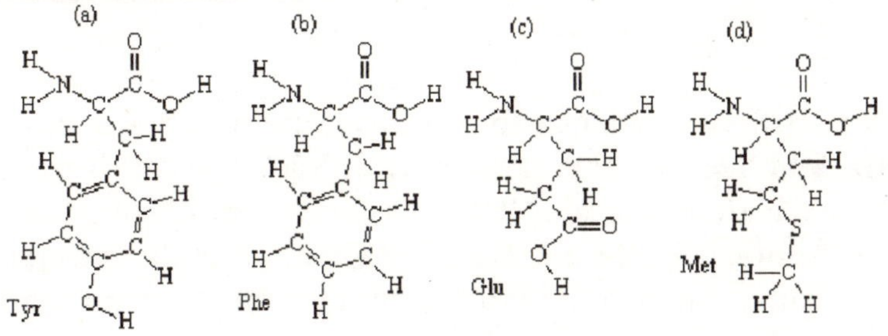

13.43   Hydrophilic side chains are characterized by the presence of N or O atoms that generate polar bonds and hydrogen-bonding capability, or an S–H bond that is polar. Among the structures shown in Problem 13.41, Tyr (O–H bond) and Glu (CO_2H group) are hydrophilic, while Phe and Met are hydrophobic.

13.45   To identify an amino acid, examine the side chain attached to the carbon atom between the N atom and the C=O bond in the amino acid backbone: (a) The side chain is just –H (hydrophobic), making this glycine, the simplest amino acid. (b) The side chain is –CH_2OH (hydrophilic side chain), so this is serine. (c) The side chain is –CH_2SH (hydrophilic side chain), so this is cysteine.

13.47   When two amino acids condense, the amino end of either molecule can link to the carboxylic acid end of the other. A water molecule is eliminated, creating a C–N bond. Three amino acids can combine in six different ways, A–B, B–A, A–C, C–A, B–C, and C–B:

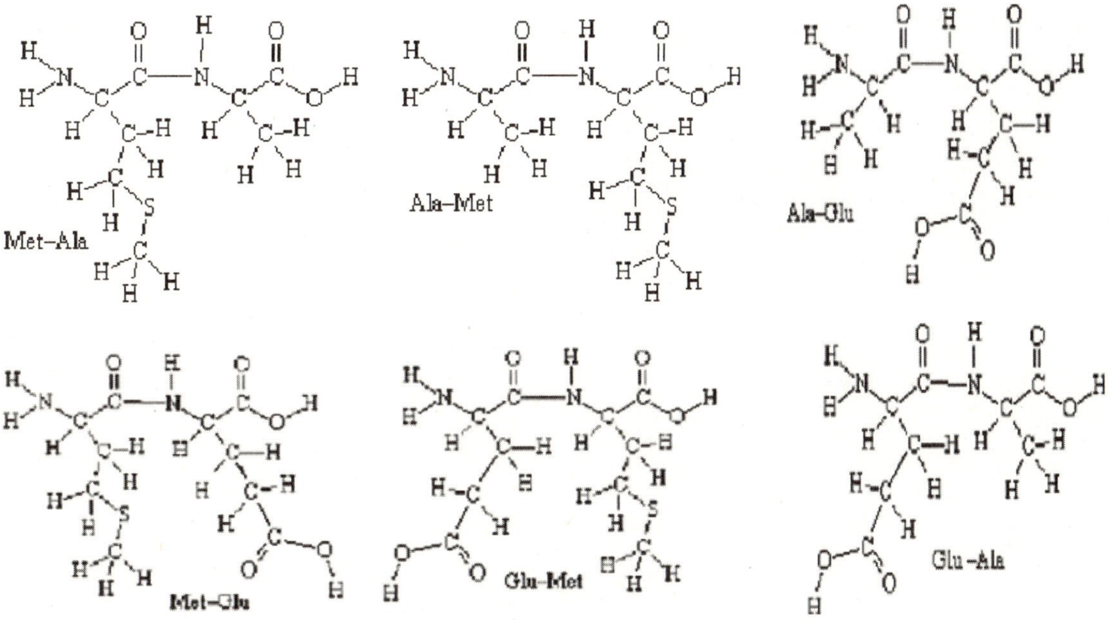

13.49   The repeat unit in polyethylene is the ethylene molecule, C_2H_4, $MM = 28.1$ g/mol, so a polymer with 744 repeat units has $MM = 744\left(\dfrac{28.1\text{ g}}{1\text{ mol}}\right) = 2.09 \times 10^4$ g/mol.

13.51   There are four possible choices for each base in a DNA strand, so the number of ways to connect 12 bases is $4^{12} = 16,777,216$ (this is an exact number, because 4 is an exact number).

13.53   (a) The monomers from which hair spray is made are substituted ethylenes, which polymerize in a free radical process in which each ethylene has an equal probability of adding to the growing chain. Thus, this polymer will have a random arrangement.

(b) The backbone of a polyethylene is a carbon chain with the substituents connected to every other carbon atom. Here is a line structure of six monomer units:

(c) The backbone of this polymer is hydrophobic, but its side chains contain highly polar C=O groups that interact with one another and with polar and hydrogen-bonding groups on hair.

13.55   A nucleotide is a combination of 1 base, 1 sugar, and 1 phosphate group. A duplex is a pair of nucleotides bound together by hydrogen-bonding interactions. The second nucleotide in a guanine duplex is cytosine.

13.57　The three steps of free radical polymerization are initiation by a free radical initiator, which generates a new free radical; propagation, in which monomer units add to the free radical end of the polymer chain; and termination, in which two free radical chains link together. The functional group on an ethylene monomer does not participate in any of these processes:

Initiation

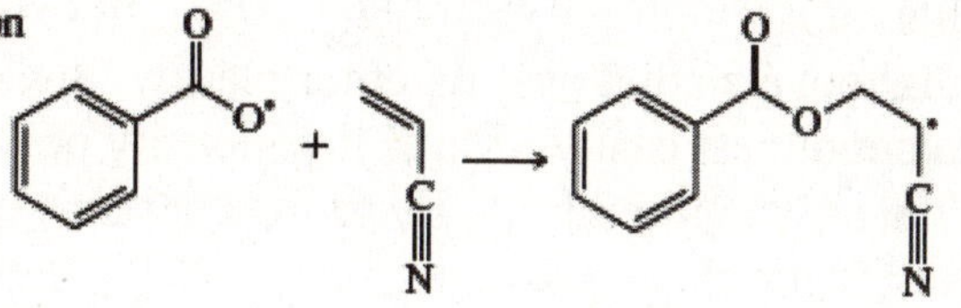

Propagation

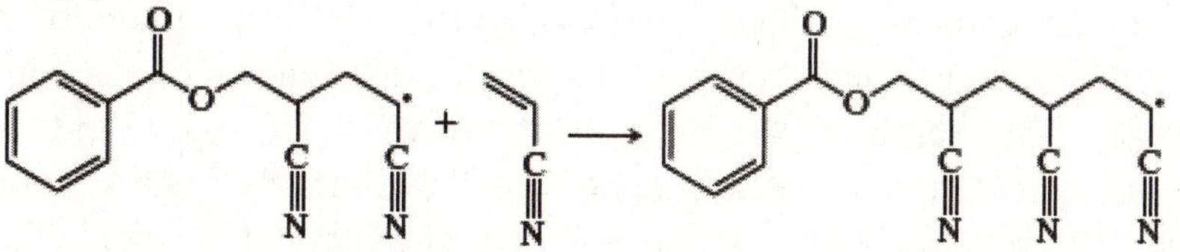

Termination:

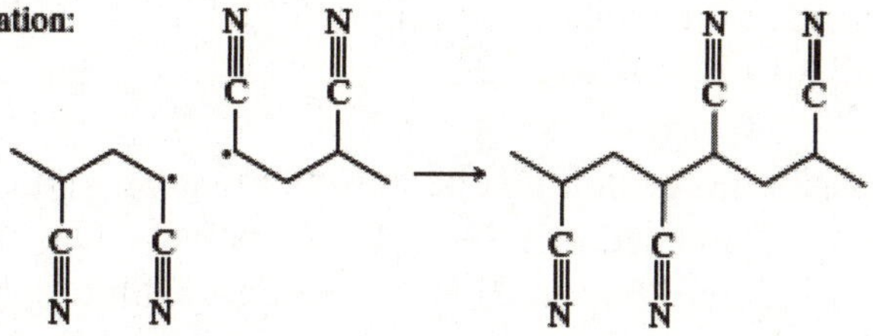

13.59　All sugar molecules have multiple –OH functional groups, each of which can participate in hydrogen bond formation with water molecules. α-glucose has 5 –OH groups; in addition, its ring O atom can participate in formation of hydrogen bonds.

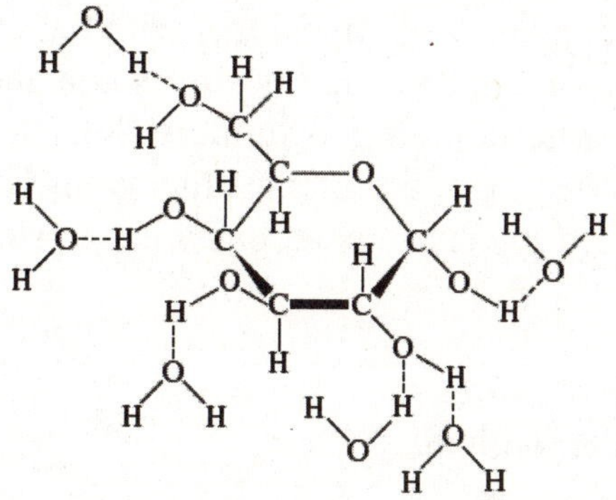

13.61　Nylons and proteins share one common feature: they form by condensation reactions between carboxylic acids and amines, so their linkages are amide groups. Otherwise, they are quite different. Nylons contain one or at most two different monomers, each of which typically contains several carbon atoms that form part of the backbone of the polymer. Proteins contain backbones that are absolutely regular repetitions of amide – C – amide bonding, but the carbon atoms in the backbone have a variety of substituent groups attached to them, generating the immense variety of different proteins (compared to only a few different nylons).

13.63 Consult your textbook for the structures of the polymers, which indicate the monomers from which they are made: (a) Kevlar is made from terephthalic acid and phenylenediamine. (b) PET is made from ethylene glycol and terephthalic acid. (c) Styrofoam is the common name for polystyrene, so it is made from styrene.

13.65 All polyethylene has the same empirical formula, $(CH_2)_n$, but whereas high-density polyethylene is all straight chains that nest together readily, low-density polyethylene has many side chains that cannot nest easily. Thus, low-density polyethylene has more open space, accounting for the lower amount of CH_2 groups per unit volume. See Figure 13-10 for a visual representation.

13.67 The Watson-Crick model of DNA requires that there be a complementary base for each base in a given strand: one G for every C, one C for every G, one A for every T, and one T for every A. This requires that the molar ratios of A to T and G to C be 1.0. Chargoff's observations indicate that this relationship holds even though DNA from different sources has different sequences, some A-T rich and others G-C rich. These differences give rise to differences in the relative amounts of A, T vs. G, C, but pairing always results in 1:1 mole ratios of A to T and G to C.

13.69 The bases of the mRNA molecule must be complementary to the bases of the template DNA on which they are modeled, meaning A generates U, C generates G, G generates C, and T generates A (remember that in RNA, U appears rather than T). Thus, the sequence generated by this strand is the same as that of Strand B except that U replaces T: 1 = cytosine, 2 = uracil, 3 = adenine, and 4 = guanine.

13.71 All proteins contain both hydrophobic and hydrophilic amino acids. When a protein is in contact with a hydrophilic medium such as aqueous solution, its hydrophilic amino acids are most stable when facing outward, in contact with the solvent, while its hydrophobic amino acids are most stable when facing inward, in contact with the protein backbone. The opposite is the case when a protein is immersed in a hydrophobic medium such as a cell wall. The tertiary structure of a protein is the manner in which the individual amino acids orient themselves, so solvent interactions are the primary determinants of this tertiary structure.

13.73 To draw the structure of a condensation product, start with the structures of the two molecules undergoing condensation and then connect them together, eliminating a small molecule such as water:

13.75   Polystyrene has the polyethylene backbone of carbon atoms, with a benzene ring attached to every second carbon atom. A divinylbenzene monomer reacts with the growing chain by means of one of its C=C double bonds, leaving the second C=C double bond available to cross-link by becoming incorporated into another polystyrene chain. Thus, the benzene rings form "bridges" between polystyrene chains:

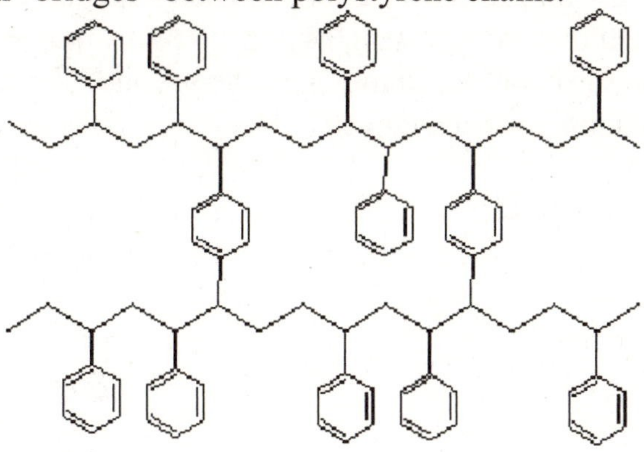

13.77   This problem is a "simple" stoichiometric problem that addresses the composition of a polymer. Use the percent by mass composition of copper to determine the mass of copper in 1 mol of the enzyme; convert to moles to obtain the molar ratio of copper to enzyme, which also gives the number of copper atoms per molecule:

$$64{,}000 \text{ g}\left(\frac{0.40\%}{100\%}\right) = 256 \text{ g of Cu in 1 mol of enzyme;}$$

$$256 \text{ g Cu}\left(\frac{1 \text{ mol}}{63.55 \text{ g}}\right) = 4.0 \text{ mol Cu/mol enzyme.}$$

There are 4 copper atoms in each molecule of this enzyme.

13.79   The description of the glucose-UDP molecule provides the information needed to construct it from the appropriate individual molecules. Uridine is uracil-ribose, and diphosphate indicates two phosphate units. Glucose is attached to the phosphate end via the specified –OH group:

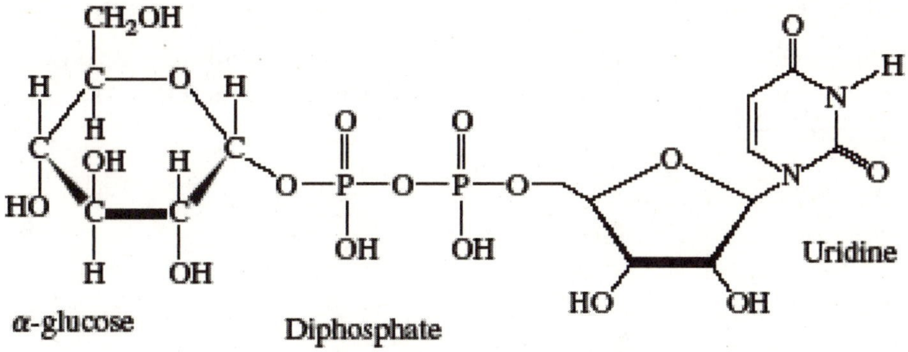

13.81 Trigonal planar geometry is best described by *sp*² hybrid orbitals. The simple Lewis structure of the peptide linkage indicates SN = 4, which normally requires *sp*³ hybrid orbitals. The trigonal planar geometry indicates that the lone pair of electrons on the N atom is delocalized in a π bonding network, which can be illustrated using the Lewis structure of the gly-gly dimer. There is a resonance structure that places a π bond between C and N. Studies of bond lengths and strengths indicate that the peptide linkage is stronger than a single bond; this along with the trigonal planar geometry about the N atom confirms this bonding arrangement:

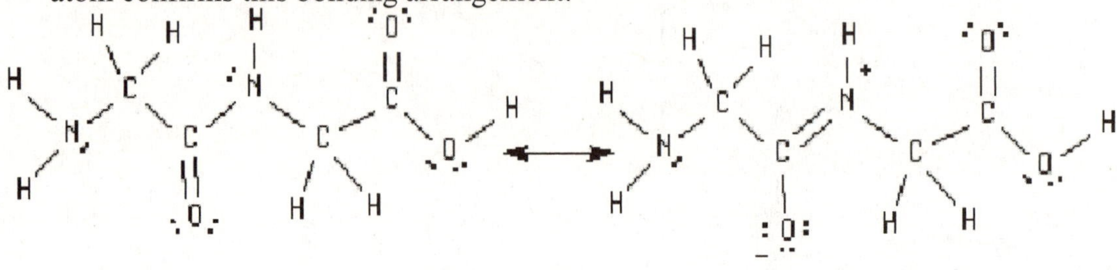

13.83 To draw the structure of a disaccharide, start with the appropriate monosaccharide and make the indicated connections. The structure of β-glucose appears in Figure 13-17.

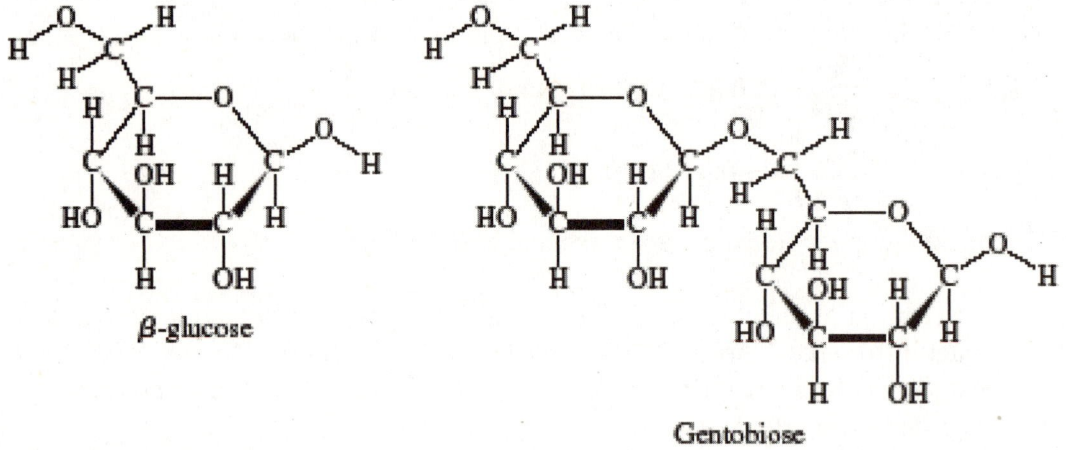

β-glucose

Gentobiose

13.85 A nucleotide consists of a base (in this case, cytosine is identified), a ribose sugar ring, and the phosphate group connected through condensation reactions at specific positions as illustrated in Figure 13-28:

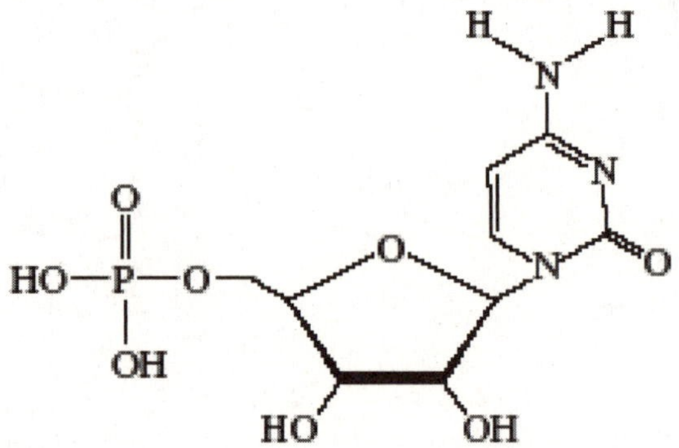

14.1   (a) A sand castle represents an organized structure constructed by a person. Waves destroy that structure, dispersing the sand grains.
   (b) Two separate liquids represent a constrained arrangement (all molecules of one kind in one container). Upon mixing, the molecules in the liquids are dispersed throughout the container.
   (c) Sticks in a bundle are constrained (all aligned in the same direction). When dropped, the sticks lose their alignment and become dispersed.
   (d) Water in a puddle is relatively constrained, as it is confined to a small volume. When the water evaporates, the molecules spread over a much larger volume and become more dispersed.

14.3   The molecules in a drop of ink are relatively constrained, because they occupy one particular part of the total volume. As they move about randomly, they spread throughout the volume of the container and become more dispersed.

14.5   (a) The air molecules in a tire are relatively constrained, because they are confined to a specific, small volume. A puncture allows gas molecules to escape from the tire and fill a much larger volume, becoming more dispersed in the process.
   (b) Like air in a tire, the fragrant molecules in a perfume bottle are relatively constrained because they are confined to a specific, small volume. When the bottle is open, molecules escape from the confined volume into a much larger space of the room, becoming more dispersed in the process.

14.7   (a) Energy dispersal requires that heat flows from high to low temperature, never in the opposite direction.
   (b) Matter dispersal requires that the spontaneous direction is toward greater dispersal, and a "sand castle" represents constrained matter.

14.9   Calculate W by determining how many ways each symbol can be placed. The first X can be placed in any of the nine compartments, the first O in any of the eight empty compartments, so $W = (9)(8) = 72$.

14.11   The entropy change accompanying a constant–temperature process is $\Delta S = \dfrac{q}{T}$.

   (a) Melting involves absorption of heat:

$$q_{ice} = n\Delta H_{fus} = 13.8 \text{ g}\left(\frac{1 \text{ mol}}{18.02 \text{ g}}\right)\left(\frac{6.01 \text{ kJ}}{1 \text{ mol}}\right)\left(\frac{10^3 \text{ J}}{1 \text{ kJ}}\right) = 4.60 \times 10^3 \text{ J}$$

$$\Delta S_{ice} = \frac{4.60 \times 10^3 \text{ J}}{273.15 \text{ K}} = 16.8 \text{ J/K}$$

   (b) The heat absorbed by the ice cube must be supplied by the pool water:

$q_{pool} = -q_{ice}$

$$\Delta S_{pool} = \frac{-4.60 \times 10^3 \text{ J}}{27.5 + 273.15 \text{ K}} = -15.3 \text{ J/K}$$

   (c) $\Delta S_{total} = \Delta S_{ice} + \Delta S_{pool} = (16.8 \text{ J/K}) + (-15.3 \text{ J/K}) = 1.5 \text{ J/K}$

14.13   In each of these spontaneous processes, $\Delta S_{total} > 0$. For system and surroundings, the sign of the entropy change will be the same as the sign of q: (a) The system (water) absorbs heat to boil, so q_{sys} is positive and ΔS_{sys} is positive. The surroundings (stove) release heat to the system, so q_{surr} is negative and ΔS_{surr} is negative. (b) The system (ice cubes) absorbs heat to melt, so q_{sys} is positive and ΔS_{sys} is positive. The surroundings (table top) release heat to the system, so q_{surr} is negative and ΔS_{surr} is negative. (c) The system (coffee) absorbs heat, so q_{sys} is positive and ΔS_{sys} is positive. The surroundings (microwave oven) release heat to the system, so q_{surr} is negative and ΔS_{surr} is negative.

14.15   The entropy change accompanying a constant–temperature process is $\Delta S = \dfrac{q}{T}$.

Condensation, the reverse of vaporization, involves release of heat:

$$q_{steam} = -n\Delta H_{vap} = -15.5 \text{ g}\left(\frac{1 \text{ mol}}{18.02 \text{ g}}\right)\left(\frac{40.79 \text{ kJ}}{1 \text{ mol}}\right)\left(\frac{10^3 \text{ J}}{1 \text{ kJ}}\right) = -3.509 \times 10^4 \text{ J}$$

$$\Delta S_{steam} = \frac{-3.509 \times 10^4 \text{ J}}{373.15 \text{ K}} = -94.0 \text{ J/K}$$

Knowing that ΔS_{total} must be positive because the overall process is spontaneous, we can say without doing further calculations that $\Delta S_{surr} > 94.0$ J/K.

14.17   The molar entropy change accompanying fusion is $\Delta S_{molar} = \dfrac{\Delta H_{fus}}{T_{fus}}$:

(a) Argon: $\Delta S_{molar} = \left(\dfrac{1.3 \text{ kJ/mol}}{83 \text{ K}}\right)\left(\dfrac{10^3 \text{ J}}{1 \text{ kJ}}\right) = 16$ J mol^{-1} K^{-1}

(b) Methane: $\Delta S_{molar} = \left(\dfrac{0.84 \text{ kJ/mol}}{90 \text{ K}}\right)\left(\dfrac{10^3 \text{ J}}{1 \text{ kJ}}\right) = 9.3$ J mol^{-1} K^{-1}

(c) Ethanol: $\Delta S_{molar} = \left(\dfrac{7.61 \text{ kJ/mol}}{156 \text{ K}}\right)\left(\dfrac{10^3 \text{ J}}{1 \text{ kJ}}\right) = 48.8$ J mol^{-1} K^{-1}

(d) Mercury: $\Delta S_{molar} = \left(\dfrac{23.4 \text{ kJ/mol}}{234 \text{ K}}\right)\left(\dfrac{10^3 \text{ J}}{1 \text{ kJ}}\right) = 1.00 \times 10^2$ J mol^{-1} K^{-1}

14.19   (a) Both are ionic solutions, but $MgCl_2$ produces three moles of ions per mole of substance, while NaCl produces two moles of ions per mole of substance, so $MgCl_2$ has the larger molar entropy.
(b) Both are ionic solids, but HgS has a higher molar mass than HgO, so HgS has the larger molar entropy.
(c) Both are diatomic molecules, but Br_2 (*l*) is liquid while I_2 (*s*) is solid, so Br_2 (*l*) has the larger molar entropy.

14.21   All these substances are small gaseous molecules. Ozone has more entropy than O_2 or H_2 because it has three atoms per molecule while the others have only two. O_2 has a higher molar mass than H_2, so it has higher entropy.

14.23   Absolute entropies are tabulated in Appendix D of your textbook. To obtain the entropy per mole of atoms, divide each value by the number of atoms in one particle:

$$He: S° = 126.153 \text{ J/mol K} \left(\frac{1 \text{ mol He}}{1 \text{ mol atom}} \right) = 126.153 \text{ J mol}^{-1} \text{ K}^{-1}$$

$$H_2: S° = 130.680 \text{ J/mol K} \left(\frac{1 \text{ mol } H_2}{2 \text{ mol atom}} \right) = 65.340 \text{ J mol}^{-1} \text{ K}^{-1}$$

$$CH_4: S° = 186.3 \text{ J/mol K} \left(\frac{1 \text{ mol } CH_4}{5 \text{ mol atom}} \right) = 37.26 \text{ J mol}^{-1} \text{ K}^{-1}$$

$$C_3H_6: S° = 226.9 \text{ J/mol K} \left(\frac{1 \text{ mol } C_3H_6}{9 \text{ mol atom}} \right) = 25.21 \text{ J mol}^{-1} \text{ K}^{-1}$$

Entropy per mole of atoms decreases as the number of atoms in a species increases, because tying together atoms into a molecule increases the constraints among those atoms (other factors, notably the atomic mass of the atoms, also play important roles).

14.25   Use Equation 14-5 to calculate entropies at pressures different from 1.00 bar:
$$S_{(p \neq 1)} = S° - R \ln p$$
Take into account amounts different from 1 mol by multiplying by the number of moles.
(a) $S = (2.50 \text{ mol})[154.843 \text{ J mol}^{-1} \text{ K}^{-1} - (8.314 \text{ J mol}^{-1} \text{ K}^{-1}) \ln (0.25 \text{ bar})] = 416 \text{ J/K}$
(b) $S = (0.75 \text{ mol})[238.9 \text{ J mol}^{-1} \text{ K}^{-1} - (8.314 \text{ J mol}^{-1} \text{ K}^{-1}) \ln (2.75 \text{ bar})] = 1.7 \times 10^2 \text{ J/K}$
(c) Calculate the entropy of each component separately, using $n_i = X_i n_{tot}$ and $p_i = X_i P_{tot}$:
$$S_{N_2} = (0.78)(0.45 \text{ mol})\{191.61 \text{ J mol}^{-1} \text{ K}^{-1}$$
$$- (8.314 \text{ J mol}^{-1} \text{ K}^{-1}) \ln [(0.78)(1.00 \text{ bar})]\} = 68 \text{ J/K}$$
$$S_{O_2} = (0.22)(0.45 \text{ mol})\{205.152 \text{ J mol}^{-1} \text{ K}^{-1}$$
$$- (8.314 \text{ J mol}^{-1} \text{ K}^{-1}) \ln [(0.22)(1.00 \text{ bar})]\} = 22 \text{ J/K}$$
$$S = S_{N_2} + S_{O_2} = 9.0 \times 10^1 \text{ J/K}$$

14.27   Standard entropy changes can be calculated using tabulated values of absolute entropies (Appendix D of your textbook) and Equation 14–6:
$$\Delta S°_{reaction} = \Sigma \text{ coeff}_{products} S°_{products} - \Sigma \text{ coeff}_{reactants} S°_{reactants}$$
(a) $\Delta S°_{reaction} = 2 \text{ mol}(192.8 \text{ J mol}^{-1} \text{ K}^{-1})$
$$- [1 \text{mol}(191.61 \text{ J mol}^{-1} \text{ K}^{-1}) + 3 \text{ mol}(130.680 \text{ J mol}^{-1} \text{ K}^{-1})] = -198.1 \text{ J/K}$$
(b) $\Delta S°_{reaction} = 2 \text{ mol}(238.9 \text{ J mol}^{-1} \text{ K}^{-1}) - 3 \text{ mol}(205.152 \text{ J mol}^{-1} \text{ K}^{-1}) = -137.7 \text{ J/K}$
(c) $\Delta S°_{reaction} = [1 \text{ mol}(64.8 \text{ J mol}^{-1} \text{ K}^{-1}) + 2 \text{ mol}(37.99 \text{ J mol}^{-1} \text{ K}^{-1})]$
$$- [1 \text{ mol}(68.7 \text{ J mol}^{-1} \text{ K}^{-1}) + 2 \text{ mol}(29.9 \text{ J mol}^{-1} \text{ K}^{-1})] = 12.3 \text{ J/K}$$
(d) $\Delta S°_{reaction} = [2 \text{ mol}(213.8 \text{ J mol}^{-1} \text{ K}^{-1}) + 2 \text{ mol}(69.95 \text{ J mol}^{-1} \text{ K}^{-1})]$
$$- [1 \text{ mol}(219.3 \text{ J mol}^{-1} \text{ K}^{-1}) + 3 \text{ mol}(205.152 \text{ J mol}^{-1} \text{ K}^{-1})] = -267.3 \text{ J/K}$$

14.29 Reactions have negative entropy changes when products are more constrained than reactants:
(a) There is a relatively large negative value because of the reduction in number of moles of gaseous substances: 4 moles of gaseous reactants are converted to 2 moles of gaseous products.
(b) There is a relatively large negative value because of the reduction in number of moles of gaseous substances: 3 moles of gaseous reactants are converted to 2 moles of gaseous products.
(c) This reaction has a near–zero entropy change because all reagents are relatively constrained solid substances.
(d) There is a relatively large negative value because of the reduction in number of moles of gaseous substances: 4 moles of gaseous reactants are converted to 2 moles of gaseous products.

14.31 The sign of $\Delta S°$ for a reaction depends on whether the atoms are more dispersed among the products or the reactants.
(a) A gas is more dispersed than a solid, so $\Delta S°$ is positive.
(b) There are gaseous molecules among the reactants but not in the products, so $\Delta S°$ is negative.
(c) Three molecules of gaseous reactants become two molecules of gaseous products, so there are more constraints and $\Delta S°$ is negative.

14.33 False statements can be made true in various ways; we choose the simplest change that corrects each statement:
(a) $\Delta G_{system} < 0$ for any spontaneous process at constant T and P.
(b) The free energy of a system decreases in any spontaneous process at constant T and P.
(c) $\Delta G = \Delta H - T\Delta S$ at constant T and P.

14.35 Standard free energy changes are calculated from Equation 14–9 using standard free energies of formation, which can be found in Appendix D of your textbook:
$$\Delta G°_{reaction} = \Sigma\ coeff_{products}\ \Delta G^{°}_{f\ products} - \Sigma\ coeff_{reactants}\ \Delta G^{°}_{f\ reactants}$$
(a) $\Delta G°_{reaction}$ = 2 mol(–16.4 kJ/mol) – [1 mol(0 kJ/mol) + 3 mol(0 kJ/mol)] = –32.8 kJ
(b) $\Delta G°_{reaction}$ = 2 mol(163.2 kJ/mol) – 3 mol(0 kJ/mol) = 326.4 kJ
(c) $\Delta G°_{reaction}$ = [1 mol(0) + 2 mol(–211.7 kJ/mol)]
           – [1 mol(–217.3 kJ/mol) + 2 mol(0 kJ/mol)] = –206.1 kJ
(d) $\Delta G°_{reaction}$ = [2 mol(–394.4 kJ/mol) + 2 mol(–237.1 kJ/mol)]
           – [1 mol(68.4 kJ/mol) + 3 mol(0 kJ/mol)] = –1331.4 kJ

14.37 Use Equation 14–11 to estimate the standard free energy change at a temperature different from 298 K: $\Delta G°_{reaction, T} \cong \Delta H°_{reaction, 298} - T\Delta S°_{reaction, 298}$. Standard entropy changes are calculated in Problem 14.27, but standard enthalpy changes need to be calculated from standard enthalpies of formation. $T = 273.15 - 85 = 188$ K:

(14.27b) $\Delta H°_{reaction} = 2$ mol(142.7 kJ/mol) – 3 mol(0 kJ/mol) = 285.4 kJ

$$\Delta G°_{reaction, T} \cong 285.4 \text{ kJ} - 188 \text{ K}(-137.7 \text{ J/K})\left(\frac{10^{-3} \text{ kJ}}{1 \text{ J}}\right) = 311.3 \text{ kJ}$$

(14.27c) $\Delta H°_{reaction} = [1$ mol(0 kJ/mol) + 2 mol(–239.7 kJ/mol)]
$\qquad\qquad$ – [1 mol(–277.4 kJ/mol) + 2 mol(0 kJ/mol)] = –202.0 kJ

$$\Delta G°_{reaction, T} \cong -202.0 \text{ kJ} - 188 \text{ K}(12.3 \text{ J/K})\left(\frac{10^{-3} \text{ kJ}}{1 \text{ J}}\right) = -204.3 \text{ kJ}$$

14.39 Use Equation 14–11 to estimate the standard free energy change at pressures different from 1 bar:

$\Delta G_{reaction} = \Delta G°_{reaction} + RT \ln Q$. $\Delta G°_{reaction}$ values are calculated in Problem 14.31.

(a) $RT \ln\left(\dfrac{p^2_{NH_3}}{p^3_{H_2} p_{N_2}}\right) = (8.314 \times 10^{-3} \text{ kJ mol}^{-1} \text{ K}^{-1})(298 \text{ K}) \ln\left(\dfrac{(15 \text{ bar})^2}{(15 \text{ bar})^3(15 \text{ bar})}\right)$
$$= -13.4 \text{ kJ}$$

$\Delta G_{reaction} = -32.8 \text{ kJ} - 13.4 \text{ kJ} = -46.2 \text{ kJ}$

(b) $RT \ln\left(\dfrac{p^2_{O_3}}{p^3_{O_2}}\right) = (8.314 \times 10^{-3} \text{ kJ mol}^{-1} \text{ K}^{-1})(298 \text{ K}) \ln\left(\dfrac{(15 \text{ bar})^2}{(15 \text{ bar})^3}\right) = -6.71 \text{ kJ}$

$\Delta G_{reaction} = 326.4 \text{ kJ} - 6.71 \text{ kJ} = 319.7 \text{ kJ}$

14.41 Use Equations 6–11, 14–6, and 14–9 to calculate standard thermodynamic values:

$\Delta H°_{reaction} = \Sigma \text{ coeff}_{products} \Delta H°_{f \, products} - \Sigma \text{ coeff}_{reactants} \Delta H°_{f \, reactants}$

$\Delta S°_{reaction} = \Sigma \text{ coeff}_{products} S°_{products} - \Sigma \text{ coeff}_{reactants} S°_{reactants}$

$\Delta G°_{reaction} = \Sigma \text{ coeff}_{products} \Delta G°_{f \, products} - \Sigma \text{ coeff}_{reactants} \Delta G°_{f \, reactants}$

$\Delta H°_{reaction} = [1$ mol(–365.6 kJ/mol) + 1 mol(–285.83 kJ/mol)]
$\qquad\qquad$ – [2 mol(–45.9 kJ/mol) + 2 mol(0 kJ/mol)] = –559.6 kJ

$\Delta S°_{reaction} = [1$ mol(151.1 J mol^{-1} K^{-1}) + 1 mol(69.95 J mol^{-1} K^{-1})]
$\qquad\qquad$ – [2 mol(192.8 J mol^{-1} K^{-1}) + 2 mol(205.152 J mol^{-1} K^{-1})] = –574.9 J/K

$\Delta G°_{reaction} = [1$ mol(–183.9 kJ/mol) + 1 mol(–237.1 kJ/mol)]
$\qquad\qquad$ – [2 mol(–16.4 kJ/mol) + 2 mol(0 kJ/mol)] = –388.2 kJ

14.43 (a) The number of molecules of gaseous products is smaller than the number of molecules of gaseous reactants, so the reaction has a negative $\Delta S°$. The reaction is a formation reaction with a positive $\Delta H_f°$ (Appendix D). Negative $\Delta S°$ combined with positive $\Delta H°$ means that $\Delta G°$ is positive at all temperatures. Thus, the reaction is not spontaneous at any temperature.

(b) Any gas–phase reaction can be made spontaneous by reducing the partial pressure of products to zero bar.

14.45 To calculate a vapor pressure, use Equation 14–14:

$$\ln vp = -\frac{\Delta H_{vap}}{RT} + \frac{\Delta S°_{vap}}{R}$$

The vaporization reaction is 4 P (s, white) $\rightarrow P_4$ (g).

$\Delta H° = 58.9 - 4(0) = 58.9$ kJ; $\Delta S° = 280.0 - 4(41.1) = 115.6$ J/K

$$\ln vp = -\left[\frac{(58.9 \text{ kJ/mol})(10^3 \text{ J/kJ})}{(8.314 \text{ J mol}^{-1}\text{ K}^{-1})(298 \text{ K})}\right] + \left(\frac{115.6 \text{ J mol}^{-1}\text{ K}^{-1}}{8.314 \text{ J mol}^{-1}\text{ K}^{-1}}\right) = -9.87$$

$vp = e^{-9.87} = 5.2$ x 10^{-5} bar

14.47 Coupled reactions share a common intermediate that connects the reaction and allows transfer of free energy between reactants and final products. In the case of the seesaw the first "reaction" would be child 1 starting on the ground and rising into the air. The second "reaction" would be child 2 starting in the air and falling to the ground. In this case child 2 would be the spontaneous reaction, child 1 would be nonspontaneous (things tend to fall down, not float up), so as child 2 is falling he will cause the first child to rise. The common intermediate is the seesaw board, which acts as a lever transferring energy from one child to the other.

14.49 The reactions that are coupled in this example are the following:
ATP + $H_2O \rightarrow$ ADP + H_3PO_4 $\qquad \Delta G° = -30.6$ kJ
$C_6H_{12}O_6$ (fructose) + $C_6H_{12}O_6$ (glucose) $\rightarrow C_{12}H_{22}O_{11}$ + H_2O $\quad \Delta G° = 23.0$ kJ
Coupled reaction:
ATP + $C_6H_{12}O_6$ (fructose) + $C_6H_{12}O_6$ (glucose) $\rightarrow$ ADP + $C_{12}H_{22}O_{11}$ + H_3PO_4
$\Delta G° = 23.0$ kJ $- 30.6$ kJ $= -7.6$ kJ

14.51 Spontaneity is determined by free energy, heat flow by enthalpy, and amount of dispersal by entropy: Use Equations 6–11, 14–6, and 14–9 to calculate standard thermodynamic values:

$\Delta H°_{reaction} = \Sigma \text{ coeff}_{products} \Delta H_f°_{products} - \Sigma \text{ coeff}_{reactants} \Delta H_f°_{reactants}$

$\Delta S°_{reaction} = \Sigma \text{ coeff}_{products} S°_{products} - \Sigma \text{ coeff}_{reactants} S°_{reactants}$

$\Delta G°_{reaction} = \Sigma \text{ coeff}_{products} \Delta G_f°_{products} - \Sigma \text{ coeff}_{reactants} \Delta G_f°_{reactants}$

(a) $\Delta G°_{reaction}$ = [2 mol(0 kJ/mol) + 3 mol(–237.1 kJ/mol)]
$- [1$ mol(–1582.3 kJ/mol) + 3 mol(0 kJ/mol)] = 871.0 kJ
The reaction is not spontaneous.

(b) $\Delta H°_{reaction} = [2 \text{ mol}(0 \text{ kJ/mol}) + 3 \text{ mol}(-285.83 \text{ kJ/mol})]$

$\qquad - [1 \text{ mol}(-1675.7 \text{ kJ/mol}) + 3 \text{ mol}(0 \text{ kJ/mol})] = 818.2 \text{ kJ}$

The reaction absorbs heat.

(c) $\Delta S°_{reaction} = [2 \text{ mol}(28.3 \text{ J mol}^{-1} \text{ K}^{-1}) + 3 \text{ mol}(69.95 \text{ J mol}^{-1} \text{ K}^{-1})]$

$\qquad - [1 \text{ mol}(50.9 \text{ J mol}^{-1} \text{ K}^{-1}) + 3 \text{ mol}(130.680 \text{ J mol}^{-1} \text{ K}^{-1})] = -176.5 \text{ J/K}$

Entropy decreases, so the products are less dispersed than the reactants.

14.53 Use Equation 14–6 to calculate the entropy change of a reaction:

$$\Delta S°_{reaction} = \Sigma \text{ coeff}_{products} \, S°_{products} - \Sigma \text{ coeff}_{reactants} \, S°_{reactants}$$

(a) $\Delta S°_{reaction} = 2 \text{ mol}(162 \text{ J mol}^{-1} \text{ K}^{-1})$

$\qquad - [2 \text{ mol}(101 \text{ J mol}^{-1} \text{ K}^{-1}) + 1 \text{ mol}(205.152 \text{ J mol}^{-1} \text{ K}^{-1})] = -83 \text{ J/K}$

(b) $\Delta S°_{reaction} = [2 \text{ mol}(87.4 \text{ J mol}^{-1} \text{ K}^{-1}) + 6 \text{ mol}(223.1 \text{ J mol}^{-1} \text{ K}^{-1})]$

$\qquad - [4 \text{ mol}(142.3 \text{ J mol}^{-1} \text{ K}^{-1}) + 3 \text{ mol}(205.152 \text{ J mol}^{-1} \text{ K}^{-1})] = 328.7 \text{ J/K}$

(c) $\Delta S°_{reaction} = [6 \text{ mol}(210.8 \text{ J mol}^{-1} \text{ K}^{-1}) + 6 \text{ mol}(69.95 \text{ J mol}^{-1} \text{ K}^{-1})]$

$\qquad - [3 \text{ mol}(121.2 \text{ J mol}^{-1} \text{ K}^{-1}) + 4 \text{ mol}(238.9 \text{ J mol}^{-1} \text{ K}^{-1})] = 365.3 \text{ J/K}$

14.55 Use Equation 14–10 to calculate the standard free energy change of a reaction:

$$\Delta G°_{reaction} = \Sigma \text{coeff}_p \, \Delta G_f^o \text{ (products)} - \Sigma \text{ coeff}_r \, \Delta G_f^o \text{ (reactants)}$$

(a) $\Delta G°_{reaction} = 2 \text{ mol}(-3 \text{ kJ/mol}) - [2 \text{ mol}(17 \text{ kJ/mol}) + 1 \text{ mol}(0 \text{ kJ/mol})] = -40 \text{ kJ}$

(b) $\Delta G°_{reaction} = [2 \text{ mol}(-742.2 \text{ kJ/mol}) + 6 \text{ mol}(0)]$

$\qquad - [4 \text{ mol}(-334.0 \text{ kJ/mol}) + 3 \text{ mol}(0 \text{ kJ/mol})] = -148.4 \text{ kJ}$

(c) $\Delta G°_{reaction} = [6 \text{ mol}(87.6 \text{ kJ/mol}) + 6 \text{ mol}(-237.1 \text{ kJ/mol})]$

$\qquad - [3 \text{ mol}(149.3 \text{ kJ/mol}) + 4 \text{ mol}(163.2 \text{ kJ/mol})] = -1997.7 \text{ kJ}$

14.57 To determine $\Delta G°$ at a temperature other than 298 K, calculate $\Delta S°$ and $\Delta H°$ using Equations 14–6 and 6–11, then use Equation 14–10. The standard entropy changes are calculated in Problem 14.53:

$$\Delta S°_{reaction} = \Sigma \text{ coeff}_{products} \, S°_{products} - \Sigma \text{ coeff}_{reactants} \, S°_{reactants}$$

$$\Delta H°_{reaction} = \Sigma \text{ coeff}_{products} \, \Delta H_f^o{}_{products} - \Sigma \text{ coeff}_{reactants} \, \Delta H_f^o{}_{reactants}$$

$$\Delta G°_{reaction, \, T} \cong \Delta H^o_{reaction, \, 298} - T \, \Delta S^o_{reaction, \, 298}$$

(a) $\Delta H°_{reaction} = 2 \text{ mol}(-104 \text{ kJ/mol}) - [2 \text{ mol}(-67 \text{ kJ/mol}) + 1 \text{ mol}(0 \text{ kJ/mol})] = -74 \text{ kJ}$

$$T\Delta S°_{reaction} = (358.2 \text{ K})\left(\frac{-83 \text{ J}}{1 \text{ K}}\right)\left(\frac{10^{-3} \text{ kJ}}{1 \text{ J}}\right) = -29.7 \text{ kJ}$$

$$\Delta G^o_{reaction, \, 358.2 \text{ K}} = (-74 \text{ kJ}) - (-29.7 \text{ kJ}) = -42 \text{ kJ}$$

(b) $\Delta H°_{reaction} = [2 \text{ mol}(-824.2 \text{ kJ/mol}) + 6 \text{ mol}(0 \text{ kJ/mol})]$

$\qquad - [4 \text{ mol}(-399.5 \text{ kJ/mol}) + 3 \text{ mol}(0 \text{ kJ/mol})] = -50.4 \text{ kJ}$

$$T\Delta S°_{reaction} = (358.2 \text{ K})\left(\frac{328.7 \text{ J}}{1 \text{ K}}\right)\left(\frac{10^{-3} \text{ kJ}}{1 \text{ J}}\right) = 117.7 \text{ kJ}$$

$$\Delta G^o_{reaction, \, 358.2 \text{ K}} = (-50.4 \text{ kJ}) - (117.7 \text{ kJ}) = -168.1 \text{ kJ}$$

(c) $\Delta H°_{reaction}$ = [6 mol(91.3 kJ/mol) + 6 mol(–285.83 kJ/mol)]

$\qquad$ – [3 mol(50.6 kJ/mol) + 4 mol(142.7 kJ/mol)] = – 1889.8 kJ

$$T\Delta S°_{reaction} = (358.2 \text{ K})\left(\frac{365.3 \text{ J}}{1 \text{ K}}\right)\left(\frac{10^{-3} \text{ kJ}}{1 \text{ J}}\right) = 130.9 \text{ kJ}$$

$$\Delta G°_{reaction, 358.2 \text{ K}} = (-1889.8 \text{ kJ}) - (130.9 \text{ kJ}) = -2020.7 \text{ kJ}$$

14.59   The figure shows a chemical reaction in which a set of diatomic molecules fragment into atoms with no change in volume or temperature.
(a) There is no volume change, so $w_{sys} = 0$.
(b) Energy must be supplied to break the chemical bonds, so $q_{sys} > 0$.
(c) Because $q_{sys} > 0$, $q_{surr} < 0$, so $\Delta S_{surr} < 0$.

14.61   A reaction is thermodynamically feasible if $\Delta G < 0$.
(a) $\Delta G°$ = [2 mol(–73.5 kJ/mol) + 1 mol(87.6 kJ/mol)]

$\qquad$ – [3 mol(51.3 kJ/mol) + 1 mol(–237.1 kJ/mol)] = 23.8 kJ

This is not thermodynamically feasible at 298 K under standard conditions.
(b) $\Delta G° = \Delta H° - T\Delta S°$

$\Delta H°$ = [2 mol(–133.9 kJ/mol) + 1 mol(91.3 kJ/mol)]

$\qquad$ – [3 mol(33.2 kJ/mol) + 1 mol(–285.83 kJ/mol)] = 9.73 kJ

$\Delta S°$ = [2 mol(266.9 J mol^{-1} K^{-1}) + 1 mol(210.8 J mol^{-1} K^{-1})]

$\qquad$ – [3 mol(240.1 J mol^{-1} K^{-1}) + 1 mol(69.95 J mol^{-1} K^{-1})] = –45.65 J K^{-1}

$\Delta G°$ = 9.73 kJ/mol – (277 K)(–45.65 J K^{-1})(10^{-3} kJ/J) = 22.4 kJ

This is not thermodynamically feasible at 277 K under standard conditions.

(c) $\Delta G = \Delta G° + RT \ln Q$ $\qquad\qquad$ $Q = \left[\dfrac{(p_{HNO_3})^2 (p_{NO})}{(p_{NO_2})^3}\right]$

$$\Delta G = 23.8 \text{ kJ} + (0.008314 \text{ kJ mol}^{-1} \text{ K}^{-1})(298 \text{ K}) \ln\left[\frac{(10^{-6})^2 (10^{-6})}{(1)^3}\right]$$

$\qquad\qquad\qquad\qquad\qquad$ = 23.8 kJ – 102.69 kJ ΔG = – 78.9 kJ

The reaction is thermodynamically feasible under these conditions.

14.63   Energy "transactions" in the body are carried out using the ATP–ADP energy–storing reaction. When a cell needs energy, it "spends" ATP; when fat or carbohydrates ("capital") are consumed, the energy is stored by converting ADP to ATP ("buying" ATP). Like money, ATP is readily transported from place to place and is readily converted into "buying power."

14.65   Spontaneity is determined by free energy, heat flow by enthalpy, and change in order by entropy. Use Equations 6–11, 14–6, and 14–10 to calculate standard thermodynamic values:

$$\Delta H°_{reaction} = \Sigma \, coeff_{products} \, \Delta H_f°_{products} - \Sigma \, coeff_{reactants} \, \Delta H_f°_{reactants}$$

$$\Delta S°_{reaction} = \Sigma \, coeff_{products} \, S°_{products} - \Sigma \, coeff_{reactants} \, S°_{reactants}$$

$$\Delta G°_{reaction} = \Sigma coeff_p \, \Delta G_f°(products) - \Sigma \, coeff_r \, \Delta G_f°(reactants)$$

(a) $\Delta G°_{reaction} = [1 \, mol(-261.905 \, kJ/mol) + 1 \, mol(-131.0 \, kJ/mol)]$
$$-1 \, mol(-384.1 \, kJ/mol) = -8.8 \, kJ$$

The reaction is spontaneous.

(b) $\Delta H°_{reaction} = [1 \, mol(-240.3 \, kJ/mol) + 1 \, mol(-167.1 \, kJ/mol)]$
$$- 1 \, mol(-411.2 \, kJ/mol) = 3.8 \, kJ$$

The reaction absorbs heat.

(c) $\Delta S°_{reaction} = [1 \, mol(58.5 \, J \, mol^{-1} \, K^{-1}) + 1 \, mol(56.5 \, J \, mol^{-1} \, K^{-1})]$
$$- 1 \, mol(72.1 \, J \, mol^{-1} \, K^{-1}) = 42.9 \, J/K$$

Entropy increases, because the ions become dispersed during this reaction.

14.67   Entropy changes for constant–temperature processes can be calculated using Equation 14–2, $\Delta S = \dfrac{q_T}{T}$, and q for water freezing can be found using $q = -n\Delta H_{fus}$:

$$n_{water} = \frac{m}{MM} = 155 \, g \left(\frac{1 \, mol}{18.02 \, g} \right) = 8.60 \, mol$$

$$T_{water} = (0.0 + 273.15 \, K) = 273.2 \, K$$

$$q_{water} = -8.60 \, mol \left(\frac{6.01 \, kJ}{1 \, mol} \right) \left(\frac{10^3 J}{1 \, kJ} \right) = -5.17 \times 10^4 \, J$$

$$T_{surr} = (-20.0 + 273.15 \, K) = 253.2 \, K$$

(a) $\Delta S_{water} = \dfrac{-5.17 \times 10^4 \, J}{273.2 \, K} = -189 \, J/K$

(b) The energy released by the water will be absorbed by the surroundings:

$$q_{surr} = -q_{water} = 5.17 \times 10^4 \, J/K$$

$$\Delta S_{surr} = \frac{5.17 \times 10^4 \, J}{253.2 \, K} = 204 \, J/K$$

$$\Delta S_{total} = \Delta S_{water} + \Delta S_{surr} = -189 + 204 = 15 \, J/K$$

(c) Cooling the ice to –20 °C is an irreversible change (because the temperatures of ice and freezer are not the same except at the end of the process), for which $\Delta S_{total} > 0$. Thus, the cooling must generate an additional entropy increase for the universe.

14.69   Ammonia is a toxic gas, while urea and ammonium nitrate both are relatively non–toxic solids. Thus, the transport and application of ammonia entail significant risks to humans. Even in aqueous solution, ammonia is highly irritating, as anyone knows who has used strong ammonia as a cleanser.

14.71   The standard free energies of combustion are given in your textbook:
$\Delta G°_{glucose} = -2870$ kJ/mol and $\Delta G°_{palmitic\ acid} = -9790$ kJ/mol
Divide molar quantities by molar masses to obtain free energy per gram:

Glucose: $\Delta G_{per\ gram} = \left(\dfrac{-2870\ kJ}{1\ mol}\right)\left(\dfrac{1\ mol}{180\ g}\right) = -15.9$ kJ/g

Palmitic acid: $\Delta G_{per\ gram} = \left(\dfrac{-9790\ kJ}{1\ mol}\right)\left(\dfrac{1\ mol}{256\ g}\right) = -38.2$ kJ/g

Palmitic acid has the higher energy content per gram.

14.73   (a) Whenever a substance cools, ΔS is negative.
(b) Whenever a gas is compressed (at constant T), ΔS is negative.
(c) Whenever two substances mix, matter is dispersed and ΔS is positive.

14.75   (a) Energy is always conserved, so $\Delta E_{total} = 0$. (b) The teaspoon warms,
so $\Delta E_{teaspoon} > 0$. (c) This is a spontaneous process, so $\Delta S_{total} > 0$. (d) The water cools,
so $\Delta S_{water} < 0$. (e) The teaspoon warms, so $q_{teaspoon} > 0$.

14.77   Entropy depends on amount, phase, temperature, and concentration. The order is:
(0.5 mol, liquid, 298 K) < (1 mol, liquid, 298 K) < (1 mol, liquid, 373 K)
< (1 mol, gas, 1 bar, 373 K) < (1 mol, gas, 0.1 bar, 373 K).

14.79   Use values from Appendix D of your textbook and Equations 6–11, 14–6, and 14–10 to
calculate standard thermodynamic values:

$$\Delta H°_{reaction} = \Sigma\ coeff_{products}\ \Delta H_f°_{products} - \Sigma\ coeff_{reactants}\ \Delta H_f°_{reactants}$$
$$\Delta S°_{reaction} = \Sigma\ coeff_{products}\ S°_{products} - \Sigma\ coeff_{reactants}\ S°_{reactants}$$
$$\Delta G°_{reaction,\ T} \cong \Delta H°_{reaction,\ 298} - T\Delta S°_{reaction,\ 298}$$

$\Delta H°_{reaction} = 2$ mol(91.3 kJ/mol) – [1 mol(0 kJ/mol) + 1 mol(0 kJ/mol)] = 182.6 kJ
$\Delta S°_{reaction} = 2$ mol(210.8 J mol^{-1} K^{-1}) – [1 mol(191.622 J mol^{-1} K^{-1})
$+ 1$ mol(205.152 J mol^{-1} K^{-1})] = 24.83 J/K

$$\Delta G°_{reaction,\ 298} = 182.6\ kJ - 298K\left(\frac{24.83\ J}{1\ K}\right)\left(\frac{10^{-3}\ kJ}{1\ J}\right) = 175.2\ kJ$$

To estimate the temperature at which the reaction becomes spontaneous,
set $\Delta G°_{reaction,\ T} = 0$ and solve for T:

$$0 = 182.6\ kJ - T\left(\frac{24.83\ J}{1\ K}\right)\left(\frac{10^{-3}\ kJ}{1\ J}\right) \qquad T = \frac{182.6\ kJ}{(24.83\ J/K)(10^{-3}\ kJ/J)} = 7350\ K$$

The temperature in an automobile engine is not anywhere near this high, but pressures are
far from standard. In particular, the pressure of NO product is very low, and this makes
the reaction spontaneous.

14.81 Vaporization at the normal boiling point is a constant temperature process, for which the entropy change can be calculated from Equation 14–2. The change takes place at constant pressure, so $q = \Delta H$:

(a) $\Delta S^\circ_{vap} = \dfrac{\Delta H_{vap}}{T} = \dfrac{(30.77 \text{ kJ/mol})(10^3 \text{ J/kJ})}{(80.1 + 273.15)} = 87.1 \text{ J/K}$

$\Delta G^\circ_{vap} = \Delta H - T\Delta S = 0$ at the normal boiling point

(b) $\Delta G^\circ_{vap} = \Delta H - T\Delta S$

$\Delta G^\circ_{vap} = (30.77 \text{ kJ/mol}) - (21 + 273.15 \text{ K})(87.1 \text{ J mol}^{-1} \text{ K}^{-1})(10^{-3} \text{ kJ/J})$

$= 5.16 \text{ kJ/mol at } 21 \text{ °C}$

(c) Use Equation 14–14 to calculate a vapor pressure:

$$\ln vp = -\dfrac{\Delta H_{vap}}{RT} + \dfrac{\Delta S^\circ_{vap}}{R}$$

$$\ln vp = -\left[\dfrac{(30.77 \text{ kJ/mol})(10^3 \text{ J/kJ})}{(8.314 \text{ J mol}^{-1} \text{ K}^{-1})(298 \text{ K})}\right] + \left(\dfrac{87.1 \text{ J mol}^{-1} \text{ K}^{-1}}{8.314 \text{ J mol}^{-1} \text{ K}^{-1}}\right) = -1.98$$

$vp = e^{-1.98} = 0.14 \text{ bar}$

14.83 The process that takes place is vaporization of the liquid bromine to form bromine vapor.

(a) At the end of the process, all the bromine is in the gas phase, with the molecules well separated, and the piston has moved back:

(b) The process is reversible, since the temperatures of system and surroundings are the same and the external and internal pressures both are 1.0 bar (bromine is at its boiling point). Thus, $\Delta S_{total} = 0$.

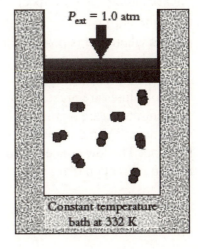

14.85 The attractive forces in CaO are substantially larger than the attractive forces in KCl, so the individual Ca^{2+} and O^{2-} ions cannot vibrate as easily in the crystal. Thus, the amount of vibrational disorder is substantially less in CaO than in KCl.

14.87 Make use of thermodynamic definitions to identify the state functions:

(a) $q_v = \Delta E$; (b) $q_p = \Delta H$; (c) $q_T = T\Delta S$.

14.89   Your molecular picture should show all three phases simultaneously present, with transfers of atoms occurring among all phases:

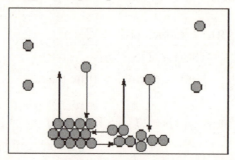

These processes are occurring at equilibrium, meaning that each is reversible, $\Delta G = 0$. For each process that leads to a more constrained phase, energy dispersal balances matter constraint, and vice–versa.

14.91   Use bond energy, intermolecular forces, and order–disorder to predict signs for ΔH and ΔS: (a) Bonds form, so energy is released and ΔH is negative. The number of independent particles decreases, so order increases and ΔS is negative. (b) When a solid melts, energy must be added; q is positive, ΔH is positive, and ΔS is positive. (c) Combustion reactions release energy, so we expect ΔH to be negative. The number of molecules of gas remains the same during this reaction, so ΔS is expected to be small, but it is not possible to predict whether it is positive or negative without doing actual calculations.

14.93   Standard conditions refers to 1 bar, 298 K, all solutes at 1 M. In a cell, $T = 37\ °C$ (310 K) and no solutes are present at 1 M. In particular, phosphoric acid and phosphate concentrations are significantly lower than 1 M. The elevated temperature makes ΔG more negative in the reaction direction that has a positive ΔS, which is the hydrolysis reaction. The low concentration of free phosphates makes ΔG more negative in the reaction direction that produces free phosphate, which also is the hydrolysis reaction.

14.95   According to your textbook, the complete oxidation of 1 mol of palmitic acid produces 130 mol of ATP. The reactions are:

$$ADP + H_3PO_4 \rightarrow ATP + H_2O \qquad \Delta G° = +30.6 \text{ kJ}$$

Palmitic acid: $C_{15}H_{31}CO_2H + 23\ O_2 \rightarrow 16\ CO_2 + 16\ H_2O \qquad \Delta G° = -9790 \text{ kJ}$

Thus, the amount of energy stored is (130)(30.6 kJ) = 3978 kJ and the efficiency is:

$$\text{Efficiency} = 100\% \left(\frac{\text{energy stored}}{\text{energy released}} \right) = 100\% \left(\frac{3978 \text{ kJ}}{9790 \text{ kJ}} \right) = 40.6\ \%$$

Metabolism of 1 mol of palmitic acid generates (9790 kJ) – (3978 kJ) = 5612 kJ of free energy. We will assume that all this free energy is converted to heat. On a per gram basis, this is:

$$q_{\text{per gram}} = \left(\frac{5612 \text{ kJ}}{1 \text{ mol}} \right) \left(\frac{1 \text{ mol}}{256.42 \text{ g}} \right) = 21.89 \text{ kJ/g}$$

Evaporation of 75 g of water requires a heat input:

$$q = n\Delta H_{\text{vap}} = 75 \text{ g} \left(\frac{1 \text{ mol}}{18.02 \text{ g}} \right) \left(\frac{40.79 \text{ kJ}}{1 \text{ mol}} \right) = 1.7 \text{ x } 10^2 \text{ kJ}$$

$$\text{Mass required} = 1.7 \text{ x } 10^2 \text{ kJ} \left(\frac{1 \text{ g}}{21.89 \text{ kJ}} \right) = 7.8 \text{ g}$$

14.97   For the first reaction, ATP is consumed and water produced, so it is likely that energy is released. There are three molecules produced from two, so ΔS is likely to be positive. The process is spontaneous, so ΔG is negative. For the second reaction, light is produced, so energy is released. There is one gaseous reactant and one gaseous product, but starting material fragments, so ΔS is positive. The process is spontaneous, so ΔG is negative.

14.99   The HCl molecule is heteronuclear, while H_2 and Cl_2 both are homonuclear. The homonuclear molecules are more symmetrical than the heteronuclear, so they have fewer possible distinguishable orientations in space. Here is an example:

Distinguishable   **Indistinguishable**

The homonuclear molecules have more possible orientations, fewer constraints, and less entropy.

15.1   In any sequence of steps, the slowest one will be rate–determining: (a) Pouring the coffee from the urn into the cup; (b) Entering the items on the cash register (if the market has a good laser scanner, paying and receiving change may be rate–determining); and (c) Preparing for the jump and passing through the door.

15.3   Every elementary reaction must depict actual molecular processes:
(a) $I_2 \rightarrow I + I$
(b) $H_2 + I_2 \rightarrow H_2I_2$
(c) $H_2 + I_2 \rightarrow H + HI_2$

15.5   A molecular picture of an elementary reaction shows the reactants, the products, and (if necessary) the intermediate collision complex:

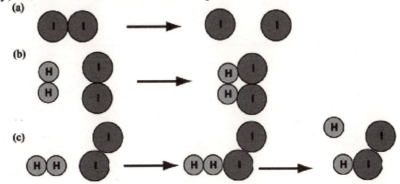

15.7   A satisfactory mechanism must consist entirely of reasonable elementary steps that sum to give the correct overall stoichiometry of the reaction:
(a) $I_2 \rightarrow I + I$
     $I + H_2 \rightarrow HI + H$
     $H + I \rightarrow HI$
(b) $H_2 + I_2 \rightarrow H_2I_2$
     $H_2I_2 \rightarrow HI + HI$
(c) $H_2 + I_2 \rightarrow H + HI_2$
     $H + HI_2 \rightarrow HI + HI$

15.9   The rate of a reaction can be expressed using the general expression:
$$\text{Rate} = -\frac{1}{a}\frac{\Delta[A]}{\Delta t}$$
(a) $\text{Rate} = -\dfrac{\Delta[Cl_2]}{\Delta t}$
(b) $-\dfrac{\Delta[Cl_2]}{\Delta t} = \dfrac{1}{2}\dfrac{\Delta[NOCl]}{\Delta t}$
(c) Use the result of (b) to calculate that NOCl appears at a rate of 94 M s^{-1}

15.11 In the reaction of NO and Cl_2, two NO molecules react for every Cl_2 that reacts, producing two NOCl molecules:

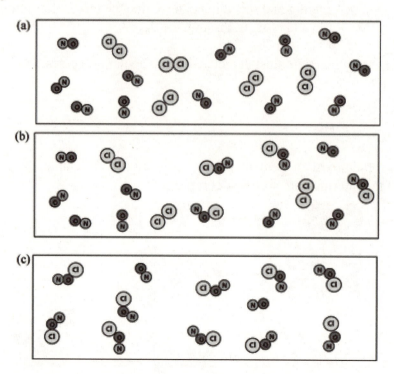

15.13 (a) To calculate the average rate of production, determine how many moles form during the time interval and divide by the time:

$$n = \frac{PV}{RT} = \frac{(0.15 \text{ atm})(5.0 \text{ L})}{(0.08206 \ \frac{\text{L atm}}{\text{mol K}})(550 + 273.15)\text{K}} = 1.11 \times 10^{-2} \text{ mol}$$

$$\frac{\Delta[CO_2]}{\Delta t} = \frac{1.11 \times 10^{-2} \text{ mol}}{5.0 \text{ min}} = 2.2 \times 10^{-3} \text{ mol/min}$$

(b) The reaction is $CaCO_3 \rightarrow CaO + CO_2$, so the amount of $CaCO_3$ decomposing is the same as the amount of CO_2 produced, 1.1×10^{-2} mol.

15.15 (a) Before the reaction begins:

(b) After 20 minutes, the amount reacted is (20 min)(0.25 molecules/min) = 5 molecules:

5.17   (a) The number of decays is directly proportional to the starting amount of the isotope in first–order kinetics, therefore there would be 6.0×10^6 decays in the 5.00 nmol case.

(b) The fraction decaying is the same in both cases:

$$\text{Fraction decaying} = \frac{\text{Number of decays}}{\text{Number of nuclei present}}$$

$$F = \frac{1.2 \times 10^6 \text{ nuclei}}{(1.00 \text{ nmol})(10^{-9} \text{ mol/nm})(6.022 \times 10^{23} \text{ nuclei/mol})} = 2.0 \times 10^{-9}$$

(c) In first–order kinetics, the *number* reacting is proportional to the amount present, but the *fraction* reacting is independent of concentrations.

15.19 (a) The rate law for an elementary step contains the product of the reactant concentrations:

Rate = $k[C][AB]$

(b) The units of a rate constant have time in the denominator along with concentrations one power less than the number of reactants: units = $(\text{conc.})^{-1}(\text{time})^{-1}$

(c) The steps of the mechanism must sum to give the observed stoichiometry for the reaction. Intermediate A can react with AB, then B and C can react:

$C + AB \rightarrow BC + A$

$A + AB \rightarrow B + A_2$

$B + C \rightarrow BC$

Net: $2\,C + 2\,AB \rightarrow 2\,BC + A_2$

15.21 (a) The rate law for an elementary step contains the product of the reactant concentrations:

Rate = $k[C][C][AB] = k[C]^2[AB]$

(b) The units of a rate constant have time in the denominator along with concentrations one power less than the number of reactants: units = $(\text{conc.})^{-2}(\text{time})^{-1}$

(c) The steps of the mechanism must sum to give the observed stoichiometry for the reaction. A bimolecular reaction between the intermediate (AC) and AB is the simplest possibility:

$2\,C + AB \rightarrow BC + AC$

$AC + AB \rightarrow BC + A_2$

Net: $2\,C + 2\,AB \rightarrow 2\,BC + A_2$

15.23 According to the stated rate law, the rate of reaction is proportional to each concentration. *A* contains 6 molecules of each type, while *B* contains 10 NO and 2 O_3:

A: Rate = $k(6)(6) = 36\,k$

B: Rate = $k(10)(2) = 20\,k$

B will react slower by a factor of $20/36 = 0.56$.

15.25 (a) One way to treat experimental data is by plotting: For first-order behavior, $\ln\left(\dfrac{[A]_0}{[A]}\right)$ vs. t is a straight line, and for second-order behavior, $\dfrac{1}{[A]} - \dfrac{1}{[A]_0}$ vs. t is a straight line. Here, the plot of $\ln[A]$ vs. t gives a straight line, so this reaction is first order.

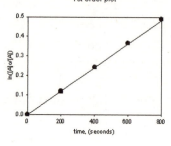

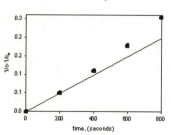

(b) For a first–order reaction, k = slope of the graph:

$$k = \text{slope} = \frac{\Delta y}{\Delta x} = \frac{0.491 - 0}{800 \text{ s} - 0 \text{ s}} = 6.14 \times 10^{-4} \text{ s}^{-1}$$

(c) To find concentration at any particular time, use $\ln[A] = \ln[A]_0 - kt$

$\ln[A] = \ln[2.50] - (6.14 \times 10^{-4} \text{ s}^{-1})(1600 \text{ s}) = -0.0661$

$[A] = e^{-0.0661} = 0.936$ atm

(d) To find the time at which concentration reaches a particular value, use

$$kt = \ln\left(\frac{[A]_0}{[A]}\right) \qquad t = \frac{\ln\left(\dfrac{[A]_0}{[A]}\right)}{k} = \frac{\ln\left(\dfrac{2.50 \text{ atm}}{0.500 \text{ atm}}\right)}{6.14 \times 10^{-4} \text{s}^{-1}} = 2.62 \times 10^3 \text{ s}$$

15.27 This is stated to be a second–order reaction, so Rate = $k[\text{NOBr}]^2$ and Equation 15–5 applies: $\dfrac{1}{[A]} - \dfrac{1}{[A]_0} = kt$

The problem states that k = 25 M^{-1} min^{-1}.

(a) $kt = \dfrac{1}{(0.010 \text{ M})} - \dfrac{1}{(0.025 \text{ M})} = 60 \text{ M}^{-1}$

$t = \dfrac{60 \text{ M}^{-1}}{25 \text{ M}^{-1} \text{ min}^{-1}} = 2.4$ min

(b) $\dfrac{1}{[A]} = \dfrac{1}{(0.025 \text{ M})} + (25 \text{ M}^{-1} \text{ min}^{-1})(125 \text{ min}) = (40 + 3125) \text{ M}^{-1} = 3165 \text{ M}^{-1}$

$[A] = 3.2 \times 10^{-4}$ M

15.29 This is stated to be a first–order reaction, so Rate = $k[\text{C}_5\text{H}_{11}\text{Br}]$ and Equation 15–3 applies:

$$kt = \ln\left(\frac{[A]_0}{[A]}\right)$$

(a) $kt = \ln\left(\dfrac{0.125}{1.25 \times 10^{-3}}\right) = 4.61$ $t = \dfrac{4.61}{0.385 \ hr^{-1}} = 12.0 \ hr$

(b) $\ln[A] = \ln[0.125] - 3.5 \ hr(0.385 \ hr^{-1}) = -3.43$

$[A] = e^{-3.43} = 0.0324 \ M$ or $3.24 \times 10^{-2} \ M$

15.31 The isolation method requires that one concentration be substantially smaller than all the others, so the experimental reaction order follows the order with respect to that one reactant. We are told that it is possible to track the concentration of N_2O_5, so this reactant has to be the one with the low concentration (otherwise the concentration would not change enough to give good data).
Do two experiments; in each $[H_2O]_0 > 100 \ [N_2O_5]_0$ but with two different values for $[H_2O]_0$. Plot $\ln\left(\dfrac{[N_2O_5]_0}{[N_2O_5]}\right)$ and $\dfrac{1}{[N_2O_5]} - \dfrac{1}{[N_2O_5]_0}$ vs. t to determine order with respect to N_2O_5, and use the ratio of slope values and H_2O concentrations to determine order with respect to H_2O.

15.33 From the data provided, we recognize this as an initial rate problem. The essential feature of the initial rate method is that we can take ratios of initial rates under different conditions. First, apply this technique to Experiments 1 and 2, which have the same initial concentration of $S_2O_8^{2-}$:

Initial rate$_1$ = 4.4×10^{-2} M/min = $k \ (0.125 \ M)^x(0.150 \ M)^y$

Initial rate$_2$ = 1.3×10^{-1} M/min = $k \ (0.375 \ M)^x(0.150 \ M)^y$

When we take the ratio of the second initial rate to the first, the rate constant and the initial concentration term for $S_2O_8^{2-}$ cancel:

$$\frac{\text{Initial rate}_2}{\text{Initial rate}_1} = \frac{1.3 \times 10^{-1} \ M/min}{4.4 \times 10^{-2} \ M/min} = \frac{k(0.375 \ M)^x (0.150 \ M)^y}{k(0.125 \ M)^x (0.150 \ M)^y} = \frac{(0.375 \ M)^x}{(0.125 \ M)^x}$$

Simplifying, we find:
$3.0 = (3.0)^x$, from which $x = 1$.
Now repeat this analysis for the third experiment and the first experiment, for which the initial concentrations of I^- are the same:

$$\frac{\text{Initial rate}_3}{\text{Initial rate}_1} = \frac{1.5 \times 10^{-2} \ M/min}{4.4 \times 10^{-2} \ M/min} = \frac{k(0.125 \ M)^x (0.050 \ M)^y}{k(0.125 \ M)^x (0.150 \ M)^y} = \frac{(0.050 \ M)^y}{(0.150 \ M)^y}$$

$0.34 = (0.33)^y$, from which $y = 1$
The reaction is first order in each reactant, so the rate law is Rate = $k[S_2O_8^{2-}][\ I^-]$
Use any of the experiments to evaluate the rate constant k:
4.4×10^{-2} M/min = $k \ (0.125 \ M)(0.150 \ M)$

$k = \dfrac{4.4 \times 10^{-2} \ M/min}{(0.125 \ M)(0.150 \ M)} = 2.3 \ M^{-1} \ min^{-1}$

15.35 The rate law should relate the rate of reaction to the concentration of the reactants. Reactive intermediates should not be shown in the rate law:

$$C + AB \rightleftharpoons BC + A, \text{ followed by}$$

$$A + AB \rightarrow B + A_2$$

$$B + C \rightarrow BC$$

Net: $2\,C + 2\,AB \rightarrow 2\,BC + A_2$

The rate law is determined by the rate–determining step: Rate = $k_2[A][AB]$. This is not satisfactory, however, because it contains the concentration of an intermediate (A). Set the rates equal for the forward and reverse first step:

$$k_1[C]\,[AB] = k_{-1}[BC][A]$$

Solve this equality for [A]: $[A] = \left(\dfrac{k_1}{k_{-1}}\right)\dfrac{[C][AB]}{[BC]}$

Substitute into the rate expression: Rate = $k_2\left(\dfrac{k_1}{k_{-1}}\right)\dfrac{[C][AB]^2}{[BC]}$

15.37 (a) The steps of a mechanism must sum to give the observed overall stoichiometry of the reaction. For ozone decomposition, this is $2\,O_3 \rightarrow 3\,O_2$. The two steps proposed by the student consume 1 O_3, produce 1 O_2, and generate an O atom, which must be consumed. Thus, the third step is $O_3 + O \rightarrow 2\,O_2$

(b) The rate law is determined by the rate–determining step: Rate = $k_2[O_5]$. This is not satisfactory, however, because it contains the concentration of an intermediate. Set the rates equal for the forward and reverse first step:

$$k_1[O_3][O_2] = k_{-1}[O_5]$$

Solve this equality for [O_5]: $[O_5] = \dfrac{k_1}{k_{-1}}[O_3][O_2]$

Substitute into the rate expression:

$$\text{Rate} = k_2\,\dfrac{k_1}{k_{-1}}\,[O_3][O_2]$$

(c) Atmospheric chemists would consider this mechanism to be molecularly unreasonable because fragmentation of O_5 in the second step (the breaking of two bonds simultaneously) is highly unlikely.

15.39 (a) The rate law is determined by the rate–determining step: Rate = $k_2[N_2O_2][O_2]$. This is not satisfactory, however, because it contains the concentration of an intermediate. Set the rates equal for the forward and reverse first step:

$$k_1[NO]^2 = k_{-1}[N_2O_2]$$

Solve this equality for [N_2O_2]: $[N_2O_2] = \left(\dfrac{k_1}{k_{-1}}\right)[NO]^2$

Substitute into the rate expression:

$$\text{Rate} = k_2\left(\dfrac{k_1}{k_{-1}}\right)[NO]^2[O_2]$$

(b) This rate expression has an overall order of $(2 + 1) = 3$, so the mechanism is consistent with third–order behavior.

(c) The intermediate species is N_2O_2. Two NO molecules could bind in several ways:

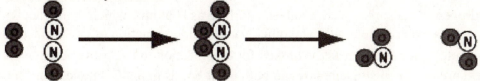

In the second step, O_2 collides with the intermediate and reacts to form two NO_2 molecules. The ONNO arrangement is the only intermediate for which the new set of bonds can easily occur:

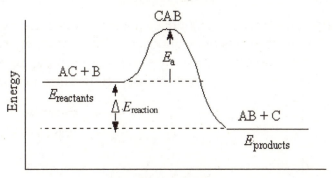

15.41   The rate constant for a reaction depends on temperature according to the Arrhenius equation (Equation 15–6): $k = Ae^{-E_a/RT}$. When $E_a = 0$, the exponent is $e^0 = 1$ and k is independent of temperature. A zero activation energy exists when a reaction can occur without first breaking any chemical bonds. The most common example is the combination of two free radicals, such as $H_3C\bullet + \bullet CH_3 \rightarrow H_3C–CH_3$.

15.43   An exothermic reaction has products lower in energy than reactants. In the activated complex A will be bonded to both B and C:

Activated complex:

CAB

AC + B

E_a

$E_{reactants}$

$\Delta E_{reaction}$

AB + C

$E_{products}$

Energy

Reaction coordinate

15.45   An exothermic reaction is "downhill" from reactants to products, and the activation energy plot should show this:

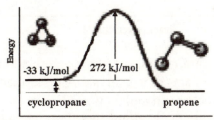

15.47 Assume that the ratio of the rate constants is proportional to the ratio of the number of flashes, then use the rearranged version of Equation 15-8 in your textbook:

$$E_a = R \ln\left(\frac{k_2}{k_1}\right)\left(\frac{1}{T_1} - \frac{1}{T_2}\right)^{-1} = (8.314 \text{ J mol}^{-1}\text{K}^{-1})\ln\left(\frac{2.7}{3.3}\right)\left(\frac{1}{(29+273)\text{K}} - \frac{1}{(23+273)\text{K}}\right)^{-1}$$

$$E_a = \left(\frac{-1.67 \text{ J mol}^{-1} \text{ K}^{-1}}{-6.7 \text{ x } 10^{-5} \text{ K}^{-1}}\right)\left(\frac{10^{-3} \text{ kJ}}{1 \text{ J}}\right) = 25 \text{ kJ/mol}$$

15.49 Hydrogen gas adsorbs on the catalyst's surface as H atoms, which then react with CO molecules much more easily (and quickly) than direct reaction of CO molecules with H_2 molecules. Bonds that must be broken for this reaction to occur (depicted by the squiggly lines) are two H–H single bonds and both the π bonds in the CO molecule. Bonds that are formed (dashed lines) are three C–H bonds and one O–H bond.

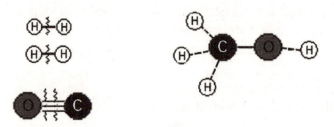

Bond Breakage   Bond Formation

15.51 (a) The effect of a catalyst on a reaction is to reduce the activation energy without changing the energy of the reactants or the products. Thus, the energies of reactants and products are the same for both curves. Only the "bump" in the activation energy diagram changes, being lower in the presence of the catalyst than in its absence:

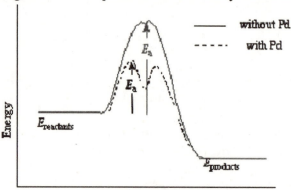

Reaction coordinate

(b) A catalyst is present at the beginning and end of the reaction but does not appear in the net reaction. Here, Pd metal acts as a catalyst. An intermediate is anything that is produced in one step of the mechanism and then used up in another step. The intermediates for this reaction are the H atoms that form when H_2 gas absorbs onto the Pd metal surface. The valley on the activation energy diagram represents this intermediate stage.

(c)

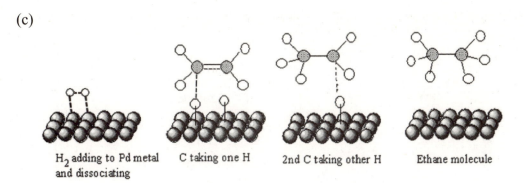

H$_2$ adding to Pd metal   C taking one H   2nd C taking other H   Ethane molecule
and dissociating

15.53 (a) The second step must use up the intermediate, O:

$O + NO \rightarrow NO_2$.

(b)

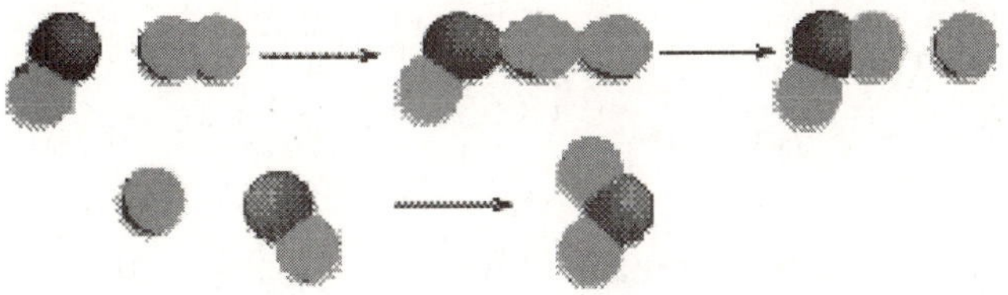

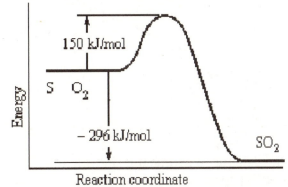

15.55 (a) A rate expression relates rate to changes: $\text{Rate} = \dfrac{\Delta[C_6H_6]}{\Delta t} = -\dfrac{1}{3}\left(\dfrac{\Delta[C_2H_2]}{\Delta t}\right)$

(b) Rate laws must always be determined experimentally. Thus, there is insufficient information to write the rate law. Experiments would have to be carried out measuring the rate as a function of $[C_2H_2]$ and the data analyzed using techniques described in your textbook.

15.57 The essential units of information needed to construct an activation energy diagram are the energy change and activation energy. Here, the reaction is a formation reaction and the change in moles of gas is zero during the reaction, so $\Delta E \cong \Delta H_f^0$:

```
         150 kJ/mol

  Energy   S   O₂


          - 296 kJ/mol
                              SO₂

         Reaction coordinate
```

15.59 When the concentration of a reactant increases by a factor of three (triples), the rate of reaction changes by 3^n, where n is the order with respect to that concentration.
(a) nine–fold increase; (b) no change; (c) rate increases by 5.2 times.

15.61 To determine the order of a reaction from a set of experimental data, prepare plots of

$$\ln\left(\frac{[A]_0}{[A]}\right) \text{ vs. } t \text{ and } \frac{1}{[A]} - \frac{1}{[A]_0} \text{ vs. } t:$$

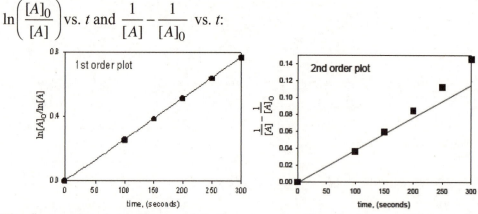

The first–order plot is linear, while the second–order plot is not. This reaction is first order. Determine the rate constant from the slope:

$$k = \text{Slope} = \frac{\Delta y}{\Delta x} = \frac{0.766 - 0}{300 \text{ s} - 0 \text{ s}} = 2.55 \times 10^{-3} \text{ s}^{-1}$$

15.63 The mechanism has the rate law, Rate = $k[X_2]$, first order in X_2 and independent of Y. a contains 5 X_2 and 8 Y, while b contains 10 X_2 and 8 Y. The rate for b will be twice that for a, because the concentration of X_2 is twice as great.

15.65 The initial concentration of N_2O_5 is much smaller than the initial concentration of H_2O in both experiments, so this is an example of the isolation technique. Assuming that the rate law has the form, Rate = $k[N_2O_5]^x[H_2O]^y$, we have

Experimental rate = $k_{obs}[N_2O_5]^x$, where $k_{obs} = k[H_2O]_0^y$

We need to plot the data: if $x = 1$, a log plot will be linear, while if $x = 2$, a $1/[N_2O_5]$ plot will be linear. Here is the first–order plot for Experiment A:

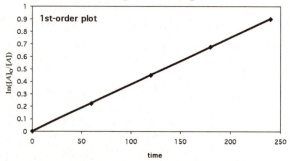

This plot is linear, with slope = $k_{obs} = 0.0038 \text{ s}^{-1}$.

We do not need to plot the data for Experiment B, in which the initial concentration of H_2O is twice as large as in Experiment A. Instead, we note that the reaction is going twice as fast: In Experiment B, it takes only 60 seconds for the concentration to fall to the value that is reached in 120 seconds in Experiment A. Doubling $[H_2O]$ leads to a doubling of the rate, so the reaction is first order in H_2O as well as first order in N_2O_5. The rate law is:

$$\text{Rate} = k[N_2O_5][H_2O]$$

To evaluate k, we can use either set of conditions. Here we use Experiment A:

$$k_{obs} = k[H_2O]_0 = 0.0038 \text{ s}^{-1} \text{ and } [H_2O]_0 = 0.025 \text{ M}$$

$$k = \frac{k_{obs}}{[H_2O]_0} = \frac{0.0038 \text{ s}^{-1}}{0.025 \text{ M}} = 0.15 \text{ M}^{-1} \text{ s}^{-1}$$

15.67   Reaction times for first–order reactions can be calculated using Equation 15–3, suitably rearranged:

$$t = \frac{\ln\left(\frac{[A]_o}{[A]}\right)}{k}$$

For 10.0% decomposition, $[A] = 0.900[A]_o$ and $\ln\left(\frac{[A]_o}{[A]}\right) = \ln\left(\frac{1.000}{0.900}\right) = 0.105$

$$t = \frac{0.105}{5.5 \times 10^{-4} s^{-1}} = 1.9 \times 10^2 \text{ s}$$

For 50.0% decomposition, $[A] = 0.500[A]_o$ and $\ln\left(\frac{[A]_o}{[A]}\right) = \ln\left(\frac{1.000}{0.500}\right) = 0.693$

$$t = \frac{0.693}{5.5 \times 10^{-4} s^{-1}} = 1.3 \times 10^3 \text{ s}$$

For 99.9% decomposition, $[A] = 0.001[A]_o$ and $\ln\left(\frac{[A]_o}{[A]}\right) = \ln\left(\frac{1.000}{0.001}\right) = 6.91$

$$t = \frac{6.91}{5.5 \times 10^{-4} s^{-1}} = 1.3 \times 10^4 \text{ s}$$

15.69   (a) False (overall reaction would give fourth–order kinetics); (b) False (rate constants must be measured for at least two different temperatures to calculate E_a); (c) True (rates of reaction increase with increasing temperature); (d) True (a unimolecular step has first–order kinetics).

15.71   Reaction orders are given by the exponents on the concentrations that appear in the rate law. Overall order is the sum of those exponents:
(a) first order in N_2O_5 and first order overall; (b) second order in NO, first order in H_2, and third order overall; and (c) first order in enzyme and first order overall.

15.73   The speed of a chemical reaction refers to how fast it proceeds. The spontaneity of a chemical reaction refers to whether or not the reaction can go in the direction written without outside intervention. A spontaneous reaction may nevertheless have a very slow speed.

15.75   Ultraviolet light causes chlorofluorocarbons to fragment, producing Cl atoms that catalyze the destruction of ozone:

$$CF_2Cl_2 \xrightarrow{h\upsilon} CF_2Cl + Cl$$

$$O_3 \xrightarrow{h\upsilon} O_2 + O$$

$$Cl + O_3 \rightarrow ClO + O_2$$

$$ClO + O \rightarrow Cl + O_2$$

Because the fourth reaction regenerates a Cl atom, the second through fourth reactions occur many hundreds of times for every CF_2Cl_2 molecule that fragments.

15.77   (a) Obtain the overall stoichiometry by adding the three steps, ignoring the reverse reaction of Step one, which leads to no net change: $2 NO + 2 H_2 \rightarrow N_2 + 2 H_2O$

(b) The rate law is determined by the rate–determining step: Rate $= k_2[N_2O_2][H_2]$. This is not satisfactory, however, because it contains the concentration of an intermediate. Set the rates equal for the forward and reverse first step: $k_1[NO]^2 = k_{-1}[N_2O_2]$

Solve this equality for $[N_2O_2]$: $[N_2O_2] = \dfrac{k_1}{k_{-1}}[NO]^2$

Substitute into the rate expression: Rate $= k_2\dfrac{k_1}{k_{-1}}[NO]^2[H_2]$

15.79   (a) A catalyst binds to one or more of the reactants in a way that weakens chemical bonds and makes it easier for bonds to rearrange to form the products. (b) When temperature increases, the average energies of the molecules increase, with the result that enough energy is present for a larger fraction of the molecules to have sufficient energy to overcome the activation energy barrier. (c) When concentration increases, the molecular density increases. There are more molecules to react, leading to a higher rate of molecular collisions. Both factors contribute to a greater rate of reaction.

15.81   Flask 1 contains 5 molecules, and Flask 2 contains 10 molecules, so the concentration is doubled. The problem states that Flask 2 reacts four times as fast. This is $(2)^2$, so the rate law is Rate $= k[A]^2$. At the molecular level, the rate–determining step could be a reaction between two A molecules. For this process, the rate increases because the higher concentration results in a higher rate of collisions.

15.83   (a) The rate law is that for an elementary bimolecular reaction: Rate $= k[H_2][X_2]$.
(b) When a first step is rate–determining, it determines the rate law: Rate $= k[X_2]$.
(c) The rate law is determined by the rate–determining step: Rate $= k_2[X][H_2]$.
Set the rates equal for the forward and reverse first step: $k_1[X_2] = k_{-1}[X]^2$

Solve this equality for $[X]$: $[X] = \left(\dfrac{k_1}{k_{-1}}\right)^{1/2}[X_2]^{1/2}$

Substitute into the rate expression: Rate $= k_2\left(\dfrac{k_1}{k_{-1}}\right)^{1/2}[H_2][X_2]^{1/2}$

15.85 (a) The net reaction can be obtained by summing the three steps, because when the reverse step occurs there is no net change: $Cl_2 + CHCl_3 \rightarrow HCl + CCl_4$

(b) Intermediates are produced in early steps and consumed in later steps: Cl and CCl_3;

(c) The rate law is determined by the rate–determining step: Rate = $k_2[CHCl_3][Cl]$. This is not satisfactory, however, because it contains the concentration of an intermediate. Set the rates equal for the forward and reverse first step: $k_1[Cl_2] = k_{-1}[Cl]^2$

Solve this equality for [Cl]: $[Cl] = \left(\dfrac{k_1}{k_{-1}}\right)^{1/2} [Cl_2]^{1/2}$

Substitute into the rate expression: Rate = $k_2\left(\dfrac{k_1}{k_{-1}}\right)^{1/2} [CHCl_3][Cl_2]^{1/2}$

15.87 Activation energies are calculated from the Arrhenius equation using Equation 15–8:

$$E_a = R\ln\left(\frac{k_2}{k_1}\right)\left(\frac{1}{T_1} - \frac{1}{T_2}\right)^{-1}$$

Development time is inversely proportional to rate constant, so $\dfrac{k_2}{k_1} = \dfrac{t_1}{t_2}$.

(a) $\dfrac{t_1}{t_2} = 2$ $T_1 = 20 + 273 = 293$ K $T_2 = 293 + 10 = 303$ K

$$\left(\frac{1}{T_1} - \frac{1}{T_2}\right)^{-1} = (3.413 \times 10^{-3} \text{ K}^{-1} - 3.300 \times 10^{-3} \text{ K}^{-1})^{-1} = 8880. \text{ K}$$

$$E_a = \left(\frac{8.314 \text{ J}}{1 \text{ mol K}}\right)\left(\frac{10^{-3} \text{ kJ}}{1 \text{ J}}\right)(\ln 2)(8880.\text{K}) = 51 \text{ kJ/mol}$$

(b) To determine the time it takes at 25 °C, use $\ln\left(\dfrac{t_1}{t_2}\right) = \dfrac{E_a}{R}\left(\dfrac{1}{T_1} - \dfrac{1}{T_2}\right)$:

$E_a = 51$ kJ/mol $t_1 = 10$ min
$T_1 = 20 + 273 = 293$ K $T_2 = 25 + 273 = 298$ K

$$\ln\left(\frac{10 \text{ min}}{t_2}\right) = \left(\frac{1}{293 \text{ K}} - \frac{1}{298 \text{ K}}\right)\left(\frac{1 \text{ mol K}}{8.314 \times 10^{-3} \text{ kJ}}\right)\left(\frac{51 \text{ kJ}}{1 \text{ mol}}\right) = 0.351$$

$\dfrac{10 \text{ min}}{t_2} = 1.42$ from which $t_2 = \dfrac{10 \text{ min}}{1.42} = 7.0$ min

15.89 (a) Prepare first–order and second–order plots and look for linear behavior:

t	s	0	1000	2000	3000	4000
c	M	0.250	0.118	0.0770	0.0572	0.0455
$\ln([A]_o/[A])$		0.000	0.751	1.18	1.47	1.70
$1/[A] - 1/[A]_o$	M^{-1}	0.00	4.47	8.99	13.5	18.0

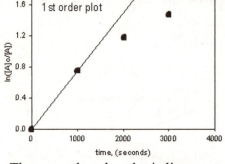
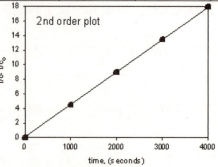

The second–order plot is linear, so Rate = $k[CH_3CHO]^2$

(b) Determine the rate constant from the slope of the second–order plot:

$$k = \text{Slope} = \frac{18.0\ M^{-1} - 0.00\ M^{-1}}{4000\ s - 0\ s} = 4.5 \times 10^{-3}\ M^{-1}\ s^{-1}$$

(c) Use Equation 15–5, suitably rearranged: $kt = \dfrac{1}{[A]} - \dfrac{1}{[A]_o}$

$$[A]_o = 0.250\ M \text{ and } [A] = 0.250\ M\left(\frac{100\% - 75\%}{100\%}\right) = 0.0625\ M$$

$$t = \frac{16.0\ M^{-1} - 4.00\ M^{-1}}{4.5 \times 10^{-3}\ M^{-1}\ s^{-1}} = 2.7 \times 10^3\ s$$

15.91 The hint for this problem suggests using the Arrhenius equation, $k = Ae^{-Ea/RT}$. Evaluate $\dfrac{E_a}{RT}$ for the catalyzed and uncatalyzed situations, using $T = 21 + 273 = 294$ K:

$$\left(\frac{E_a}{RT}\right)_{uncat} = \frac{125\ kJ\ mol^{-1}}{(8.314 \times 10^{-3}\ kJ\ mol^{-1}\ K^{-1})(294\ K)} = 51.1$$

$$\left(\frac{E_a}{RT}\right)_{cat} = \frac{46\ kJ\ mol^{-1}}{(8.314 \times 10^{-3}\ kJ\ mol^{-1}\ K^{-1})(294\ K)} = 18.8$$

$k_{uncat} = Ae^{-51.1}$ and $k_{cat} = Ae^{-18.8}$

Divide one of these by the other to eliminate A and find the ratio of rate constants:

$$\frac{k_{cat}}{k_{uncat}} = e^{(51.1 - 18.8)} = e^{32.3} = 1.1 \times 10^{14}$$

15.93 Neither intermediates nor catalysts appear in the overall stoichiometry of the reaction, so any species that appears in the mechanism but not in the overall stoichiometry is either an intermediate or a catalyst. Catalysts are consumed in early steps and regenerated in later steps, while intermediates are produced in early steps and consumed in later steps.

15.95 The only true statements are (d) and (f). Statements (a) and (b) are false because $\Delta E_{\text{reaction}} = C - A$. Here is an energy diagram showing the quantities involved:

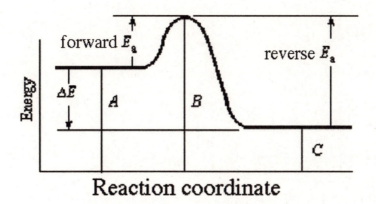

16.1   The reaction proceeds, converting *cis*–butene into *trans*–butene, but as the concentration of *trans*–butene builds up, the reverse reaction becomes increasingly important until, when the concentration ratio is 3, the rates are equal. Then there is no net change. A plot of concentration vs. time appears similar to Figure 16–2, with 0.75 bar for the final pressure of *trans*–butene and 0.25 bar for the final pressure of *cis*–butene:

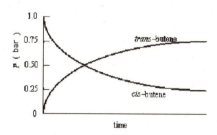

16.3   (a) According to the principle of reversibility, every elementary reaction can run in either direction, so as the system approaches equilibrium, the reverse reactions become important:

$$Cl^- \,(aq) + ClO_2^- \,(aq) \xrightarrow{\;k_{-1}\;} ClO^- \,(aq) + ClO^- \,(aq)$$

$$Cl^- \,(aq) + ClO_3^- \,(aq) \xrightarrow{\;k_{-2}\;} ClO^- \,(aq) + ClO_2^- \,(aq)$$

(b) The overall reaction is $3\,ClO^- \,(aq) \rightleftharpoons 2\,Cl^- \,(aq) + ClO_3^- \,(aq)$. By inspection, the equilibrium constant expression is $K_{eq} = \dfrac{[ClO_3^-]_{eq}[Cl^-]^2_{eq}}{[ClO^-]^3_{eq}}$

(c) The overall reaction is the sum of the two elementary steps, so the equilibrium constant is the product of the rate ratios of the two steps: $K_{eq} = \dfrac{k_1 k_2}{k_{-1} k_{-2}}$

16.5   A molecular picture of an elementary reaction shows the reactants, the products, and (if necessary) the intermediate collision complex.

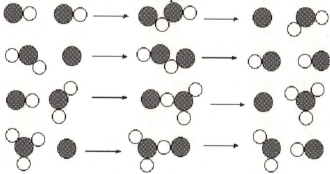

16.7   To test the reversibility of a reaction, set up a system containing the products and observe if reactants form. Here, a solution containing Cl^- and ClO_3^- ions should react to form some ClO^- ions. (If the reaction of ClO^- to form Cl^- and ClO_3^- goes virtually to completion, this experiment may not succeed.)

16.9   Equilibrium constant expressions can be written by inspection of the overall stoichiometry, remembering that pure liquids and solids do not appear in the expression. Equilibrium constant expressions contain product concentrations over reactant concentrations raised to their stoichiometric coefficient:

(a) $K_{eq} = \dfrac{(p_{IF_5})^2_{eq}}{(p_{F_2})^5_{eq}}$ (b) $K_{eq} = \dfrac{1}{\left(p_{O_2}\right)^5_{eq}}$ (c) $K_{eq} = (p_{CO})^2_{eq}$

(d) $K_{eq} = \dfrac{1}{(p_{CO})_{eq}(p_{H_2})^2_{eq}}$ (e) $K_{eq} = \dfrac{[H_3O^+]^3_{eq}[PO_4^{3-}]_{eq}}{[H_3PO_4]_{eq}}$

16.11   Equilibrium constant expressions can be written by inspection of the overall stoichiometry, remembering that pure liquids and solids do not appear in the expression:

(a) $2IF_5\,(g) \rightleftharpoons I_2\,(s) + 5\,F_2\,(g)$ $K_{eq} = \dfrac{(p_{F_2})^5_{eq}}{(p_{IF_5})^2_{eq}}$

(b) $P_4O_{10}\,(s) \rightleftharpoons P_4\,(s) + 5\,O_2\,(g)$ $K_{eq} = \left(p_{O_2}\right)^5_{eq}$

(c) $BaO\,(s) + 2\,CO\,(g) \rightleftharpoons BaCO_3\,(s) + C\,(s)$ $K_{eq} = \dfrac{1}{(p_{CO})^2_{eq}}$

(d) $CH_3OH\,(l) \rightleftharpoons CO\,(g) + 2\,H_2\,(g)$ $K_{eq} = (p_{CO})_{eq}(p_{H_2})^2_{eq}$

(e) $PO_4^{3-}\,(aq) + 3\,H_3O^+\,(aq) \rightleftharpoons H_3PO_4\,(aq) + 3\,H_2O\,(l)$ $K_{eq} = \dfrac{[H_3PO_4]_{eq}}{[H_3O^+]^3_{eq}[PO_4^{3-}]_{eq}}$

16.13   The standard states of gases are gases at 1 bar, the standard states of solutes are solutions at 1 M, and the standard states of solvents and pure liquids and solids are unit mole fractions, $X = 1$:
(a) $p = 1$ bar for F_2 and IF_5, $X = 1$ for I_2
(b) $p = 1$ bar for O_2, $X = 1$ for P_4 and P_4O_{10}
(c) $p = 1$ bar for CO, $X = 1$ for others
(d) $p = 1$ bar for CO and H_2, $X = 1$ for CH_3OH
(e) $c = 1$ M for H_3PO_4, H_3O^+, and PO_4^{3-}, $X = 1$ for H_2O

16.15   Your views should be like those in Figure 16–4, showing two different size samples with equal sized portions of each highlighted to show that equal volumes contain equal numbers of atoms:

16.17 To determine an equilibrium constant at standard temperature from thermodynamic tables, calculate $\Delta G^o{}_{\text{reaction}}$ from tabulated values for ΔG_f^o and then use Equation 16–3:

$$\Delta G^o = -RT \ln K_{eq}$$

The solubility reaction is $LiCl\ (s) \rightleftharpoons Li^+\ (aq)\ + Cl^-\ (aq)$ $K_{eq} = K_{sp}$

$\Delta G^o{}_{\text{reaction}} = [1\ (-293.31\ \text{kJ/mol}) + 1\ (-131.0\ \text{kJ/mol})] - [1\ (-384.4\ \text{kJ/mol})]$

$$= -39.91\ \text{kJ/mol}$$

$$\ln K_{sp} = -\frac{-3.991 \times 10^4\ \text{J mol}^{-1}}{(8.314\ \text{J mol}^{-1}\ \text{K}^{-1})(298\ \text{K})} = 16.11 \qquad K_{sp} = e^{16.11} = 9.9 \times 10^6$$

16.19 To determine an equilibrium constant at standard temperature from thermodynamic tables, calculate $\Delta G^o{}_{\text{reaction}}$ from tabulated values for ΔG_f^o and then use Equation 16–3:

$$\Delta G^o = -RT \ln K_{eq}$$

(a) $\Delta G^o{}_{\text{reaction}} = [1\ (0\ \text{kJ/mol}) + 1\ (-394.4\ \text{kJ/mol})]$

$\qquad - [1\ (-137.2\ \text{kJ/mol}) + 1\ (-237.1\ \text{kJ/mol})] = -20.1\ \text{kJ/mol}$

$$\ln K_{eq} = -\frac{-2.01 \times 10^4\ \text{J mol}^{-1}}{(8.314\ \text{J mol}^{-1}\ \text{K}^{-1})(298\ \text{K})} = 8.11 \qquad K_{eq} = e^{8.11} = 3.3 \times 10^3$$

(b) $\Delta G^o{}_{\text{reaction}} = 2\ (-394.4\ \text{kJ/mol}) - [1\ (0\ \text{kJ/mol}) + 2\ (-137.2\ \text{kJ/mol})]$

$$= -514.4\ \text{kJ/mol}$$

$$\ln K_{eq} = -\frac{-5.144 \times 10^5\ \text{J mol}^{-1}}{(8.314\ \text{J mol}^{-1}\ \text{K}^{-1})(298\ \text{K})} = 207.6 \qquad K_{eq} = e^{207.6} = 1 \times 10^{90}$$

(c) $\Delta G^o{}_{\text{reaction}} = [1\ (-520.3\ \text{kJ/mol}) + 2\ (-137.2\ \text{kJ/mol})]$

$\qquad - [1\ (0\ \text{kJ/mol}) + 1\ (-1134.4\ \text{kJ/mol})] = 339.7\ \text{kJ/mol}$

$$\ln K_{eq} = -\frac{3.397 \times 10^5\ \text{J mol}^{-1}}{(8.314\ \text{J mol}^{-1}\ \text{K}^{-1})(298\ \text{K})} = -137.1 \qquad K_{eq} = e^{-137.1} = 3 \times 10^{-60}$$

(d) $\Delta G^o{}_{\text{reaction}} = [3\ (-237.1\ \text{kJ/mol}) + 1\ (74.62\ \text{kJ/mol})]$

$\qquad - [6\ (0\ \text{kJ/mol}) + 3\ (-137.2\ \text{kJ/mol})] = -225.1\ \text{kJ/mol}$

$$\ln K_{eq} = -\frac{-2.251 \times 10^5\ \text{J mol}^{-1}}{(8.314\ \text{J mol}^{-1}\ \text{K}^{-1})(298\ \text{K})} = 90.86 \qquad K_{eq} = e^{90.86} = 2.9 \times 10^{39}$$

16.21 To estimate the equilibrium constant at a temperature different from 298 K, calculate $\Delta H^o{}_{\text{reaction}}$ and $\Delta S^o{}_{\text{reaction}}$ at 298 K and then use Equations 14–10 and 16–3:

$$\Delta G^o = \Delta H^o - T\Delta S^o \qquad\qquad \Delta G^o = -RT \ln K_{eq}$$

16.19 (b) $\Delta H^o{}_{\text{reaction}} = 2\ (-393.5\ \text{kJ/mol}) - [1\ (0\ \text{kJ/mol}) + 2\ (-110.5\ \text{kJ/mol})]$

$$= -566.0\ \text{kJ/mol}$$

$\Delta S^o{}_{\text{reaction}} = 2\ (213.8\ \text{J/mol K}) - [1\ (205.15\ \text{J/mol K}) + 2\ (197.7\ \text{J/mol K})]$

$$= -173.0\ \text{J/mol K}$$

$\Delta G^o{}_{\text{reaction, 250K}} \cong (-566.0\ \text{kJ/mol}) - (250\ \text{K})(-0.1730\ \text{kJ/mol K}) = -522.8\ \text{kJ/mol}$

$$\ln K_{eq} = -\frac{-5.228 \times 10^5 \text{ J mol}^{-1}}{(8.314 \text{ J mol}^{-1} \text{ K}^{-1})(250 \text{ K})} = 251.5 \qquad K_{eq} = e^{251.5} = 2 \times 10^{109}$$

16.19 (c) $\Delta H°_{reaction} = 1 \ (-548.0 \text{ kJ/mol}) + 2 \ (-110.5 \text{ kJ/mol})$
$$- [1 \ (0 \text{ kJ/mol}) + 1(-1213.0 \text{ kJ/mol})] = 444.0 \text{ kJ/mol}$$

$\Delta S°_{reaction} = [1 \ (72.1 \text{ J/mol K}) + 2 \ (197.7 \text{ J/mol K})]$
$$-[1 \ (5.7 \text{ J/mol K}) + 1 \ (112.1 \text{ J/mol K})] = 349.7 \text{ J/mol K}$$

$\Delta G°_{reaction, 250K} \cong (444.0 \text{ kJ/mol}) - (250 \text{ K})(0.3497 \text{ J/mol K}) = 356.6 \text{ kJ/mol}$

$$\ln K_{eq} = -\frac{356.6 \times 10^3 \text{ J mol}^{-1}}{(8.314 \text{ J mol}^{-1} \text{ K}^{-1})(250 \text{ K})} = -171.6 \qquad K_{eq} = e^{-171.6} = 3 \times 10^{-75}$$

16.23   To estimate the equilibrium constant at a temperature different from 298 K, calculate $\Delta H°_{reaction}$ and $\Delta S°_{reaction}$ at 298 K and then use Equations 14–10 and 16–3:

$$\Delta G° = \Delta H° - T\Delta S° \qquad\qquad \Delta G° = -RT \ln K_{eq}$$

16.19 (a)
$$\Delta H°_{reaction} = [1 \ (-393.5 \text{ kJ/mol}) + 1 \ (0 \text{ kJ/mol})]$$
$$- [1 \ (-110.5 \text{ kJ/mol}) + 1 \ (-241.83 \text{ kJ/mol})] = -41.2 \text{ kJ/mol}$$

$$\Delta S°_{reaction} = [1 \ (213.8 \text{ J/mol K}) + 1 \ (130.680 \text{ J/mol K})]$$
$$- [1 \ (197.7 \text{ J/mol K}) + 1 \ (188.835 \text{ J/mol K})] = -42.1 \text{ J/mol K}$$

$\Delta G°_{reaction, 395K} \cong (-41.2 \text{ kJ/mol}) - (395 \text{ K})(-0.0421 \text{ kJ/mol K}) = -24.6 \text{ kJ/mol}$

$$\ln K_{eq} = -\frac{-2.46 \times 10^4 \text{ J mol}^{-1}}{(8.314 \text{ J mol}^{-1} \text{ K}^{-1})(395 \text{ K})} = 7.49 \qquad K_{eq} = e^{7.49} = 1.8 \times 10^3$$

16.19 (d)
$$\Delta H°_{reaction} = 1 \ (20.0 \text{ kJ/mol}) + 3 \ (-241.83 \text{ kJ/mol})$$
$$- [3 \ (-110.5 \text{ kJ/mol}) + 6(0 \text{ kJ/mol})] = -374.0 \text{ kJ/mol}$$

$$\Delta S°_{reaction} = [1 \ (226.9 \text{ J/mol K}) + 3 \ (188.835 \text{ J/mol K})]$$
$$-[3 \ (197.7 \text{ J/mol K}) + 6 \ (130.680 \text{ J/mol K})] = -583.8 \text{ J/mol K}$$

$\Delta G°_{reaction, 395K} \cong (-374.0 \text{ kJ/mol}) - (395 \text{ K})(-0.5838 \text{ J/mol K}) = -143.4 \text{ kJ/mol}$

$$\ln K_{eq} = -\frac{-1.434 \times 10^5 \text{ J mol}^{-1}}{(8.314 \text{ J mol}^{-1} \text{ K}^{-1})(395 \text{ K})} = 43.67 \qquad K_{eq} = e^{43.67} = 9.2 \times 10^{18}$$

16.25   Use Le Châtelier's principle to determine the effect on an equilibrium position caused by adding one reagent:
In reactions (a), (b), and (d), CO is a reactant, so adding CO causes the reaction to go to the right, forming products; in reaction (c), CO is a product, so adding CO causes the reaction to proceed to the left, forming reactants.

16.27   Use Le Châtelier's principle to determine the effect on an equilibrium position caused by a change in conditions:
(a) Solid reactants do not appear in the equilibrium constant expression, so adding $PbCl_2$ has no effect.

(b) Addition of water dilutes the solution, reducing the concentrations of the product ions, so more of the solid will dissolve.

(c) Addition of NaCl increases the concentration of one of the product ions, Cl^-, so some solid will precipitate.

(d) Addition of KNO_3 does not change the concentrations in the equilibrium constant expression, so adding this solid has no effect.

16.29 Use Le Châtelier's principle to determine what changes in conditions will drive an equilibrium in any given direction. This reaction can be driven to the left by removing SO_2, removing Cl_2, adding SO_2Cl_2, or increasing the temperature.

16.31 To calculate an equilibrium constant when amounts are available, we generally must convert amounts data into concentrations or pressures using stoichiometric reasoning, then complete an amounts table and substitute into the equilibrium constant expression. In this problem, however, the equilibrium constant expression contains p^2 in the numerator and denominator, so units cancel and amounts can be used directly. Set up an amounts table, using the fact that the change is 0.49 mol for CO:

Reaction:	H_2 +	CO_2 $\rightleftharpoons$	H_2O +	CO
Initial amount (mol)	1.00	1.00	0	0
Change in amount (mol)	–0.49	–0.49	+ 0.49	+ 0.49
Equilibrium amount (mol)	0.51	0.51	0.49	0.49

Now substitute into the equilibrium constant expression and evaluate K:

$$K_{eq} = \frac{(p_{H_2O})_{eq}(p_{CO})_{eq}}{(p_{H_2})_{eq}(p_{CO_2})_{eq}} = \frac{(n_{H_2O})_{eq}(n_{CO})_{eq}}{(n_{H_2})_{eq}(n_{CO_2})_{eq}} = \frac{(0.49)^2}{(0.51)^2} = 0.92$$

16.33 To calculate an equilibrium constant when experimental data concerning amounts are available, identify the reaction, convert amounts data into concentration using stoichiometric reasoning, then complete a concentration table and substitute into the equilibrium constant expression. For solutes, concentrations must be in mol/L.

$$[C_2H_5CO_2H]_{initial} = \frac{0.0500 \text{ mol}}{0.500 \text{ L}} = 0.100 \text{ M}$$

To complete the concentration table, use stoichiometric reasoning and the fact that the change is 1.15×10^{-3} M for H_3O^+ (The concentration of water is not needed, because water is the solvent):

Reaction:	H_2O +	$C_2H_5CO_2H$ $\rightleftharpoons$	$C_2H_5CO_2^-$ +	H_3O^+
Initial conc. (M)		0.100	0	0
Change in conc. (M)		-1.15×10^{-3}	$+ 1.15 \times 10^{-3}$	$+1.15 \times 10^{-3}$
Equilibrium conc. (M)		0.099	1.15×10^{-3}	1.15×10^{-3}

Now substitute into the equilibrium constant expression and evaluate K:

$$K_a = \frac{[C_2H_5CO_2^-]_{eq}[H_3O^+]_{eq}}{[C_2H_5CO_2H]_{eq}} = \frac{(1.15 \times 10^{-3})^2}{(0.099)} = 1.3 \times 10^{-5}$$

16.35 To calculate concentrations at equilibrium from initial conditions, set up a concentration table. For a gas phase reaction, concentrations must be expressed in bar. Let x = change in p_{H_2} :

Reaction:	H_2 +	$Br_2 \rightleftharpoons$	2 HBr
Initial pressure (bar)	0	0	10.0
Change in pressure (bar)	$+x$	$+x$	$-2x$
Equilibrium pressure (bar)	x	x	$10.0 - 2x$

Now substitute into the equilibrium constant expression and solve for x:

$$K_{eq} = \frac{(p_{HBr})_{eq}^2}{(p_{H_2})_{eq}(p_{Br_2})_{eq}} = \frac{(10.0 - 2x)^2}{(x)(x)} = 1.6 \times 10^5$$

To simplify, assume that $2x \ll 10.0$:

$$1.6 \times 10^5 = \frac{(10.0)^2}{(x)^2} \qquad so \qquad x^2 = \frac{100}{1.6 \times 10^5} = 6.25 \times 10^{-4}$$

$x = 2.5 \times 10^{-2} = (p_{H_2})_{eq} = (p_{Br_2})_{eq}$

$(p_{HBr})_{eq} = 10.0 - 2(2.5 \times 10^{-2}) = 10.0$ bar

$2(2.5 \times 10^{-2}) = 0.050 \ll 10.0$, so the approximation is valid.

16.37 To calculate concentrations at equilibrium from initial conditions, set up a concentration table. For a gas phase reaction, concentrations must be expressed in bar. Let x = change in p_{CO_2} :

Reaction:	FeO (s) +	CO (g) $\rightleftharpoons$	CO_2 (g) +	Fe (s)
Initial pressure (bar)	excess	5.0	0	——
Change in pressure (bar)	——	$-x$	$+x$	——
Equilibrium pressure (bar)	——	$5.0 - x$	x	——

Now substitute into the equilibrium constant expression and solve for x:

$$K_{eq} = \frac{(p_{CO_2})_{eq}}{(p_{CO})_{eq}} = \frac{x}{(5.0 - x)} = 0.403$$

$x = (0.403)(5.0 - x) = 2.015 - 0.403 x \qquad so \qquad 1.403 x = 2.015$

$x = 1.436 \qquad (p_{CO_2})_{eq} = 1.4$ bar $\qquad (p_{CO})_{eq} = 5.0 - 1.4 = 3.6$ bar

16.39 To identify species in solution, first identify the nature of the solute. Strong acids, strong bases, and salts generate ions, while all other substances remain molecular:
(a) weak acid, major species are H_2O and CH_3CO_2H; (b) salt, major species are H_2O, NH_4^+ and Cl^-; (c) salt, major species are H_2O, K^+ and Cl^-; (d) salt, major species are H_2O, Na^+ and $CH_3CO_2^-$; and (e) strong base, major species are H_2O, Na^+ and OH^-.

16.41   The equilibria among major species depend on the nature of the species. The proton transfer reaction of water always plays a role:

$$H_2O \ (l) + H_2O \ (l) \rightleftharpoons H_3O^+ \ (aq) + OH^- \ (aq)$$

(a) weak acid equilibrium:

$$CH_3CO_2H \ (aq) + H_2O \ (l) \rightleftharpoons CH_3CO^- \ (aq) + H_3O^+ \ (aq)$$

(b) ammonium ion is the weak conjugate acid of NH_3:

$$NH_4^+ \ (aq) + H_2O \ (l) \rightleftharpoons NH_3 \ (aq) + H_3O^+ \ (aq)$$

(c) There are no equilibria other than the water equilibrium.
(d) acetate ion is the weak conjugate base of acetic acid:

$$CH_3CO_2^- \ (aq) + H_2O \ (l) \rightleftharpoons CH_3CO_2H \ (aq) + OH^- \ (aq)$$

(e) There are no equilibria other than the water equilibrium.

16.43   To identify species in solution, first identify the nature of the solute. Strong acids, strong bases, and salts generate ions, while all other substances remain molecular:
(a) major species are $(CH_3)_2CO$ (acetone) and H_2O; (b) salt, major species are H_2O, K^+, and Br^-; (c) strong base, major species are H_2O, Li^+, and OH^-; and (d) strong acid, major species are H_2O, H_3O^+, and HSO_4^-.

16.45   Equilibrium constant expressions have the concentrations of the products in the numerator and the concentrations of the reactants in the denominator with each concentration raised to the power of its stoichiometric coefficient. Remember to omit liquids and solids from the expressions. If the reaction involves a weak acid (or base) reacting with water, K_{eq} will be related to K_a (or K_b if a base). If the reaction involves a solid, K_{eq} will be related to K_{sp}:

(a) $K_{eq} = \dfrac{[ClO_2^-][H_3O^+]}{[HClO_2]} = K_a$

(b) $K_{eq} = \dfrac{1}{[Fe^{3+}][OH^-]^3} = \dfrac{1}{K_{sp}}$

(c) $K_{eq} = \dfrac{[HCN]}{[CN^-][H_3O^+]} = \dfrac{1}{K_a}$

16.47   To identify the spectator ions, first determine the reaction (if any) that occurs when the solutions are mixed. Species that do not participate in the reaction are spectator ions:

(a) $CH_3CO_2H + OH^- \rightleftharpoons CH_3CO_2^- + H_2O$; spectator ion: Na^+

(b) $3 \ Ca^{2+} + 2 \ PO_4^{3-} \rightleftharpoons Ca_3(PO_4)_2$; spectator ions: Cl^- and K^+

(c) $H_3O^+ + OH^- \rightleftharpoons 2 \ H_2O$; spectator ions: K^+ and NO_3^-

16.49 Equilibrium constant expressions are determined by the stoichiometry of the overall reaction, with pure liquids, solids, and solvent omitted from the expression. Equilibrium constant expressions have the concentrations of the products in the numerator and the concentrations of the reactants in the denominator, with each concentration raised to the power of its stoichiometric coefficient:

(a) $K_{eq} = (p_{CO_2})_{eq}(p_{H_2O})_{eq}$

(b) $K_{eq} = \dfrac{(p_{NH_3})_{eq}^4 (p_{O_2})_{eq}^3}{(p_{N_2})_{eq}^2}$

(c) $K_{eq} = \dfrac{(p_{CH_3CHO})_{eq}^2}{(p_{C_2H_4})_{eq}^2 (p_{O_2})_{eq}}$

(d) $K_{eq} = [Ag^+]_{eq}^2 [SO_4^{2-}]_{eq}$

(e) $K_{eq} = (p_{H_2S})_{eq}(p_{NH_3})_{eq}$

16.51 This problem describes an equilibrium reaction. We are asked to determine the equilibrium pressures of all the gases. To calculate pressures at equilibrium from initial conditions, set up a concentration table, write the K_{eq} expression, and solve for the pressures. For a gas phase reaction, concentrations must be expressed in bar.

$$\frac{p_1}{T_1} = \frac{p_2}{T_2} \qquad p_2 = \frac{p_1 T_2}{T_1} = \frac{(3.00\ \text{bar})(273 + 352)\ \text{K}}{(298\ \text{K})} = 6.29\ \text{bar}$$

Reaction:	$COCl_2 \rightleftharpoons$	$CO\ +$	Cl_2
Initial pressure (bar)	6.29	0	0
Change in pressure (bar)	$-x$	$+x$	$+x$
Equilibrium pressure (bar)	$6.29 - x$	x	x

Now substitute into the equilibrium constant expression and solve for x:

$$K_{eq} = 8.3 \times 10^{-4} = \frac{(p_{CO})_{eq}(p_{Cl_2})_{eq}}{(p_{COCl_2})_{eq}} = \frac{(x)^2}{6.29 - x}\ ; \text{Assume that } x \ll 6.29:$$

$$8.3 \times 10^{-4} = \frac{x^2}{6.29} \qquad\qquad x^2 = (6.29)(8.3 \times 10^{-4}) = 5.22 \times 10^{-3}$$

$x = 7.2 \times 10^{-2}$ bar $= (p_{CO})_{eq} = (p_{Cl_2})_{eq}$ $(p_{COCl_2})_{eq} = 6.29 - 7.2 \times 10^{-2} = 6.22$ bar

7.2×10^{-2} is 1.1% of 6.29, so x is $\ll 6.29$ and the approximation is valid.

16.53 At equilibrium, a molecular picture should show the presence of both reactants and products, in relative amounts that are determined by the value of the equilibrium constant. Set up a "concentration" table to determine how many of each species are present at equilibrium:

Species:		$\rightleftharpoons$		
Initial	12	12	0	0
Change	$-x$	$-x$	$+x$	$+x$
Equilibrium	$12 - x$	$12 - x$	x	x

Substitute in the equilibrium constant expression and solve for x:

$K_{eq} = 25 = \dfrac{(x)^2}{(12 - x)^2}$; Taking the square root of each side gives $5 = \dfrac{x}{(12 - x)}$

$(60 - 5\,x) = x$ $\qquad\qquad\qquad 6\,x = 60$ and $x = 10$

The molecular picture should show $(12 - 10) = 2$ of each reactant and 10 of each product:

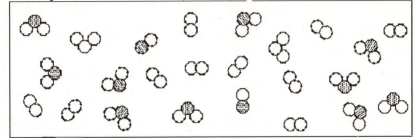

16.55 At equilibrium between a solute and solution, the rate at which molecules leave the surface of the solute equals the rate at which molecules are captured from solution at the surface of the solute. Adding more solute to a solution that is saturated increases the total surface area, so the overall rate at which molecules enter the solution becomes greater. However, the rate at which molecules are captured also becomes greater. The rate per unit surface area is unchanged, so the concentration of chloroform in the aqueous solution remains the same.

16.57 To calculate equilibrium pressures, set up a concentration table. In this problem, the table can be short, because two of the three equilibrium pressures are provided:

Reaction:	Br_2 (g) +	I_2 (g) $\rightleftharpoons$	2 IBr (g)
Equilibrium pressure (bar)	0.512	0.327	x

Substitute into the equilibrium constant expression and solve for x:

$322 = \dfrac{(p_{IBr})^2_{eq}}{(p_{Br_2})_{eq}(p_{I_2})_{eq}} = \dfrac{x^2}{(0.512)(0.327)}$

$x^2 = (322)(0.512)(0.327) = 53.91$

$x = 7.34$ $\qquad\qquad (p_{IBr})_{eq} = 7.34$ bar

16.59 To determine an equilibrium constant at standard temperature from thermodynamic tables, calculate $\Delta G^\circ_{reaction}$ from tabulated values for ΔG°_f. To estimate the equilibrium constant at a temperature different from 298 K, calculate $\Delta H^\circ_{reaction}$ and $\Delta S^\circ_{reaction}$ at 298K and then use Equations 14–10 and 16–3:

$\qquad\qquad \Delta G^\circ = \Delta H^\circ - T\Delta S^\circ \qquad\qquad \Delta G^\circ = -RT \ln K_{eq}$

At 298 K:

$\qquad\qquad \Delta G^\circ_{reaction} = 1\,(99.8\ kJ/mol) - 2\,(51.3\ kJ/mol) = -2.8\ kJ/mol$

$\qquad\qquad\qquad \ln K_{eq} = -\dfrac{-2.8 \times 10^3\ J\ mol^{-1}}{(8.314\ J\ mol^{-1}\ K^{-1})(298\ K)} = 1.1$

$\qquad\qquad\qquad\qquad K_{eq} = e^{1.1} = 3.0$

At 525 K:

$$\Delta H^{\circ}_{reaction} = 1\,(11.1\ kJ/mol) - 2\,(33.2\ kJ/mol)\ = -55.3\ kJ/mol$$

$$\Delta S^{\circ}_{reaction} = 1\,(304.4\ J/mol\ K) - 2\,(240.1\ J/mol\ K)\ = -175.8\ J/mol\ K$$

$$\Delta G^{\circ}_{reaction} \cong (-55.3\ kJ/mol) - (525\ K)\left(\frac{-175.8\ J}{1\ mol\ K}\right)\left(\frac{10^{-3}\ kJ}{1\ J}\right) = 37.0\ kJ/mol$$

$$\ln K_{eq} = -\frac{3.70 \times 10^{4}\ J\ mol^{-1}}{(8.314\ J\ mol^{-1}\ K^{-1})(525\ K)} = -8.48 \qquad K_{eq} = e^{-8.48} = 2.1 \times 10^{-4}$$

Notice that the exothermic reaction has a smaller K_{eq} at higher temperature.

16.61   (a) The general weak base reaction is $B + H_2O \rightleftharpoons BH^+ + OH^-$, and $B = (CH_3)_3N$:

$$(CH_3)_3N + H_2O \rightleftharpoons (CH_3)_3NH^+ + OH^- \qquad K_b = \frac{[(CH_3)_3NH^+]_{eq}[OH^-]_{eq}}{[(CH_3)_3N]_{eq}}$$

(b) The general weak acid reaction is $HA + H_2O \rightleftharpoons A^- + H_3O^+$, and $HA = HF$:

$$HF + H_2O \rightleftharpoons F^- + H_3O^+ \qquad K_a = \frac{[F^-]_{eq}[H_3O^+]_{eq}}{[HF]_{eq}}$$

(c) A solubility reaction involves solid dissolving to produce ions:

$$CaSO_4\,(s) \rightleftharpoons Ca^{2+}(aq) + SO_4^{2-}(aq) \qquad K_{sp} = [Ca^{2+}]_{eq}[SO_4^{2-}]_{eq}$$

16.63   To predict effects of changes on equilibrium position, apply Le Châtelier's principle: The system will respond in the direction that reduces the effect of the change. The reaction in Example 16–14 is: $2\ NO\,(g) + O_2\,(g) \rightleftharpoons 2\ NO_2\,(g)$:

(a) NO_2 is a product, so reducing its pressure causes the reaction to proceed to the right. (b) There are more moles of gas on the reactant side, doubling the volume causes the reaction to proceed to the left. (c) Adding Ar does not change any of the pressures in the equilibrium expression, so this change has no effect.

16.65   To estimate the equilibrium constant at a temperature different from 298 K, calculate $\Delta H^{\circ}_{reaction}$ and $\Delta S^{\circ}_{reaction}$ at 298 K and then use Equations 14–10 and 16–3:

$$\Delta G^{\circ} = \Delta H^{\circ} - T\Delta S^{\circ} \qquad\qquad \Delta G^{\circ} = -RT \ln K_{eq}$$

The solubility reaction is $LiCl\,(s) \rightleftharpoons Li^+\,(aq)\ + Cl^-\,(aq)$

$\Delta H^{\circ}_{reaction} = [1\,(-167.1\ kJ/mol) + 1\,(-278.47\ kJ/mol)]$

$$- [1\,(-408.6\ kJ/mol)] = -36.97\ kJ/mol$$

$\Delta S^{\circ}_{reaction} = [1\,(56.5\ J/mol\ K) + 1\,(12.2\ J/mol\ K)]$

$$- [1\,(59.3\ J/mol\ K)] = 9.4\ J/mol\ K$$

$\Delta G^{\circ}_{reaction,\ 373K} \cong (-36.97\ kJ/mol) - (373\ K)(-0.0094\ kJ/mol\ K) = -40.5\ kJ/mol$

$$\ln K_{eq} = - \frac{-4.05 \times 10^4 \text{ J mol}^{-1}}{(8.314 \text{ J mol}^{-1} \text{ K}^{-1})(373 \text{ K})} = 13.06 \qquad K_{eq} = e^{13.06} = 4.7 \times 10^5$$

From Problem 16.17, at 298 K, $K_{eq} = 9.9 \times 10^6$; LiCl is less soluble at high temperature.

16.67 To calculate an equilibrium constant from initial and equilibrium conditions, set up a concentration table:

Reaction:	$CCl_4 (g) \rightleftharpoons$	$2 Cl_2 (g) +$	$C (s)$
Initial pressure (bar)	1.00	0	——
Change in pressure (bar)	$-x$	$+ 2x$	——
Equilibrium pressure (bar)	$1.00 - x$	$2x$	——

The problem gives the total equilibrium pressure, which is the sum of partial pressures:
1.35 bar $= (1.00 - x) + 2x = 1.00 + x$, from which $x = (1.35 - 1.00) = 0.35$
Substitute into the equilibrium constant expression and calculate K:

$$K_{eq} = \frac{(p_{Cl_2})_{eq}^2}{(p_{CCl_4})_{eq}} = \frac{[(2)(0.35)]^2}{(1.00 - 0.35)} = \frac{0.70^2}{0.65} = 0.75$$

16.69 To determine an equilibrium constant at standard temperature from thermodynamic tables, calculate $\Delta G^o_{reaction}$ from tabulated values for ΔG^o_f; to estimate the equilibrium constant at a temperature different from 298 K, calculate $\Delta H^o_{reaction}$ and $\Delta S^o_{reaction}$ at 298 K and then use Equations 14–10 and 16–3:

$$\Delta G^o = \Delta H^o - T\Delta S^o \qquad\qquad \Delta G^o = -RT \ln K_{eq}$$

(a) $\Delta G^o_{reaction} = 1 (-210.7 \text{ kJ/mol})$
$$- [1 (31.8 \text{ kJ/mol}) + 1 (-178.6 \text{ kJ/mol})] = -63.9 \text{ kJ/mol}$$

$$\ln K_{eq} = - \frac{-6.39 \times 10^4 \text{ J mol}^{-1}}{(8.314 \text{ J mol}^{-1} \text{ K}^{-1})(298 \text{ K})} = 25.8 \qquad K_{eq} = e^{25.8} = 1.6 \times 10^{11}$$

(b) When the equilibrium pressure is 1.00 bar, $K_{eq} = 1$, $\ln(K_{eq}) = 0$ and $\Delta G^o_{reaction} = 0$

$$0 = \Delta H^o - T\Delta S^o \qquad T\Delta S^o = \Delta H^o \quad \text{and} \quad T = \frac{\Delta H^o}{\Delta S^o}$$

$\Delta H^o_{reaction} = 1 (-265.4 \text{ kJ/mol})$
$$- [1 (61.4 \text{ kJ/mol}) + 1 (-224.3 \text{ kJ/mol})] = -102.5 \text{ kJ/mol}$$
$\Delta S^o_{reaction} = 1 (191.6 \text{ J/mol K})$
$$- [1 (175.0 \text{ J/mol K}) + 1 (146.0 \text{ J/mol K})] = -129.4 \text{ J/mol K}$$

$$T = \left(\frac{-102.5 \text{ kJ}}{1 \text{ mol}} \right) \left(\frac{10^3 \text{ J}}{1 \text{ kJ}} \right) \left(\frac{1 \text{ mol K}}{-129.4 \text{ J}} \right) = 792 \text{ K}$$

(c) $\Delta G^{\circ}_{\text{reaction, 1050 K}} \cong (-102.5 \text{ kJ/mol}) - (1050 \text{ K}) \left(\dfrac{-129.4 \text{ J}}{1 \text{ mol K}} \right) \left(\dfrac{10^{-3} \text{ kJ}}{1 \text{ J}} \right) = 33.4 \text{ kJ/mol}$

$\ln K_{\text{eq}} = -\dfrac{3.34 \times 10^4 \text{ J mol}^{-1}}{(8.314 \text{ J mol}^{-1} \text{ K}^{-1})(1050 \text{ K})} = -3.83 \qquad K_{\text{eq}} = e^{-3.83} = 2.2 \times 10^{-2}$

Notice that K_{eq} decreases as T increases for this exothermic reaction.

16.71 (a) The equilibrium constant expression for a reaction can be written by inspection of the stoichiometry of the reaction, omitting pure liquids, solids, and solvents:

$$K_{\text{eq}} = \dfrac{[CO_3^{2-}]_{\text{eq}}}{(p_{CO_2})_{\text{eq}}[OH^-]^2_{\text{eq}}}$$

(b) Use Le Châtelier's Principle to predict the effect of changes on a system at equilibrium. Dissolving Na_2CO_3 leads to an increase in the concentration of CO_3^{2-}, so the equilibrium shifts to the left and the pressure of CO_2 increases.

(c) At first glance, it may appear that HCl will not affect this equilibrium, but recall that HCl is a strong acid which generates H_3O^+ in solution. This, in turn, will react with OH^-, reducing the concentration of a reactant. Again the equilibrium shifts to the left and the pressure of CO_2 increases.

16.73 To find an equilibrium total pressure, it is necessary to calculate equilibrium partial pressures of all gaseous participants. First determine the initial pressures of the gases, using the ideal gas equation, converting from atm to bar:

$$p = \dfrac{nRT}{V} = \dfrac{(0.494 \text{ mol})(0.08206 \tfrac{\text{L atm}}{\text{mol K}})(1020 + 273 \text{ K})}{1.00 \text{ L}} \left(\dfrac{1.013 \text{ bar}}{1 \text{ atm}} \right) = 53.1 \text{ bar}$$

Set up a concentration table, write the equilibrium expression and solve for the pressure:

Reaction:	$C_{(s)}$ +	$CO_{2\,(g)}$ ⇌	$2\,CO_{(g)}$
Initial pressure (bar)	——	53.1	53.1
Change in pressure (bar)	——	$-x$	$+2x$
Equilibrium pressure (bar)	——	$53.1 - x$	$53.1 + 2x$

Substitute into the equilibrium constant expression and solve for x:

$167.5 = \dfrac{(53.1 + 2x)^2}{(53.1 - x)} \qquad\qquad (167.5)(53.1 - x) = (53.1 + 2x)^2$

$8894 - 167.5\,x = 2820 + 212.4\,x + 4x^2 \qquad\qquad 4x^2 + 379.9\,x - 6074 = 0$

$x = \dfrac{-b \pm \sqrt{b^2 - 4ac}}{2a} = \dfrac{-379.9 \pm \sqrt{(379.9)^2 - 4(4)(-6074)}}{2(4)} = \dfrac{-379.9 \pm 491.4}{8} = 13.9$

Rule out the negative value, which would give a negative pressure:

$(p_{CO_2})_{\text{eq}} = 53.1 - 13.9 = 39.2 \text{ bar} \qquad\qquad (p_{CO})_{\text{eq}} = 53.1 + 2(13.9) = 80.9 \text{ bar}$

$P_{\text{total}} = 39.2 + 80.9 = 1.20 \times 10^2 \text{ bar}$

16.75   To determine the volume that will contain a single molecule of SnH_4 we first need to find the equilibrium pressure of the gas. To calculate equilibrium pressures, set up a concentration table:

Reaction:	Sn (s) +	2 H₂ (g) ⇌	SnH₄ (g)
Initial pressure (bar)	——	200	0
Change in pressure (bar)	——	–2 x	+ x
Equilibrium pressure (bar)	——	200 – 2 x	x

Because K_{eq} is very small, we assume that $2x << 200$:

$$1.1 \times 10^{-33} = \frac{[SnH_4]_{eq}}{[H_2]^2_{eq}} = \frac{x}{(200)^2} \qquad x = (1.1 \times 10^{-33})(200)^2 = 4.4 \times 10^{-29}$$

$[SnH_4]_{eq} = 4.4 \times 10^{-29}$ bar (a very small pressure)

To calculate the volume that would be expected to contain a single molecule, use

$$pV = nRT \qquad\qquad \# = nN_A$$

$$\frac{V}{\#} = \frac{RT}{pN_A}$$

First, convert the pressure from bar to atm:

$$p = (4.4 \times 10^{-29} \text{ bar})\left(\frac{1 \text{ atm}}{1.013 \text{ bar}}\right) = 4.3 \times 10^{-29} \text{ atm}$$

$$V/1 \text{ molecule} = \frac{RT}{pN_A} = \frac{(0.08206 \frac{L \text{ atm}}{mol \text{ K}})(298 \text{ K})}{(6.022 \times 10^{23} \text{ mol}^{-1})(4.3 \times 10^{-29} \text{ bar})} = 9.4 \times 10^5 \text{ L}$$

16.77   This molecular picture illustrates starting conditions and equilibrium conditions. The symbols indicate BG_3 as the starting material and G_2 and GB as products. Count numbers of symbols to determine initial and equilibrium concentrations. Set up a concentration table using numbers of symbols:

Reaction			
Initial	15	0	0
Change	–12	+12	+12
Equilibrium	3	12	12

(a) From the amounts of change, the stoichiometry is 1:1, so the net reaction is

;

(b) Use numbers of symbols at equilibrium to calculate the equilibrium constant:

$$K_{eq} = \frac{[GB]_{eq}[G_2]_{eq}}{[BG_3]_{eq}} = \frac{(12)(12)}{(3)} = 48$$

17.1   HBr is a strong acid that will transfer its hydrogen atom to a water molecule, generating a hydronium cation and a bromide anion:

17.3 $HClO_4$ is a strong acid that dissolves in water to generate H_3O^+ cations and ClO_4^- anions. Since there are no bases present for the hydronium ion to react with, the only equilibrium occurring is the proton transfer reaction between water molecules:

$$H_2O + H_2O \rightleftharpoons OH^- + H_3O^+ \qquad K_w = 1.00 \times 10^{-14}$$

To determine the concentrations of hydroxide and hydronium ions, set up a concentration table, write the equilibrium expression, and solve for the final concentrations.

The initial concentration of hydronium ions is the same as the concentration of perchloric acid, $[H_3O^+] = 1.25 \times 10^{-3}$ M. Here is the concentration table:

Reaction:	H_2O +	H_2O $\rightleftharpoons$	H_3O^+ +	OH^-
Initial concentration (M)	solvent	solvent	1.25×10^{-3}	0
Change in concentration (M)	solvent	solvent	$+x$	$+x$
Final concentration (M)	solvent	solvent	$1.25 \times 10^{-3} + x$	x

$K_w = [H_3O^+][OH^-]$ $1.00 \times 10^{-14} = [0.00125 + x][x]$

Assume that $x \ll 0.00125$ M: $1.00 \times 10^{-14} = [0.00125][x]$

$x = 8.00 \times 10^{-12}$ M $= [OH^-]$ The assumption that $x \ll 0.00125$ is valid.

$[H_3O^+] = 1.25 \times 10^{-3}$ M $+ 8.00 \times 10^{-12}$ M $= 1.25 \times 10^{-3}$ M

17.5   We are asked to determine the final concentrations of all the ions in a final solution. Begin by analyzing the chemistry. HCl is strong acid that dissolves in water to generate H_3O^+ cations and Cl^- anions. Any water solution always has OH^- and H_3O^+ ions with the equilibrium:

$$H_2O + H_2O \rightleftharpoons OH^- + H_3O^+ \qquad K_w = 1.00 \times 10^{-14}$$

Therefore, the ions present in this solution are H_3O^+, OH^-, and Cl^-.

The process involves dilution, so the first step is to determine the concentration of HCl in the flask after the dilution:

$M_i V_i = M_f V_f$

$$M_f = \frac{M_i V_i}{V_f} = \frac{(12.1 \text{ M})(1.00 \text{ mL})}{100. \text{ mL}} = 0.121 \text{ M HCl in final solution.}$$

Since Cl^- is a spectator ion, its concentration is the same as that of HCl, $[Cl^-] = 0.121$ M. The rest of the ion concentrations are determined by the equilibrium. Set up a concentration table, write the equilibrium expression, and solve for the final concentrations.

Reaction:	H_2O +	H_2O $\rightleftharpoons$	H_3O^+ +	OH^-
Initial concentration (M)	solvent	solvent	0.121	0
Change in concentration (M)	solvent	solvent	$+x$	$+x$
Final concentration (M)	solvent	solvent	$0.121 + x$	x

$$K_w = [H_3O^+][OH^-] \qquad\qquad 1.00 \times 10^{-14} = x(0.121 + x)$$

Aassume that $x \ll 0.121$: $\qquad\qquad 1.00 \times 10^{-14} = x(0.121)$

$x = [OH^-] = 8.26 \times 10^{-14}$ M $\qquad$ The assumption is valid.

$[H_3O^+] = 0.121$ M $+ 8.26 \times 10^{-14}$ M $= 0.121$ M

Here are the final concentrations: $[Cl^-] = [H_3O^+] = 0.121$ M $\qquad [OH^-] = 8.26 \times 10^{-14}$ M

17.7   We are asked to determine the concentrations of hydroxide and hydronium ions in the solution. Begin by analyzing the chemistry. HCl is a strong acid that dissolves in water to generate H_3O^+ cations and Cl^- anions. Any water solution always has OH^- and H_3O^+ ions with the equilibrium:

$$H_2O + H_2O \rightleftharpoons OH^- + H_3O^+ \qquad\qquad K_w = 1.00 \times 10^{-14}$$

The first step is to determine the initial number of moles of HCl gas and convert that to concentration of HCl dissolved in the solution:

$MM_{HCl} = 1.008$ g/mol $+ 35.453$ g/mol $= 36.461$ g/mol

$$0.488 \text{ g}\left(\frac{1 \text{ mol}}{36.461 \text{ g}}\right) = 0.01338 \text{ mol HCl dissolved}$$

$$[HCl] = \frac{0.01338 \text{ mol}}{0.325 \text{ L}} = 0.0412 \text{ M}$$

To determine the concentrations of the ions, construct a concentration table, write the equilibrium expression, and solve for the final concentrations. The initial concentration of hydronium ions will be the same as the concentration of HCl, $[H_3O^+] = 0.0412$ M.

Reaction:	H_2O +	$H_2O \rightleftharpoons$	H_3O^+ +	OH^-
Initial concentration (M)	solvent	solvent	0.0412	0
Change in concentration (M)	solvent	solvent	$+x$	$+x$
Final concentration (M)	solvent	solvent	$0.0412 + x$	x

$K_w = [H_3O^+][OH^-]$ $\qquad\qquad\qquad 1.00 \times 10^{-14} = (0.0412 + x)x$

Assume that $x \ll 0.0412$: $\qquad\qquad\qquad 1.00 \times 10^{-14} = (0.0412)x$

$x = 2.43 \times 10^{-13}$ M $= [OH^-]$ $\qquad$ The assumption is valid.

$[H_3O^+] = 0.0412$ M $+ 2.43 \times 10^{-13}$ M $= 0.0412$ M

17.9   Conversion from hydronium ion molarity to pH is accomplished by taking the logarithm to base ten and changing sign, pH $= -(\log[H_3O^+])$:
(a) -0.60; (b) 5.426; (c) 2.32; and (d) 3.593

17.11   Because pH + pOH = 14.00, pH = 14.00 − pOH. Convert from hydroxide ion molarity to pOH by taking the logarithm to base ten and changing sign, pOH $= -(\log[OH^-])$.
Thus, pH $= 14 + \log[OH^-]$:
(a) 14.60 ; (b) 8.574; (c) 11.68; and (d) 10.407

17.13   Take 10^{-pH} to convert pH into hydronium ion concentration:
(a) 0.22 M; (b) 1.4×10^{-8} M; (c) 2.1×10^{-4} M; and (d) 4.7×10^{-15} M

17.15 Use pH + pOH = 14.00 to convert pH to pOH. Then take 10^{-pOH} to convert pOH into hydroxide ion concentration:
(a) pOH = 13.34, $[OH^-]$ = 4.6 x 10^{-14} M; (b) pOH = 6.15, $[OH^-]$ = 7.1 x 10^{-7} M;
(c) pOH = 10.32, $[OH^-]$ = 4.8 x 10^{-11} M; and (d) pOH = – 0.33, $[OH^-]$ = 2.1 M

17.17 To calculate the pH of a solution, it is necessary to determine either the hydronium ion concentration or the hydroxide ion concentration.
Determine the nature and initial concentration of each species, construct a concentration table, write the equilibrium expression, and solve for the concentrations:
(a) Weak base, carry out an equilibrium calculation to determine $[OH^-]$:

Reaction:	H_2O +	C_5H_5N ⇌	$C_5H_5NH^+$ +	OH^-
Initial concentration (M)		1.5	0	0
Change in concentration (M)		$-x$	$+x$	$+x$
Final concentration (M)		$1.5 - x$	x	x

$$K_b = 1.7 \times 10^{-9} = \frac{[C_5H_5NH^+]_{eq}[OH^-]_{eq}}{[C_5H_5N]_{eq}} = \frac{x^2}{1.5 - x}; \text{ Assume that } x \ll 1.5:$$

$1.7 \times 10^{-9} = \dfrac{x^2}{1.5}$, from which $x^2 = 2.55 \times 10^{-9}$ and $x = 5.0 \times 10^{-5}$; assumption is valid.

$[OH^-] = 5.0 \times 10^{-5}$ M   pOH = $-\log(5.0 \times 10^{-5})$ = 4.30   pH = 14.00 – 4.30 = 9.70

(b) Weak base, carry out an equilibrium calculation to determine $[OH^-]$:

Reaction:	H_2O +	NH_2OH ⇌	NH_3OH^+ +	OH^-
Initial concentration (M)		1.5	0	0
Change in concentration (M)		$-x$	$+x$	$+x$
Final concentration (M)		$1.5 - x$	x	x

$$K_b = 8.7 \times 10^{-9} = \frac{[NH_3OH^+]_{eq}[OH^-]_{eq}}{[NH_2OH]_{eq}} = \frac{x^2}{1.5 - x} \qquad \text{Assume that } x \ll 1.5$$

$8.7 \times 10^{-9} = \dfrac{x^2}{1.5}$, from which $x^2 = 1.31 \times 10^{-8}$ and $x = 1.1 \times 10^{-4}$; assumption is valid.

$[OH^-] = 1.1 \times 10^{-4}$ M   pOH = $-\log(1.1 \times 10^{-4})$ = 3.96
pH = 14.00 – 3.96 = 10.04

(c) Weak acid, carry out an equilibrium calculation to determine $[H_3O^+]$:

Reaction:	H_2O +	HCO_2H ⇌	HCO_2^- +	H_3O^+
Initial concentration (M)		1.5	0	0
Change in concentration (M)		$-x$	$+x$	$+x$
Final concentration (M)		$1.5 - x$	x	x

$$K_a = 1.8 \times 10^{-4} = \frac{[HCO_2^-]_{eq}[H_3O^+]_{eq}}{[HCO_2H]_{eq}} = \frac{x^2}{1.5 - x} \qquad \text{Assume that } x \ll 1.5$$

$1.8 \times 10^{-4} = \dfrac{x^2}{1.5}$, from which $x^2 = 2.7 \times 10^{-4}$ and $x = 1.6 \times 10^{-2}$; assumption is valid.

$[H_3O^+] = 1.6 \times 10^{-2}$ M   pH = $-\log(1.6 \times 10^{-2})$ = 1.80

17.19 $HONH_2$ is a weak base that will accept a proton from water to form a hydroxide ion:

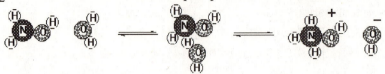

HCO_2H is a weak acid that will donate a proton to water to form a hydronium ion:

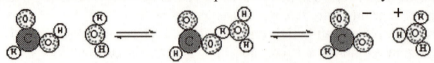

17.21 Follow standard procedures for dealing with equilibrium problems:
(a) HN_3 is a weak acid. Major species: HN_3 and H_2O, Minor species: N_3^-, H_3O^+, and OH^-
(b) Construct a concentration table, write the equilibrium expression, and solve for the concentrations:

Reaction:	H_2O +	$HN_3 \rightleftharpoons$	N_3^- +	H_3O^+
Initial concentration (M)		1.50	0	0
Change in concentration (M)		$-x$	$+x$	$+x$
Final concentration (M)		$1.50 - x$	x	x

$$K_a = 2.5 \times 10^{-5} = \frac{[N_3^-]_{eq}[H_3O^+]_{eq}}{[HN_3]_{eq}} = \frac{x^2}{1.50 - x} \qquad \text{Assume that } x \ll 1.50$$

$$2.5 \times 10^{-5} = \frac{x^2}{1.50} \qquad\qquad x^2 = 3.75 \times 10^{-5} \text{ and } x = 6.1 \times 10^{-3}$$

$[H_3O^+] = [N_3^-] = 6.1 \times 10^{-3}$ M, and $[HN_3] = 1.50$ M $- 6.1 \times 10^{-3}$ M $= 1.50$ M

$$[OH^-] = \frac{1.0 \times 10^{-14}}{6.1 \times 10^{-3}} = 1.6 \times 10^{-12} \text{ M}$$

(c) pH $= -\log (6.1 \times 10^{-3}) = 2.21$
(d) The dominant equilibrium is proton transfer from water to HN_3:

17.23 Determine the percent ionization using Equation 17–3 in your textbook:

$$\% \text{ HA ionized} = 100\% \frac{\left[H_3O^+\right]_{eq}}{[HA]_{initial}}$$

The equilibrium concentration of hydronium ions was calculated in Problem 17.21, and the initial concentration of the weak acid is given in the problem:

$$\% \text{ HA ionized} = 100\% \frac{\left(6.1 \times 10^{-3} \text{ M}\right)}{(1.50 \text{ M})} = 0.41 \%$$

17.25 Follow standard procedures for dealing with equilibrium problems:
(a) $N(CH_3)_3$ is a weak base. Major species: $N(CH_3)_3$, H_2O, Minor species: $HN(CH_3)_3^+$, OH^-, and H_3O^+
(b) Construct a concentration table, write the equilibrium expression, and solve for the concentrations:

Reaction: H_2O +	$N(CH_3)_3 \rightleftharpoons$	OH^- +	$HN(CH_3)_3^+$
Initial concentration (M)	0.350	0	0
Change in concentration (M)	$-x$	$+x$	$+x$
Final concentration (M)	$0.350 - x$	x	x

$$K_b = 6.5 \times 10^{-5} = \frac{[HN(CH_3)_3^+]_{eq}[OH^-]_{eq}}{[N(CH_3)_3]_{eq}} = \frac{x^2}{0.350 - x} \quad \text{Assume that } x \ll 0.350$$

$$6.5 \times 10^{-5} = \frac{x^2}{0.350} \quad\quad x^2 = 2.28 \times 10^{-5} \text{ and } x = 4.8 \times 10^{-3}; \text{ assumption is valid.}$$

$$[OH^-] = [HN(CH_3)_3^+] = 4.8 \times 10^{-3} \text{ M} \quad\quad [N(CH_3)_3] = 0.350 - 4.8 \times 10^{-3} = 0.345 \text{ M}$$

$$[H_3O^+] = \frac{1.0 \times 10^{-14}}{4.8 \times 10^{-3}} = 2.1 \times 10^{-12} \text{ M}$$

(c) pH = $-\log(2.1 \times 10^{-12}) = 11.68$
(d) The dominant equilibrium is proton transfer from water to trimethylamine:

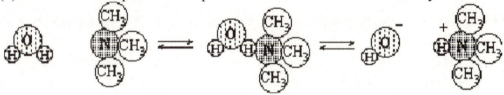

17.27 Determine the percent ionization using Equation 17–3 in your textbook:

$$\% \text{ HA ionized} = 100\% \frac{[H_3O^+]_{eq}}{[HA]_{initial}}$$

Follow standard procedures for dealing with equilibrium problems:
Acetic acid is a weak acid. Major species: CH_3CO_2H and H_2O
Construct a concentration table, write the equilibrium expression, and solve for the concentration of hydronium ions:

Reaction: H_2O +	$CH_3CO_2H \rightleftharpoons$	$CH_3CO_2^-$ +	H_3O^+
Initial concentration (M)	0.75	0	0
Change in concentration (M)	$-x$	$+x$	$+x$
Final concentration (M)	$0.75 - x$	x	x

Find K_a in Appendix E of your textbook:

$$K_a = 1.8 \times 10^{-5} = \frac{[CH_3CO_2^-]_{eq}[H_3O^+]_{eq}}{[CH_3CO_2H]_{eq}} = \frac{x^2}{0.75 - x} \quad\quad \text{Assume that } x \ll 0.75$$

$$1.8 \times 10^{-5} = \frac{x^2}{0.75} \quad\quad x^2 = 1.35 \times 10^{-5} \text{ and } x = 3.7 \times 10^{-3}$$

$[H_3O^+] = 3.7 \times 10^{-3}$ M

Substitute this result into Equation 17–3:

$$\% \text{ HA ionized} = 100\% \frac{(3.7 \times 10^{-3} \text{ M})}{(0.75 \text{ M})} = 0.49\%$$

17.29   Examine the chemical formula to determine the nature of a compound: (a) weak base; (b) weak acid; (c) weak acid; and (d) strong base.

17.31   Identify each substance using the standard color code. The first is $C_2H_5CO_2H$, propanoic acid, a carboxylic acid. The second is H_3PO_4, phosphoric acid. The third is CH_3NH_2, methyl amine, a base. Draw the conjugate base of an acid by removing an acidic H^+, leaving a negative charge. Draw the conjugate acid of a base by adding H^+ to an atom that has a lone pair:

17.33   The conjugate base will have one less H and one lower charge, a conjugate acid will have one more H and one higher charge:

C_5H_5N conjugate acid is $C_5H_5NH^+$ $HONH_2$ conjugate acid is $HONH_3^+$

HCO_2H conjugate base is HCO_2^-

17.35   Conjugate pairs are connected. Any aqueous solution has OH^- and H_3O^+ ions with the equilibrium:

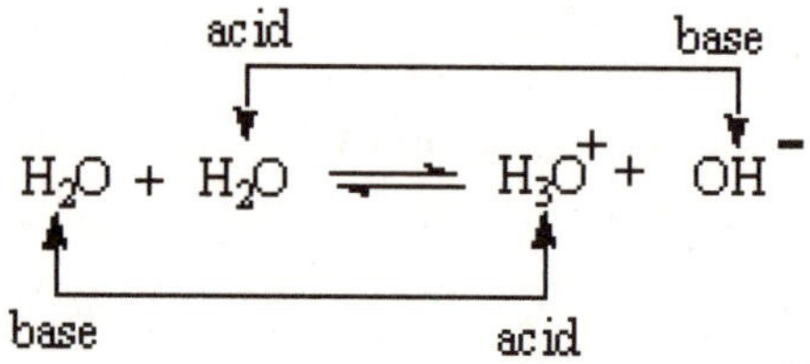

$$\underset{\text{base}}{\underset{\downarrow}{\overset{\text{acid}}{\overset{\downarrow}{H_2O}}}} + \underset{}{H_2O} \rightleftharpoons \overset{\text{base}}{\underset{\text{acid}}{H_3O^+}} + \overset{\text{base}}{OH^-}$$

acid   base

$H_2O + H_2O \rightleftharpoons H_3O^+ + OH^-$

base   acid

(a) NH_3 is a weak base:

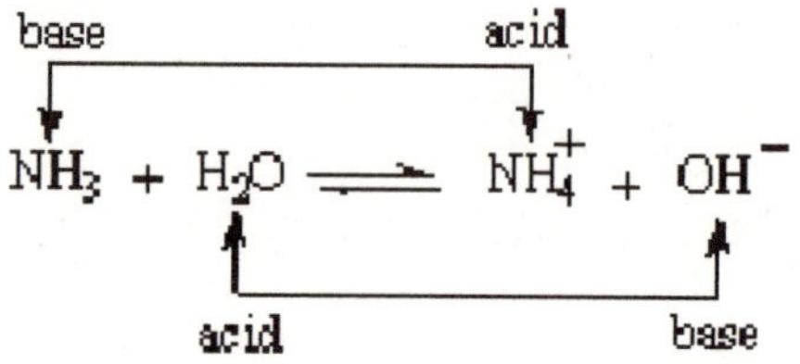

base   acid

$NH_3 + H_2O \rightleftharpoons NH_4^+ + OH^-$

acid   base

(b) HCNO is a weak acid:

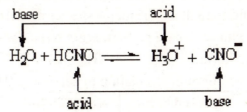

(c) HClO is a weak acid:

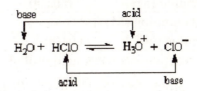

(d) Ba(OH)$_2$ is a strong base:

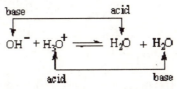

17.37 Follow standard procedures for dealing with equilibrium problems:
(a) The compound is a salt, so major species are Na^+, SO_3^{2-}, and H_2O.
(b) The species with acid–base properties are SO_3^{2-} (a weak base) and H_2O, so the dominant equilibrium is H_2O (*l*) + SO_3^{2-} (*aq*) $\rightleftharpoons$ HSO_3^- (*aq*) + OH^-(*aq*).
(c) Construct a concentration table, write the equilibrium expression, and carry out an equilibrium calculation to determine [OH^-]:

Reaction: H_2O +	SO_3^{2-} $\rightleftharpoons$	HSO_3^- +	OH^-
Initial concentration (M)	0.45	0	0
Change in concentration (M)	$-x$	$+x$	$+x$
Final concentration (M)	$0.45 - x$	x	x

The equilibrium reaction is a weak base proton transfer. HSO_3^- is the species resulting from the gain of a proton by SO_3^{2-}, so use K_{a2} to evaluate K_b:

$$K_{a2} = 6.3 \times 10^{-8} \qquad K_b = \frac{K_w}{K_{a2}} = \frac{1.0 \times 10^{-14}}{6.3 \times 10^{-8}} = 1.6 \times 10^{-7}$$

$$K_b = 1.6 \times 10^{-7} = \frac{[HSO_3^-]_{eq}[OH^-]_{eq}}{[SO_3^{2-}]_{eq}} = \frac{x^2}{0.45 - x} \qquad \text{Assume } x \ll 0.45$$

$1.6 \times 10^{-7} = \dfrac{x^2}{0.45}$, from which $x^2 = 7.2 \times 10^{-8}$ and $x = 2.7 \times 10^{-4}$; assumption is valid.

[OH^-] = 2.7×10^{-4} M   pOH = $-\log(2.7 \times 10^{-4}) = 3.57$
pH = $14.00 - 3.57 = 10.43$

17.39 Follow standard procedures for dealing with equilibrium problems:

(a) The compound is a salt, so the major species are NH_4^+, NO_3^-, and H_2O.

(b) The species with acid–base properties are NH_4^+ (a weak acid) and H_2O, so the dominant equilibrium is $H_2O\ (l) + NH_4^+\ (aq) \rightleftharpoons NH_3\ (aq) + H_3O^+(aq)$.

(c) Carry out an equilibrium calculation to determine $[H_3O^+]$:

Reaction: H_2O +	$NH_4^+ \rightleftharpoons$	NH_3 +	H_3O^+
Initial concentration (M)	0.0100	0	0
Change in concentration (M)	$-x$	$+x$	$+x$
Final concentration (M)	$0.0100 - x$	x	x

The equilibrium reaction is a weak acid proton transfer, NH_3 is the species resulting from the loss of a proton from NH_4^+, so use K_b to determine K_a:

$$K_b = 1.5 \times 10^{-5} \qquad K_a = \frac{K_w}{K_b} = \frac{1.0 \times 10^{-14}}{1.8 \times 10^{-5}} = 5.6 \times 10^{-10}$$

$$K_a = 5.6 \times 10^{-10} = \frac{[NH_4^+]_{eq}[H_3O^+]_{eq}}{[NH_3]_{eq}} = \frac{x^2}{0.0100 - x}$$

Assume that $x \ll 0.0100$: $\qquad 5.6 \times 10^{-10} = \dfrac{x^2}{0.0100}$

$x^2 = 5.6 \times 10^{-12}$ and $x = 2.37 \times 10^{-6}$; assumption is valid.

$[H_3O^+] = 2.37 \times 10^{-6}$ M $\qquad\qquad$ pH $= -\log(2.37 \times 10^{-6}) = 5.63$

17.41 To determine the acid–base properties of a salt solution, examine the species for their acidic or basic character:

(a) Species are H_2O (acid and base), Na^+ (neither), and HS^- (conjugate base of H_2S); Solution is basic with pH determined by
$$H_2O\ (l) + HS^-\ (aq) \rightleftharpoons OH^-\ (aq) + H_2S\ (aq)$$

(b) Species are H_2O (acid and base), Na^+ (neither), and OI^- (conjugate base of HOI); Solution is basic with pH determined by
$$H_2O\ (l) + OI^-\ (aq) \rightleftharpoons OH^-\ (aq) + HOI\ (aq)$$

(c) Species are H_2O (acid and base), Li^+ (neither), and ClO_4^- (conjugate base of a strong acid, thus neither); Solution is neutral, with pH determined by the water equilibrium:
$$H_2O\ (l) + H_2O\ (l) \rightleftharpoons OH^-\ (aq) + H_3O^+\ (aq)$$

(d) Species are H_2O (acid and base), $HC_5H_5N^+$ (conjugate acid of pyridine, a weak base), and Cl^- (conjugate base of a strong acid, thus neither); Solution is acidic with pH determined by $H_2O\ (l) + HC_5H_5N^+\ (aq) \rightleftharpoons H_3O^+\ (aq) + C_5H_5N\ (aq)$

17.43 The pH of an aqueous solution of a salt is determined by the acid–base characteristics of the cation and anion. Because Na^+ has no acid–base tendencies, the anions in these compounds determine the pH of their solutions. Solution pH increases with the strength of the basic anion, which in turn is inversely proportional to the strength of the parent weak acid. Here is the order for these compounds:

NaI (neutral) < NaF ($K_a = 6.3 \times 10^{-4}$) < $NaC_6H_5CO_2$ ($K_a = 6.3 \times 10^{-5}$)

< Na_3PO_4 ($K_{a3} = 4.8 \times 10^{-13}$) < NaOH (strong base)

17.45 (a) H_2SO_4 is stronger because anions are poorer proton donors than neutral species.

(b) HClO is stronger because Cl is a more electronegative atom than I. A higher electronegativity means that Cl attracts more of the electron density around it than I, weakening the H–X bond and making it easier to break (hence a better proton donor).

(c) $HClO_2$ is stronger. O atoms are highly electronegative and attract electron density around them. Having two O atoms, $HClO_2$ will have less electron density in the H–O bond than HClO, thus, making the bond easier to break.

17.47 Use arrows of different sizes to show differences in electron density shifts.

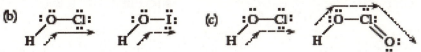

17.49 The phenolate anion can be drawn with several resonance structures, placing the negative charge at different locations around the benzene ring instead of on the oxygen atom:

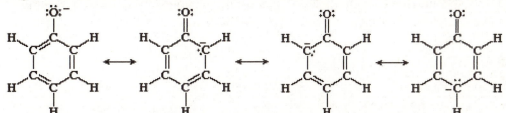

Acid strength increases as the stability of the conjugate base increases. The Lewis structures show that the phenolate anion can distribute its negative charge around the benzene ring, increasing its stability compared to that of the localized O^- that results from removal of a proton from ethanol. That is why phenol is a weak acid, whereas alcohols such as ethanol are not acidic.

17.51 This problem asks for the concentration of all ionic species present in the solution. Begin by analyzing the chemistry. $NaC_2H_3O_2$ is a salt that dissolves in solution to form Na^+ and $C_2H_3O_2^-$. Acetate anion is a weak base:

$$H_2O + C_2H_3O_2^- \rightleftharpoons C_2H_3O_2H + OH^- \qquad K_b = 5.6 \times 10^{-10}$$

Every aqueous solution has the water equilibrium:

$$H_2O + H_2O \rightleftharpoons H_3O^+ + OH^- \qquad K_w = 1.0 \times 10^{-14}$$

Thus, the ionic species present in the solution are: Na^+, $C_2H_3O_2^-$, H_3O^+, and OH^-

Na^+ is a spectator ion and will have the same concentration as the initial salt:

$$[Na^+] = 0.250 \text{ M}$$

For the remaining ions, set up concentration tables, write the equilibrium expressions, and solve for the ionic concentrations.

Reaction: H_2O +	$C_2H_3O_2^- \rightleftharpoons$	$C_2H_3O_2H$ +	OH^-
Initial concentration (M)	0.250	0	0
Change in concentration (M)	$-x$	$+x$	$+x$
Final concentration (M)	$0.250 - x$	x	x

$$K_b = 5.6 \times 10^{-10} = \frac{x^2}{0.250 - x} \qquad \text{Assume that } x \ll 0.250$$

$$5.6 \times 10^{-10} = \frac{x^2}{0.250} \qquad x^2 = 1.4 \times 10^{-10} \qquad x = 1.2 \times 10^{-5}; \text{ assumption is valid.}$$

$$[C_2H_3O_2H] = [OH^-] = 1.2 \times 10^{-5} \text{ M}$$

$$[C_2H_3O_2^-] = 0.250 \text{ M} - 1.2 \times 10^{-5} \text{ M} = 0.250 \text{ M}$$

Use a second concentration table to determine the concentrations of hydronium ions:

Reaction: H_2O +	$H_2O \rightleftharpoons$	H_3O^+ +	OH^-
Initial concentration (M)	solvent	0	1.2×10^{-5}
Change in concentration (M)	solvent	$+x$	$+x$
Final concentration (M)	solvent	x	$1.2 \times 10^{-5} + x$

$$K_w = 1.00 \times 10^{-14} = (x)(1.2 \times 10^{-5} + x) \qquad \text{Assume that } x \ll 1.2 \times 10^{-5}$$

$$1.00 \times 10^{-14} = (x)(1.2 \times 10^{-5})$$

$$x = 8.3 \times 10^{-10} \text{ M} = [H_3O^+]; \quad \text{assumption is valid.}$$

$$[OH^-] = 1.2 \times 10^{-5} \text{ M}$$

The ionic concentrations are:

$$[Na^+] = [C_2H_3O_2^-] = 0.250 \text{ M} \qquad [OH^-] = 1.2 \times 10^{-5} \text{ M} \qquad [H_3O^+] = 8.3 \times 10^{-10} \text{ M}$$

17.53 This problem asks for the concentration of all ionic species present in the solution. Begin by analyzing the chemistry. H_2CO_3 is a diprotic acid with equilibria:

$$H_2O + H_2CO_3 \rightleftharpoons HCO_3^- + H_3O^+ \qquad K_{a1} = 4.5 \times 10^{-7}$$

$$H_2O + HCO_3^- \rightleftharpoons CO_3^{-2} + H_3O^+ \qquad K_{a2} = 4.7 \times 10^{-11}$$

Every aqueous solution has the water equilibrium:

$$H_2O + H_2O \rightleftharpoons H_3O^+ + OH^- \qquad K_w = 1.00 \times 10^{-14}$$

Thus, the ionic species present in the solution are: HCO_3^-, CO_3^{-2}, H_3O^+, and OH^-

Set up concentration tables, write equilibrium expressions, and solve for the ionic concentrations.

Reaction: H_2O +	$H_2CO_3 \rightleftharpoons$	HCO_3^- +	H_3O^+
Initial concentration (10^{-2} M)	1.55	0	0
Change in concentration (10^{-2} M)	$-x$	$+x$	$+x$
Final concentration (10^{-2} M)	$1.55-x$	x	x

$$K_{a1} = 4.5 \times 10^{-7} = \frac{x^2}{1.55 \times 10^{-2} - x} \qquad \text{Assume that } x \ll 1.55 \times 10^{-2}$$

$$4.5 \times 10^{-7} = \frac{x^2}{1.55 \times 10^{-2}}$$

from which $x^2 = 6.98 \times 10^{-9}$ and $x = 8.4 \times 10^{-5}$; assumption is valid.

$[H_3O^+] = [HCO_3^-] = 8.4 \times 10^{-5}$ M $[H_2CO_3] = 1.55 \times 10^{-2}$ M

Set up a concentration table for the second equilibrium:

Reaction: H_2O +	$HCO_3^- \rightleftharpoons$	CO_3^{-2} +	H_3O^+
Initial concentration (10^{-5} M)	8.4	0	8.4
Change in concentration (10^{-5} M)	$-x$	$+x$	$+x$
Final concentration (10^{-5} M)	$8.4-x$	x	$8.4 + x$

$$K_{a2} = 4.7 \times 10^{-11} = \frac{x(8.4 \times 10^{-5} + x)}{8.4 \times 10^{-5} - x} \qquad \text{Assume that } x \ll 8.4 \times 10^{-5}$$

$$x = 4.7 \times 10^{-11} \text{M} = [CO_3^{2-}]$$

Use K_w to determine the concentration of hydroxide ions:

$$K_w = 1.00 \times 10^{-14} = (8.4 \times 10^{-5})[OH^-] \qquad [OH^-] = 1.2 \times 10^{-10} \text{ M}$$

ionic concentrations:

$[H_3O^+] = [HCO_3^-] = 8.4 \times 10^{-5}$ M

$[CO_3^{2-}] = 4.7 \times 10^{-11}$ M $[OH^-] = 1.2 \times 10^{-10}$ M

17.55 In order to determine concentrations of all ions, we need to consider more than one equilibrium. This is done in stages, starting with the dominant equilibrium. The problem asks us for the concentrations of the ions in aqueous sodium carbonate, in which the major species are Na^+, CO_3^{2-} and H_2O. The sodium ion is a spectator ion.

The carbonate anion undergoes proton transfer with equilibrium constant K_{b2}:

$$H_2O \,(l) + CO_3^{2-} \,(aq) \rightleftharpoons OH^- \,(aq) + HCO_3^- \,(aq) \qquad K_{b2} = \frac{[OH^-]_{eq} \, [HCO_3^-]_{eq}}{[CO_3^{2-}]_{eq}}$$

Table 17–2 provides the value of the acid equilibrium constant: $K_{a2} = 4.7 \times 10^{-11}$

$$K_{b2} = \frac{K_w}{K_{a2}} = \frac{1.0 \times 10^{-14}}{4.7 \times 10^{-11}} = 2.1 \times 10^{-4}$$

This is much larger than K_w, so this is the dominant equilibrium, which we use to begin our calculations.

Because carbonic acid is diprotic, a second proton transfer equilibrium has an effect on the ion concentrations, and the water equilibrium also plays a secondary role.

Now we are ready to organize the data and the unknowns and do the calculations.

The spectator ion is easiest to deal with: $[Na^+] = 2\,[CO_3^{2-}] = 2(0.055\ M) = 0.11\ M$

There are multiple equilibria affecting ion concentrations, so we must work with more than one concentration table, starting with the dominant equilibrium. Set up a concentration table to determine concentrations of the ions generated by this reaction:

Reaction:	$H_2O\ +$	$CO_3{}^{2-} \rightleftharpoons$	$OH^-\ +$	HCO_3^-
Initial concentration (M)		0.055	0	0
Change in concentration (M)		$-x$	$+x$	$+x$
Final concentration (M)		$0.055 - x$	x	x

Substitute the equilibrium concentrations into the equilibrium constant expression and solve for x. We cannot make an approximation, so use the quadratic formula:

$$K_{b2} = 2.1 \times 10^{-4} = \frac{(x)(x)}{(0.055 - x)}$$

$$x^2 = (2.1 \times 10^{-4})(0.055 - x) = 1.16 \times 10^{-5} - 2.1 \times 10^{-4}\,x$$

$$0 = x^2 + 2.1 \times 10^{-4}\,x - 1.16 \times 10^{-5}$$

$$x = \frac{-b \pm \sqrt{b^2 - 4ac}}{2a} = \frac{-(2.1 \times 10^{-4}) \pm \sqrt{(2.1 \times 10^{-4})^2 - 4(-1.16 \times 10^{-5})}}{2} = 3.2 \times 10^{-3}$$

$[OH^-] = [HCO_3^-] = 3.2 \times 10^{-3}\ M$, and $[CO_3^-] = 0.055 - 0.0032 = 0.052\ M$

Take into account the proton transfer equilibrium involving HCO_3^- with a second concentration table, using concentrations calculated for the first equilibrium:

Reaction:	$H_2O\ +$	$HCO_3^- \rightleftharpoons$	$OH^-\ +$	H_2CO_3
Initial concentration (M)		3.2×10^{-3}	3.2×10^{-3}	0
Change in concentration (M)		$-x$	$+x$	$+x$
Final concentration (M)		$3.2 \times 10^{-3} - x$	$3.2 \times 10^{-3} + x$	x

Substitute equilibrium concentrations into the equilibrium constant expression and solve for x, making the approximation that $x \ll 3.2 \times 10^{-3}$:

$$K_{b1} = \frac{K_w}{K_{a1}} = \frac{1.0 \times 10^{-14}}{4.5 \times 10^{-7}} = 2.2 \times 10^{-8} = \frac{[OH^-]_{eq}\,[H_2CO_3]_{eq}}{[HCO_3^-]_{eq}}$$

$$2.2 \times 10^{-8} = \frac{(x)(3.2 \times 10^{-3} + x)}{3.2 \times 10^{-3} - x} \cong \frac{(x)(3.2 \times 10^{-3})}{3.2 \times 10^{-3}} = x$$

This value is too small to cause a measurable change in the concentrations already calculated, but it does tell us the concentration of carbonic acid in the solution:

$[H_2CO_3] = 2.2 \times 10^{-8}\ M$

One more ion remains, H_3O^+, generated from the water equilibrium. Apply the water equilibrium expression directly:

$$[H_3O^+] = \frac{K_w}{[OH^-]} = \frac{(1.0 \times 10^{-14})}{(3.2 \times 10^{-3})} = 3.1 \times 10^{-12}\ M$$

17.57 (a) An acid–base equilibrium reaction involves proton transfer, in this case from boric acid to water:

(b) To calculate the pH of a solution, follow the standard procedure for equilibrium calculations:

Reaction:	H_2O +	H_3BO_3 ⇌	$H_2BO_3^-$ +	H_3O^+
Initial concentration (M)		0.050	0	0
Change in concentration (M)		$-x$	$+x$	$+x$
Final concentration (M)		$0.050 - x$	x	x

$$K_a = 5.4 \times 10^{-10} = \frac{[H_3BO_2^-]_{eq}[H_3O^+]_{eq}}{[H_3BO_3]_{eq}} = \frac{x^2}{0.050 - x} \qquad \text{Assume that } x \ll 0.050$$

$$5.4 \times 10^{-10} = \frac{x^2}{0.050} \qquad x^2 = 2.7 \times 10^{-11} \text{ and } x = 5.2 \times 10^{-6}; \text{ assumption is valid.}$$

$[H_3O^+] = 5.2 \times 10^{-6} \text{ M}$ $\qquad\qquad$ $pH = -\log(5.2 \times 10^{-6}) = 5.28$

17.59 There is much interesting chemical information provided in the statement of this problem, but the calculation is a straightforward equilibrium determination for a solution of a weak base. Follow the standard procedures to determine the pH. Begin by constructing a concentration table, write the equilibrium expression, and solve for the concentration of hydroxide ions.

Reaction:	H_2O +	LSD ⇌	$LSDH^+$ +	OH^-
Initial concentration (M)		0.55	0	0
Change in concentration (M)		$-x$	$+x$	$+x$
Final concentration (M)		$0.55 - x$	x	x

$$K_b = 7.6 \times 10^{-7} = \frac{[LSDH^+]_{eq}[OH^-]_{eq}}{[LSD]_{eq}} = \frac{x^2}{0.55 - x} \qquad \text{Assume that } x \ll 0.55$$

$$7.6 \times 10^{-7} = \frac{x^2}{0.55}, \text{ from which } x^2 = 4.2 \times 10^{-7} \text{ and } x = 6.5 \times 10^{-4}; \text{ assumption is valid.}$$

$[OH^-] = 6.5 \times 10^{-4} \text{ M}$

$pOH = -\log(6.5 \times 10^{-4}) = 3.19$, and $pH = 14.00 - 3.19 = 10.81$

17.61 Follow the standard procedure for dealing with equilibrium calculations:
The major species are Na^+, A^-, and H_2O, and the acid–base equilibrium is:
A^- (aq) + H_2O (l) ⇌ HA (aq) + OH^- (aq)

Set up a concentration table and use it to determine K_b:

pOH = 14.00 – pH = 3.00 $[OH^-] = 10^{-3.00} = 1.00 \times 10^{-3}$ M

Reaction: H_2O +	$A^- \rightleftharpoons$	HA +	OH^-
Initial concentration (M)	0.0100	0	0
Change in concentration (M)	–0.00100	+0.00100	+0.00100
Final concentration (M)	0.0090	0.00100	0.00100

$$K_b = \frac{[HA]_{eq}[OH^-]_{eq}}{[A^-]_{eq}} = \frac{(1.00 \times 10^{-3})^2}{(0.0090)} = 1.1 \times 10^{-4}$$

$$K_a = \frac{K_w}{K_b} = \frac{1.00 \times 10^{-14}}{1.1 \times 10^{-4}} = 9.1 \times 10^{-11}$$

17.63   Determine the percent ionization using Equation 17–3 in your textbook:

$$\% \text{ HA ionized} = 100\% \frac{[H_3O^+]_{eq}}{[HA]_{initial}}$$

The equilibrium concentration of hydronium ions was calculated in Problem 17.57: $[H_3O^+] = 5.2 \times 10^{-6}$ M, and the initial concentration of the weak acid is given in the problem, 0.050 M:

$$\% \text{ HA ionized} = 100\% \frac{\left(5.2 \times 10^{-6} \text{ M}\right)}{(0.050 \text{ M})} = 0.010 \%$$

17.65   There are two amine groups, one at either end of the molecule, each of which can accept a proton from a water molecule. Convert the line structure to a Lewis structure using the standard procedures, then show the transfer of one proton to each N atom:

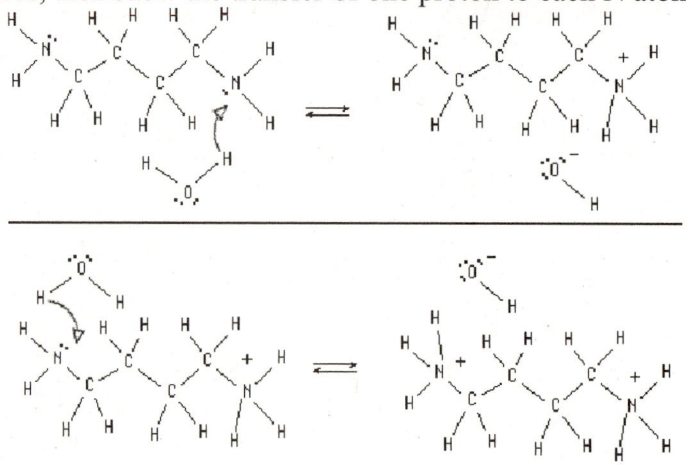

17.67   Follow the standard procedure for solving equilibrium problems:

1.) Major species: H_2O, Na^+, and F^-

2.) Dominant acid–base equilibrium: $H_2O + F^- \rightleftharpoons HF + OH^-$

3.) From Table 17–3, $K_a = 6.3 \times 10^{-4}$; $K_b = \dfrac{K_w}{K_a} = \dfrac{1.0 \times 10^{-14}}{6.3 \times 10^{-4}} = 1.6 \times 10^{-11}$

4.) Set up a concentration table:

Reaction: H_2O +	F^- $\rightleftharpoons$	HF +	OH^-
Initial concentration (M)	0.250	0	0
Change in concentration (M)	$-x$	$+x$	$+x$
Final concentration (M)	$0.250 - x$	x	x

$K_b = 1.6 \times 10^{-11} = \dfrac{[HF]_{eq}[OH^-]_{eq}}{[F^-]_{eq}} = \dfrac{x^2}{0.250 - x}$ Assume that $x \ll 0.250$

$1.6 \times 10^{-11} = \dfrac{x^2}{0.250}$, from which $x^2 = 4.0 \times 10^{-12}$ and $x = 2.0 \times 10^{-6}$; assumption is valid.

$[OH^-] = 2.0 \times 10^{-6}$ M $pOH = -\log(2.0 \times 10^{-6}) = 5.70$ $pH = 14.00 - 5.70 = 8.30$

17.69   To determine concentrations of species in a solution, follow the standard procedure:

1.) This is a strong acid. Major species are H_2O, H_3O^+, and HSO_4^-.

2.) The dominant acid–base equilibrium is $H_2O + HSO_4^- \rightleftharpoons SO_4^{2-} + H_3O^+$.

3.) From Table 17–2, $K_{a2} = 1.0 \times 10^{-2}$.

4.) In this solution, there are hydronium ions from the strong acid present initially:

Reaction: H_2O +	HSO_4^- $\rightleftharpoons$	SO_4^{2-} +	H_3O^+
Initial concentration (M)	2.00	0	2.00
Change in concentration (M)	$-x$	$+x$	$+x$
Final concentration (M)	$2.00 - x$	x	$2.00 + x$

$K_{a2} = 1.0 \times 10^{-2} = \dfrac{[SO_4^{2-}]_{eq}[H_3O^+]_{eq}}{[HSO_4^-]_{eq}} = \dfrac{x(2.00 + x)}{2.00 - x}$ Assume that $x \ll 2.00$

$1.0 \times 10^{-2} = \dfrac{(2.00)(x)}{(2.00)}$, from which $x = 1.0 \times 10^{-2}$; assumption is valid.

$[H_3O^+] = 2.00 + 0.010 = 2.01$ M $[SO_4^{2-}] = 1.0 \times 10^{-2}$ M

$[HSO_4^-] = 2.00 - x = 1.99$ M

17.71   (a) H_2SO_4 is a strong acid, so the major species in solution are H_2O, HSO_4^-, and H_3O^+.
        The hydrogen sulfate ion is a weak acid, so the equilibrium reaction that determines
        the pH is $H_2O + HSO_4^- \rightleftharpoons SO_4^{2-} + H_3O^+$

(b) Na_2SO_4 is a salt, so the major species in solution are H_2O, SO_4^{2-}, and Na^+. The
        sulfate ion is the conjugate base of a weak acid, so the equilibrium reaction that
        determines pH is $SO_4^{2-} + H_2O \rightleftharpoons HSO_4^- + OH^-$

(c) NaHSO$_4$ is a salt, so the major species in solution are H$_2$O, HSO$_4^-$, and Na$^+$. The hydrogen sulfate ion is a weak acid, so the equilibrium reaction that determines the pH is   H$_2$O + HSO$_4^-$ ⇌ SO$_4^{2-}$ + H$_3$O$^+$

(d) NH$_4$Cl is a salt, so the major species in solution are H$_2$O, Cl$^-$, and NH$_4^+$. The ammonium ion is the conjugate acid of a weak base, so the equilibrium reaction that determines pH is   H$_2$O + NH$_4^+$ ⇌ NH$_3$ + H$_3$O$^+$

17.73   Tabulated equilibrium constants for acid–base reactions always refer to reactions in which H$_2$O is one of the reactants. The reaction in this problem is the reverse of a base reaction:

$$HPO_4^{2-} (aq) + OH^- (aq) \rightleftharpoons PO_4^{3-} (aq) + H_2O (l)$$

Table 17–2 lists K_a values for phosphoric acid:

$$HPO_4^{2-} (aq) + H_2O (l) \rightleftharpoons PO_4^{3-} (aq) + H_3O^+ (aq) \qquad K_{a3} = 4.8 \times 10^{-13}$$

K_a and K_b for a conjugate acid–base pair are related through $K_a K_b = K_w$:

$$K_b = \frac{1.0 \times 10^{-14}}{4.8 \times 10^{-13}} = 2.1 \times 10^{-2}$$

$$K_{eq} = \frac{1}{K_b} = 48$$

17.75   Molecular pictures must show the correct relative numbers of the various species in the solution. From the starting condition (six molecules of oxalic acid), make appropriate changes and then draw new pictures:

(a) Hydroxide ions react with oxalic acid to form water and hydrogen oxalate ions:
H$_2$C$_2$O$_4$ + OH$^-$ ⇌ H$_2$O + HC$_2$O$_4^-$ The picture shows 2 molecules of oxalic acid and 4 each of water and hydrogen oxalate:

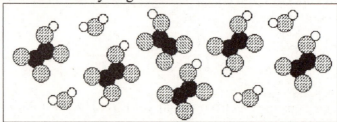

(b) When all oxalic acid has reacted, hydroxide ions react with hydrogen oxalate ions to form water and oxalate ions:
HC$_2$O$_4^-$ + OH$^-$ ⇌ H$_2$O + C$_2$O$_4^{2-}$. The picture shows 4 hydrogen oxalate ions, 8 water molecules, and 2 oxalate ions:

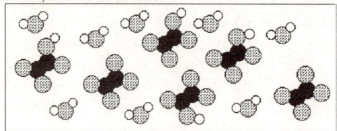

(c) NH_3, a weak base, accepts a proton from oxalic acid, a weak acid:

$$H_2C_2O_4 + NH_3 \rightleftharpoons NH_4^+ + HC_2O_4^-$$

The picture shows 2 oxalic acid molecules and four each ammonium and hydrogen oxalate ions:

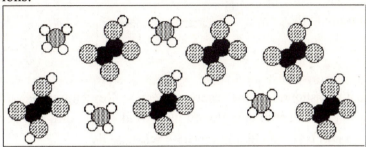

17.77   The chemical reaction that occurs is:

$$P_4O_{10} + 6\,H_2O \rightarrow 4\,H_3PO_4$$

(a) The major species present are H_2O and H_3PO_4

(b) To determine the ranking of the minor species, consider what reactions generate them. H_3PO_4 undergoes proton transfer with water to form $H_2PO_4^-$ and H_3O^+ in equal concentrations, but $H_2PO_4^-$ undergoes further proton transfer with water to form HPO_4^{2-} and H_3O^+. HPO_4^{2-}, in turn, generates a tiny amount of PO_4^{3-}, and the water equilibrium generates a tiny amount of OH^-. The minor species are (in order of highest concentration to lowest concentration):

$$H_3O^+,\ H_2PO_4^-,\ HPO_4^{2-},\ OH^-,\ \text{and}\ PO_4^{3-}$$

(c) The dominant equilibrium that determines the pH is:

$$H_3PO_4 + H_2O \rightleftharpoons H_2PO_4^- + H_3O^+$$

Set up a concentration table, solve for hydronium ion concentration, then calculate the pH. Determine the initial concentration using standard stoichiometric procedures:

$$n(H_3PO_4) = 3.5\ \text{g}\ P_4O_{10} \left(\frac{1\ \text{mol}}{283.88\ \text{g}} \right) \left(\frac{4\ \text{mol}\ H_3PO_4}{1\ \text{mol}\ P_4O_{10}} \right) = 0.0493\ \text{mol}$$

$$[H_3PO_4] = \frac{0.0493\ \text{mol}}{1.50\ \text{L}} = 0.033\ \text{M}$$

Reaction: H_2O +	$H_3PO_4 \rightleftharpoons$	$H_2PO_4^-$ +	H_3O^+
Initial concentration (M)	0.033	0	0
Change in concentration (M)	$-x$	$+x$	$+x$
Final concentration (M)	$0.033 - x$	x	x

Now substitute into the equilibrium constant expression and solve for x:

$$K_{a1} = 0.0069 = \frac{[H_2PO_4^-]_{eq}[H_3O^+]_{eq}}{[H_3PO_4]_{eq}} = \frac{x^2}{0.033 - x}\ ;\ \text{solve by the quadratic equation}$$

$$x = [H_3O^+] = 0.012\ \text{M} \qquad\qquad pH = -\log(0.012) = 1.92$$

17.79 Proton transfer occurs from the carboxylic acid O–H (shown screened below) and the amino nitrogen atom:

17.81 Net ionic equations show only the reacting species. Remember that strong acids generate H_3O^+ in solution and react to completion with weak bases, and strong bases generate OH^- in solution and react to completion with weak acids:

(a) strong base reacting with weak acid: $OH^- + C_6H_5CO_2H \rightleftharpoons H_2O + C_6H_5CO_2^-$

(b) strong acid reacting with weak base: $H_3O^+ + (CH_3)_3N \rightleftharpoons H_2O + (CH_3)_3NH^+$

(c) weak base reacting with weak acid: $SO_4^{2-} + CH_3CO_2H \rightleftharpoons HSO_4^- + CH_3CO_2^-$

HSO_4^-, $pK_a = 1.99$; CH_3CO_2H, $pK_a = 4.75$; HSO_4^- is stronger, so this reaction proceeds to a small extent.

(d) strong base reacting with weak acid: $OH^- + NH_4^+ \rightleftharpoons H_2O + NH_3$

(e) weak base reacting with weak acid: $HPO_4^{2-} + NH_3 \rightleftharpoons PO_4^{3-} + NH_4^+$

HPO_4^{2-}, $pK_a = 12.32$; NH_4^+, $pK_a = 9.25$; NH_4^+ is stronger, so this reaction proceeds to a small extent.

18.1 A buffer solution must contain both a weak acid and its conjugate weak base. Begin each analysis by calculating the initial amounts of acid and base.

(a) The strong base reacts with NH_4^+, a weak acid, to generate the conjugate weak base.

Moles $(NH_4^+) = 0.25$ M$(0.100$ L$) = 0.025$ mol

Moles $(OH^-) = 0.25$ M$(0.150$ L$) = 0.0375$ mol

There is excess strong base (0.0375 mol compared with 0.025 mol), so no weak acid remains after mixing, this solution does not have buffer properties.

(b) Initial amounts:

Moles $(NH_4^+) = 0.25$ M$(0.100$ L$) = 0.025$ mol

Moles $(OH^-) = 0.25$ M$(0.050$ L$) = 0.0125$ mol

The strong base reacts completely with the weak acid to generate the conjugate weak base:

Reaction:	NH_4^+ +	OH^- ⇌	NH_3 +	H_2O
Start (mol)	0.025	0.0125	0	---
Change (mol)	−0.0125	−0.0125	+0.0125	---
Final (mol)	0.0125	0.0000	0.0125	solvent

Since both acid and conjugate base exist, this solution is an NH_4^+/ NH_3 buffer. Use the buffer equation, Equation 18-1, with amounts to determine the pH:

$$pH = pK_a + \log\left\{\frac{(mol\ A^-)}{(mol\ HA)}\right\} \qquad \text{For } NH_4^+, pK_a = 9.25$$

$$pH = 9.25 + \log\left\{\frac{(0.0125\ mol)}{(0.0125\ mol)}\right\} = 9.25$$

(c) NH_4^+ is a weak acid and HCl is a strong acid. There is no conjugate base present in high concentration, so this solution does not have buffer properties.

(d) NH_3 is a weak base and HCl is a strong acid. The strong acid reacts completely with the weak base to generate the conjugate weak acid.

Initial amounts:

Moles$(NH_3) = 0.25$ M$(0.100$ L$) = 0.025$ mol

Moles$(H_3O^+) = 0.25$ M$(0.050$ L$) = 0.0125$ mol

Reaction:	NH_3 +	H_3O^+ ⇌	NH_4^+ +	H_2O
Start (mol)	0.025	0.0125	0	---
Change (mol)	−0.0125	−0.0125	+0.0125	---
Final (mol)	0.0125	0.0000	0.0125	solvent

This is an NH_4^+/ NH_3 buffer. Use the buffer equation, Equation 18-1:

$$pH = pK_a + \log\left\{\frac{(mol\ A^-)}{(mol\ HA)}\right\} \qquad \text{where } pK_a = 9.25$$

$$pH = 9.25 + \log\left\{\frac{(0.0125\ mol)}{(0.0125\ mol)}\right\} = 9.25$$

18.3   The two solutions that are buffered are (b) and (d). The concentration tables for these solutions show that they are identical in acid/base composition, despite being prepared differently. Thus only one calculation is required. The added strong acid reacts completely with the weak base of the buffer to generate the conjugate weak acid.

Reaction:	NH_3 +	H_3O^+ $\rightleftharpoons$	NH_4^+ +	H_2O
Start (mol)	0.0125	0.0050	0.0125	---
Change	−0.0050	−0.0050	+0.0050	---
Final	0.0075	0.0000	0.0175	solvent

Use the buffer equation, Equation 18-1, with the new amounts:

$$pH = pK_a + \log\left\{\frac{(mol\ A^-)}{(mol\ HA)}\right\} = 9.25 + \log\left\{\frac{(0.0075\ mol)}{(0.0175\ mol)}\right\} = 9.25 - 0.37 = 8.88$$

18.5   The two solutions that are buffered are (b) and (d). The concentration tables for these solutions show that they are identical in acid/base composition, despite being prepared differently. Thus only one calculation is required. Addition of strong base consumes weak acid, produces weak base, and increases the pH of the buffer. First use the buffer equation to calculate the acid/base ratio in a solution whose pH is greater by 0.10 unit than the initial buffer solution. Then set up a concentration ratio, using x = the amount of added base, and solve for x: new pH = 9.25 + 0.10 = 9.35

$[NH_3] = 0.0125 + x$ $[NH_4^+] = 0.0125 - x$

$$9.35 = 9.25 + \log\left\{\frac{(0.0125 + x)}{(0.0125 - x)}\right\} \qquad \log\left\{\frac{(0.0125 + x)}{(0.0125 - x)}\right\} = 0.10$$

$$\left\{\frac{(0.0125 + x)}{(0.0125 - x)}\right\} = 1.26 \qquad (0.0125 + x) = 1.26(0.0125 - x) = 0.0157 - 1.26\,x$$

$0.0032 = 2.26\,x$, and $x = 1.4 \times 10^{-3}$ mol = amount of base that must be added

18.7   base consumes added hydronium ions:

The acid is HCO_3^-, and its conjugate base is CO_3^{2-}.

Consumption of hydronium ions: CO_3^{2-} (aq) + H_3O^+ (aq) → HCO_3^- (aq) + H_2O (l)

Consumption of hydroxide ions: HCO_3^- (aq) + OH^- (aq) → CO_3^{2-} (aq) + H_2O (l)

18.9   Buffer solutions must contain a weak acid and its conjugate base. A weak acid and its salt, a strong acid and a weak base, or a strong base and a weak acid can meet these requirements. To determine if a pair of reactants can generate a buffer solution, identify the acid-base properties of the pair:

(a) $NaHCO_3$ is both a weak base and a weak acid, and NaOH is a strong base. This pair generates a buffer solution containing HCO_3^- and CO_3^{2-}.

(b) NaOH is a strong base, and NH_3 is a weak base; no buffer possibilities.

(c) H_3PO_4 is a weak acid, and HCl is a strong acid; no buffer possibilities.

(d) HCl is a strong acid, and Na_2CO_3 is a weak base. This pair generates a buffer solution containing HCO_3^- and CO_3^{2-}.

18.11   A buffer solution requires a conjugate acid/base pair whose pK_a is within 1 pH unit of the desired pH. For pH = 3.50, the HCO_2H/HCO_2^- system, pK_a = 3.75, would be most suitable. Sodium formate, $NaHCO_2$, could be used along with HCl solution, which would protonate some formate anions to form the conjugate weak acid. For pH = 12.60, the HPO_4^{2-}/PO_4^{3-} system, pK_a = 12.32, would be most suitable. Potassium phosphate, K_3PO_4, could be used along with HCl solution, which would protonate some phosphate anions to form the conjugate weak acid.

18.13   This is a NH_3/NH_4^+ buffer. Use the buffer equation, Equation 18-1, to determine the amount of ammonium chloride needed:

$MM(NH_4Cl)$ = 14.01 g/mol + 4(1.01 g/mol) + 35.45 g/mol = 53.50 g/mol

$$\log\left\{\frac{\text{mol } NH_3}{\text{mol } NH_4^+}\right\} = pH - pK_a = 8.90 - 9.25 = -0.35$$

$$\frac{\text{mol } NH_3}{\text{mol } NH_4^+} = 10^{-0.35} = 0.447$$

$$\text{mol } NH_4^+ = \frac{\text{mol } NH_3}{0.447} = \frac{(1.25 \text{ L})(0.25 \text{ M})}{0.447} = 0.699 \text{ mol} = \text{moles of } NH_4Cl$$

$$m = 0.699 \text{ mol}\left(\frac{53.50 \text{ g}}{1 \text{ mol}}\right) = 37 \text{ g}$$

18.15   The question asks how much strong acid is required to lower the pH from 8.90 to 8.65. Use the buffer equation to determine the number of moles of acid required to cause this change, with x = moles of acid added:

$$\log\left\{\frac{0.3125 - x}{0.699 + x}\right\} = pH - pK_a = 8.65 - 9.25 = -0.60$$

$$\frac{0.3125 - x}{0.699 + x} = 10^{-0.60} = 0.251 \qquad\qquad x = 0.110 \text{ moles HCl}$$

$$V = 0.110 \text{ mol}\left(\frac{1 \text{ L}}{2.0 \text{ mol}}\right) = 0.055 \text{ L or 55 mL}$$

18.17   To prepare the buffer solution, dissolve the appropriate amount of Na_2CO_3 in water, add sufficient 0.500 M HCl solution, and make up to 1.5 L with additional water. Calculate the carbonate/hydrogen carbonate ratio using the buffer equation:

$$\log\left\{\frac{[CO_3^{2-}]}{[HCO_3^-]}\right\} = pH - pK_a = 9.85 - 10.33 = -0.48, \quad \left\{\frac{[CO_3^{2-}]}{[HCO_3^-]}\right\} = 10^{-0.48} = 0.33$$

Thus, $[CO_3^{2-}] = 0.33 [HCO_3^-]$ $[CO_3^{2-}] + [HCO_3^-] = 0.35$ M

$0.33 [HCO_3^-] + [HCO_3^-] = 0.35$ M

$1.33 [HCO_3^-] = 0.35$ M and $[HCO_3^-] = 0.263$ M

$[CO_3^{2-}] = 0.33[HCO_3^-] = 0.087$ M

$$n_{HCO_3^-} = (0.263 \text{ M})(1.5 \text{ L}) = 0.3945 \text{ mol}$$

$$n_{CO_3^{2-}} = (0.087 \text{ M})(1.5 \text{ L}) = 0.1305 \text{ mol}$$

All the carbonate and hydrogen carbonate must come from sodium carbonate:

$$n_{Na_2CO_3} = 0.3945 + 0.1305 = 0.525 \text{ mol}$$

$$m_{Na_2CO_3} = n \, MM = (0.525 \text{ mol})(106 \text{ g/mol}) = 55.7 \text{ g}$$

The 0.3945 mol of hydrogen carbonate in the buffer solution is generated by adding strong acid:

$$V_{HCl} = \frac{n}{M} = \frac{0.3945 \text{ mol}}{0.500 \text{ mol/L}} = 0.789 \text{ L}$$

Dissolve 55.7 g sodium carbonate in about 1 L of water, add 0.789 L 0.500 M HCl and enough water to make 1.5 L (two significant figure accuracy would be sufficient for the final volume, but the reagents should be measured to three significant figures).

18.19   The question asks how much strong base is required to raise the pH by 0.15 units, from 9.85 to 10.00. Let x be the moles of added base, and use the buffer equation with a ratio in moles rather than concentration to determine the number of moles required to cause this change:
From the solution to Problem 18.17, the original buffer solution contains

$$n_{HCO_3^-} = (0.263 \text{ M})(1.5 \text{ L}) = 0.3945 \text{ mol}$$

$$n_{CO_3^{2-}} = (0.087 \text{ M})(1.5 \text{ L}) = 0.1305 \text{ mol}$$

$$\log\left\{\frac{0.1305 + x}{0.3945 - x}\right\} = pH - pK_a = 10.00 - 10.33 = -0.33$$

$$\left\{\frac{0.1305 + x}{0.3945 - x}\right\} = 10^{-0.33} = 0.468$$

$$(0.468)(0.3945 - x) = 0.1305 + x$$

$$0.184 - 0.468 \, x = 0.1305 + x$$

$$0.0535 = 1.468 \, x$$

$$x = 0.0364 \text{ moles NaOH}$$

$$m_{NaOH} = n \, MM = (0.0364 \text{ moles})(40. \text{ g/mol}) = 1.5 \text{ g}$$

We round to two significant figures because that is the typical accuracy of the buffer equation. The resulting solution is still buffered, because the ratio of acid to conjugate base remains in the buffer region, between 0.1 and 10.

18.21   To determine the approximate pH at the stoichiometric point of an acid-base titration, examine the equilibrium that determines the pH in the vicinity of the stoichiometric point:
(a) This is a weak base-strong acid titration. At the stoichiometric point, major acid-base species are NH_4^+ and H_2O, and the dominant equilibrium is

$$NH_4^+ \, (aq) + H_2O \, (l) \rightleftharpoons NH_3 \, (aq) + H_3O^+ \, (aq)$$

Hydronium ions are produced in this reaction, so pH < 7 at the stoichiometric point.

(b) This is a strong acid-strong base titration. The only acid-base species present at the stoichiometric point is H_2O, and the dominant equilibrium is

$$H_2O \; (l) + H_2O \; (l) \rightleftharpoons OH^- \; (aq) + H_3O^+ \; (aq)$$

Thus, pH = 7 at the stoichiometric point.

(c) This is a weak base-strong acid titration. At the stoichiometric point, major acid-base species are CH_3CO_2H and H_2O, and the dominant equilibrium is

$$CH_3CO_2H \; (aq) + H_2O \; (l) \rightleftharpoons CH_3CO_2^- \; (aq) + H_3O^+ \; (aq)$$

Hydronium ions are produced in this reaction, so pH < 7 at the stoichiometric point.

18.23 This problem describes an acid-base titration. The major species differ at various points during a titration, so there are different dominant equilibria that must be identified before doing an equilibrium calculation to determine pH.

(a) Before titration begins, the major species are a weak acid, aspirin (HA) and H_2O, and the dominant equilibrium is $HA + H_2O \rightleftharpoons A^- + H_3O^+$, for which $K_a = 3.0 \times 10^{-4}$.

Reaction: H_2O +	HA $\rightleftharpoons$	A^- +	H_3O^+
Initial concentration (M)	10^{-2}	0	0
Change in concentration (M)	$-x$	$+x$	$+x$
Final concentration (M)	$10^{-2} - x$	x	x

Now substitute into the equilibrium constant expression and solve for x:

$$K_a = 3.0 \times 10^{-4} = \frac{[A^-]_{eq}[H_3O^+]_{eq}}{[HA]_{eq}} = \frac{x^2}{(10^{-2} - x)}$$

$$x^2 = (3.0 \times 10^{-4})(10^{-2} - x) = (3.0 \times 10^{-6}) - (3.0 \times 10^{-4})x$$

$$x^2 + (3.0 \times 10^{-4})x - (3.0 \times 10^{-6}) = 0$$

$$x = \frac{-b \pm \sqrt{b^2 - 4ac}}{2a} = \frac{-(3.0 \times 10^{-4}) \pm \sqrt{(3.0 \times 10^{-4})^2 - 4(-3.0 \times 10^{-6})}}{2} = 1.6 \times 10^{-3}$$

$$[H_3O^+] = 1.6 \times 10^{-3} \text{ M} \qquad pH = -\log(1.6 \times 10^{-3}) = 2.8$$

(b) At the stoichiometric point, the major acid-base species present are A^- and H_2O and the dominant equilibrium is:

$$A^- + H_2O \rightleftharpoons HA + OH^- \qquad K_b = \frac{K_w}{K_a} = \frac{1.00 \times 10^{-14}}{3.0 \times 10^{-4}} = 3.3 \times 10^{-11}$$

Reaction: H_2O +	A^- $\rightleftharpoons$	HA +	OH^-
Initial concentration (M)	10^{-2}	0	0
Change in concentration (M)	$-x$	$+x$	$+x$
Final concentration (M)	$10^{-2} - x$	x	x

Now substitute into the equilibrium constant expression and solve for x:

$$K_b = 3.3 \times 10^{-11} = \frac{[HA]_{eq}[OH^-]_{eq}}{[A^-]_{eq}} = \frac{x^2}{(10^{-2} - x)}; \text{ Assume } x \ll 10^{-2}$$

$$x^2 = (10^{-2})(3.3 \times 10^{-11}) = 3.3 \times 10^{-13} \qquad x = 5.7 \times 10^{-7}; \text{ the assumption is valid.}$$

$$[OH^-]_{eq} = 5.7 \times 10^{-7} \text{ M}$$

$$pOH = -\log(5.7 \times 10^{-7}) = 6.24 \qquad pH = 14.00 - pOH = 14.00 - 6.24 = 7.8$$

(c) At the midpoint of the titration, half of the weak acid has been converted into its conjugate base, and $pH = pK_a = -\log(3.0 \times 10^{-4}) = 3.5$.

18.25 The best indicator for a titration is one for which $pK_{In} = pH_{\text{stoichiometric point}}$; the pH at the stoichiometric point was calculated in Problem 18.23 to be 7.8. The best indicator is phenol red, $pK_{In} = 7.9$.

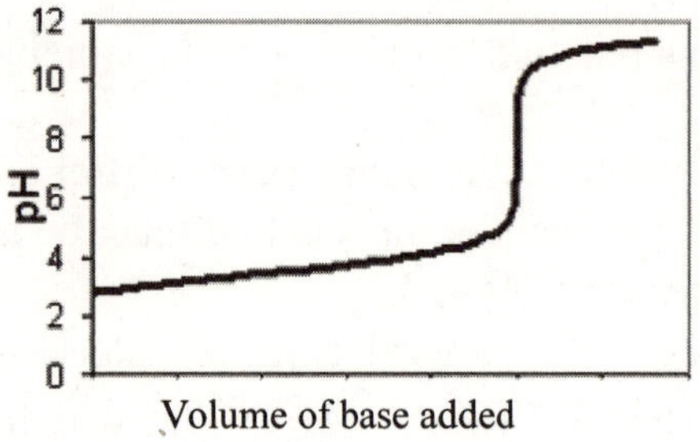
12
10
8
pH
6
4
2
0
Volume of base added

18.27 This is a titration type problem for a diprotic acid asking for information at the second stoichiometric point. At the second stoichiometric point, all the weak acid (H_2A) has been converted into A^{2-}, so the major acid-base species present are A^{2-} and H_2O and the dominant equilibrium is

$$A^{2-}(aq) + H_2O(l) \rightleftharpoons HA^-(aq) + OH^-(aq)$$

$$K_{b1} = \frac{K_w}{K_{a2}} = \frac{1.00 \times 10^{-14}}{3.9 \times 10^{-6}} = 2.6 \times 10^{-9}$$

Begin by determining the initial amount of A^{2-} (we can assume that dilution effects to be negligible). Then construct an equilibrium table, write the equilibrium expression, and solve for the hydroxide concentration to determine the pH:

At the second stoichiometric point, $[A^{2-}]_{\text{initial}} = [H_2A]_{\text{before titration}} = 1.50 \times 10^{-2}$ M:

Here is the completed equilibrium table:

Reaction: H_2O +	A^{2-} $\rightleftharpoons$	HA^- +	OH^-
Initial concentration (M)	0.0150	0	0
Change in concentration (M)	$-x$	$+x$	$+x$
Final concentration (M)	$0.0150 - x$	x	x

Now substitute into the equilibrium constant expression and solve for x:

$$K_{b1} = 2.6 \times 10^{-9} = \frac{[HP^-]_{eq}[OH^-]_{eq}}{[P^{2-}]_{eq}} = \frac{x^2}{0.0150 - x} \quad \text{Assume that } x \ll 0.0150$$

$x^2 = (2.6 \times 10^{-9})(0.0150) = 3.90 \times 10^{-11}$

$x = 6.24 \times 10^{-6}$

$[OH^-] = 6.24 \times 10^{-6}$ M $\quad$ $pOH = -\log(6.24 \times 10^{-6}) = 5.20$

$pH = 14.00 - 5.20 = 8.80$; a suitable indicator for this titration is thymol blue, $pK_{in} = 8.9$.

(If dilution is taken into account, $[OH^-] = 6.14 \times 10^{-6}$ M, $pOH = 5.21$)

18.29 The major species present are different at various points during a titration, so there are different dominant equilibria that must be identified before doing a calculation to determine pH. We also need to know the initial concentration of the solution:

$$n = \frac{m}{MM} = \frac{1.51 \text{ g}}{120. \text{ g/mol}} = 0.0126 \text{ mol} \qquad M = \frac{n}{V} = \frac{0.0126 \text{ mol}}{0.100 \text{ L}} = 0.126 \text{ M}$$

(a) Before titration begins, the major species are a weak acid, dihydrogen phosphate (HA) and H_2O, and the dominant equilibrium is $HA + H_2O \rightleftharpoons A^- + H_3O^+$, for which $K_a = 6.2 \times 10^{-8}$ and $pK_a = 7.21$:

Reaction: H_2O +	HA $\rightleftharpoons$	A^- +	H_3O^+
Initial concentration (M)	0.121	0	0
Change in concentration (M)	$-x$	$+x$	$+x$
Final concentration (M)	$0.126 - x$	x	x

Now substitute into the equilibrium constant expression and solve for x:

$$K_a = 6.2 \times 10^{-8} = \frac{[A^-]_{eq}[H_3O^+]_{eq}}{[HA]_{eq}} = \frac{x^2}{(0.126 - x)} \qquad \text{Assume that } x << 0.126$$

$x^2 = (6.2 \times 10^{-8})(0.126) = 7.8 \times 10^{-9}$, from which $x = 8.8 \times 10^{-5}$

$[H_3O^+] = 8.8 \times 10^{-5} \text{ M} \qquad \qquad pH = -\log(8.8 \times 10^{-5}) = 4.05$

(b) At the first midpoint, both acid and conjugate base of dihydrogen phosphate are present in equal concentrations, the solution is buffered, and the pH can be calculated using the buffer equation:

$$pH = pK_{a1} + \log 1 = 7.21$$

(c) At the first stoichiometric point, Equation 18-2 from your textbook applies:

$$[H_3O^+]_{eq, \text{1st stoichiometric point}} = \sqrt{K_{a1} K_{a2}}$$

$$[H_3O^+] = \sqrt{\left(6.2 \times 10^{-8}\right)\left(4.8 \times 10^{-13}\right)} = \sqrt{\left(2.976 \times 10^{-20}\right)} = 1.7 \times 10^{-10} \text{ M}$$

$$pH = -\log(1.7 \times 10^{-10}) = 9.76$$

(d) At the second midpoint, both acid and conjugate base of hydrogenphosphate are present in equal concentrations, the solution is buffered, and the pH can be calculated using the buffer equation: $\qquad pH = pK_{a2} + \log 1 = 12.32$

(e) At the second stoichiometric point, all the hydrogenphosphate has been converted into its conjugate base, so the major acid-base species present are phosphate (A^-) and H_2O and the dominant equilibrium is:

$$A^- + H_2O \rightleftharpoons HA + OH^-, \quad K_b = \frac{K_w}{K_{a2}} = \frac{1.0 \times 10^{-14}}{4.8 \times 10^{-13}} = 2.1 \times 10^{-2}$$

Reaction: H_2O +	A^- $\rightleftharpoons$	HA +	H_3O^+
Initial concentration (M)	0.126	0	0
Change in concentration (M)	$-x$	$+x$	$+x$
Final concentration (M)	$0.126 - x$	x	x

Now substitute into the equilibrium constant expression and solve for x:

$$K_a = 2.1 \times 10^{-2} = \frac{[HA]_{eq}[OH^-]_{eq}}{[A^-]_{eq}} = \frac{x^2}{(0.126 - x)}$$

$$x^2 = (2.1 \times 10^{-2})(0.126 - x) = (2.6 \times 10^{-3}) - (2.1 \times 10^{-2})x$$

$$x^2 + (2.1 \times 10^{-2})x - (2.6 \times 10^{-3}) = 0$$

$$x = \frac{-b \pm \sqrt{b^2 - 4ac}}{2a} = \frac{-(2.1 \times 10^{-2}) \pm \sqrt{(2.1 \times 10^{-2})^2 - 4(-2.6 \times 10^{-3})}}{2} = 4.2 \times 10^{-2}$$

$[OH^-]_{eq} = 4.2 \times 10^{-2}$ M

pOH = -log(4.2×10^{-2}) = 1.38, and pH = 14.00 − pOH = 14.00 −1.38 = 12.62

18.31   To write a solubility product equilibrium expression, first determine the chemical formula and stoichiometry of the salt:

(a) $AgCl\,(s) \rightleftharpoons Ag^+\,(aq) + Cl^-\,(aq)$ $K_{sp} = [Ag^+]_{eq}[Cl^-]_{eq}$

(b) $BaSO_4\,(s) \rightleftharpoons Ba^{2+}\,(aq) + SO_4{}^{2-}\,(aq)$ $K_{sp} = [Ba^{2+}]_{eq}[SO_4^{2-}]_{eq}$

(c) $Fe(OH)_2\,(s) \rightleftharpoons Fe^{2+}\,(aq) + 2\,OH^-\,(aq)$ $K_{sp} = [Fe^{2+}]_{eq}[OH^-]_{eq}^2$

(d) $Ca_3(PO_4)_2\,(s) \rightleftharpoons 3Ca^{2+}\,(aq) + 2PO_4{}^{3-}\,(aq$ $K_{sp} = [Ca^{2+}]_{eq}^3[PO_4^{3-}]_{eq}^2$

18.33   "Determine the mass that dissolves" means "calculate the amount present in solution at equilibrium". Initial amounts are zero, so it is easy to complete a concentration table. Calculate molarity from the solubility product expression, then convert to mass using standard stoichiometric methods. Let x be the number of moles of salt dissolving in 1 L of solution.

(a)

Reaction:	$AgCl\,(s) \rightleftharpoons$	$Ag^+\,(aq) +$	$Cl^-\,(aq)$
Initial concentration (M)	solid	0	0
Change in concentration (M)	solid	+x	+x
Final concentration (M)	solid	x	x

$$K_{sp} = 1.8 \times 10^{-10} = [Ag^+]_{eq}[Cl^-]_{eq} = x^2 \qquad x = 1.34 \times 10^{-5}\ M$$

$$m_{AgCl} = 0.475\ L \left(\frac{1.34 \times 10^{-5}\ mol}{1\ L} \right) \left(\frac{143.32\ g}{1\ mol} \right) = 9.1 \times 10^{-4}\ g$$

(b)

Reaction:	$BaSO_4\,(s) \rightleftharpoons$	$Ba^{2+}\,(aq) +$	$SO_4{}^{2-}\,(aq)$
Initial concentration (M)	solid	0	0
Change in concentration (M)	solid	+x	+x
Final concentration (M)	solid	x	x

$$K_{sp} = 1.1 \times 10^{-10} = [Ba^{2+}]_{eq}[SO_4^{2-}]_{eq} = x^2 \qquad x = 1.05 \times 10^{-5}\ M$$

$$m_{BaSO_4} = 0.475\ L \left(\frac{1.05 \times 10^{-5}\ mol}{1\ L} \right) \left(\frac{233.39\ g}{1\ mol} \right) = 1.2 \times 10^{-3}\ g$$

(c)

Reaction:	$Fe(OH)_2\ (s) \rightleftharpoons$	$Fe^{2+}\ (aq)+$	$2\ OH^-\ (aq)$
Initial concentration (M)	solid	0	0
Change in concentration (M)	solid	$+x$	$+2x$
Final concentration (M)	solid	x	$2x$

$$K_{sp} = 4.8 \times 10^{-17} = [Fe^{2+}]_{eq}[OH^-]_{eq}^2 = 4x^3$$

$$x = 2.3 \times 10^{-6}\ M$$

$$m_{Fe(OH)_2} = 0.475\ L \left(\frac{2.3 \times 10^{-6}\ mol}{1\ L} \right) \left(\frac{89.87\ g}{1\ mol} \right) = 9.8 \times 10^{-5}\ g$$

(d)

Reaction:	$Ca_3(PO_4)_2\ (s) \rightleftharpoons$	$3\ Ca^{2+}\ (aq)+$	$2\ PO_4^{3-}\ (aq)$
Initial concentration (M)	solid	0	0
Change (M)	solid	$+3x$	$+2x$
Final concentration (M)	solid	$3x$	$2x$

$$K_{sp} = 2.0 \times 10^{-33} = [Ca^{2+}]_{eq}^3[PO_4^{3-}]_{eq}^2 = 108\ x^5$$

$$x = 1.1 \times 10^{-7}\ M$$

$$m_{Ca_3(PO_4)_2} = 0.475\ L \left(\frac{1.1 \times 10^{-7}\ mol}{1\ L} \right) \left(\frac{310.18\ g}{1\ mol} \right) = 1.6 \times 10^{-5}\ g$$

18.35 To calculate a solubility product from the mass that dissolves in solution, convert to molarity of ions and then evaluate the equilibrium constant expression. Begin by writing the chemical reaction and the equilibrium expression:

$$CaC_2O_4\ (s) \rightleftharpoons Ca^{2+}\ (aq) + C_2O_4^{2-}\ (aq) \qquad K_{sp} = [Ca^{2+}]_{eq}[C_2O_4^{2-}]_{eq}$$

$$n = \frac{m}{MM} = (6.1\ mg) \left(\frac{10^{-3}\ g}{1\ mg} \right) \left(\frac{1\ mol}{128.1\ g} \right) = 4.76 \times 10^{-5}\ mol$$

Since the stoichiometry is 1:1, the moles of calcium ions and oxalate ions are the same as the moles of calcium oxalate:

$$[Ca^{2+}]_{eq} = [C_2O_4^{2-}]_{eq} = \frac{n}{V} = \frac{4.76 \times 10^{-5}\ mol}{1.0\ L} = 4.76 \times 10^{-5}\ mol/L$$

$$K_{sp} = [Ca^{2+}]_{eq}[C_2O_4^{2-}]_{eq} = (4.76 \times 10^{-5})^2 = 2.3 \times 10^{-9}$$

18.37 This problem describes a solution containing a common ion. To determine concentrations of ions at equilibrium, follow the five-step procedure for working equilibrium problems:

1.) Species present initially are Pb^{2+}, NO_3^-, and $PbCl_2$

2.) Reaction is a solubility process: $PbCl_2\ (s) \rightleftharpoons Pb^{2+}\ (aq) + 2\ Cl^-\ (aq)$

3.) $K_{sp} = [Pb^{2+}]_{eq}[Cl^-]_{eq}^2 = 1.7 \times 10^{-5}$

4.) Set up and complete a concentration table, letting x be the increase in $[Pb^{2+}]$:

Reaction:	$PbCl_2$ (s) $\rightleftharpoons$	Pb^{2+} (aq) +	$2\,Cl^-$ (aq)
Initial concentration (M)	solid	0.650	0
Change in concentration (M)	solid	$+x$	$+2x$
Final concentration (M)	solid	$0.650 + x$	$2x$

5.) Substitute into the equilibrium constant expression and solve for x:

$1.7 \times 10^{-5} = (0.650 + x)(2x)^2$; Assume $x << 0.650$

$$4x^2 = \frac{1.7 \times 10^{-5}}{0.650}$$

$x^2 = 6.54 \times 10^{-6}$ $x = 2.56 \times 10^{-3}$

$2.56 \times 10^{-3} < 5\%$ of 0.650, so the approximation is valid.

The increase in lead ions is $\Delta\,[Pb^{2+}] = 2.56 \times 10^{-3}$ M

To convert to moles Pb^{2+} that dissolve, multiply by the solution volume:

$n = MV = (2.56 \times 10^{-3}$ M$)(0.750$ L$) = 1.92 \times 10^{-3}$ mol

The amount of $PbCl_2$ that dissolves is determined by the amount of increase in Pb^{2+}; convert to mass by multiplying by the molar mass of $PbCl_2$:

$$m = n\,MM = 1.92 \times 10^{-3} \text{ mol}\left(\frac{278.1 \text{ g}}{1 \text{ mol}}\right) = 0.53 \text{ g}$$

18.39   The chemical reactions are the following:

$$CaSO_3\,(s) \rightleftharpoons Ca^{2+}\,(aq) + SO_3^{2-}\,(aq) \qquad\qquad K_{sp}$$

$$\underline{SO_3^{2-}\,(aq) + H_3O^+\,(l) \rightleftharpoons HSO_3^-\,(aq) + H_2O\,(l) \qquad\qquad 1/K_{a2}}$$

$$CaSO_3\,(s) + H_3O^+\,(aq) \rightleftharpoons Ca^{2+}\,(aq) + HSO_3^-\,(aq) + H_2O\,(l)$$

$$K_{eq} = \frac{[Ca^{2+}][HSO_3^-]}{[H_3O^+]} = [Ca^{2+}][SO_3^{2-}]\frac{[HSO_3^-]}{[H_3O^+][SO_3^{2-}]} = \frac{K_{sp}}{K_{a_2}} = \frac{1.0 \times 10^{-4}}{6.3 \times 10^{-8}} = 1.6 \times 10^3$$

18.41   This problem describes a common ion effect solubility reaction. We are asked to determine the final concentration of calcium ions in solution. Begin by analyzing the chemistry. The starting materials are HCl (aq), a strong acid, and $CaSO_3$ (s), an insoluble salt. In addition to water, the major species in solution are H_3O^+, Cl^-, and H_2O.

Hydronium ions will react with calcium sulfite (see problem 18.39):

$$CaSO_3\,(s) + H_3O^+\,(aq) \rightleftharpoons Ca^{2+}\,(aq) + HSO_3^-\,(aq) + H_2O\,(l)$$

$$K_{eq} = \frac{[Ca^{2+}][HSO_3^-]}{[H_3O^+]} = \frac{K_{sp}}{K_{a2}} = 1.6 \times 10^3$$

Because the K_{eq} for the reaction is large, assume that the reaction goes to completion and then do the equilibrium calculations:

At completion, all of the hydronium has reacted to form Ca^{2+} and HSO_3^-:

$[Ca^{2+}] = [HSO_3^-] = 0.125$ M

Here is the concentration table for the return from completion to equilibrium:

Reaction	$CaSO_3$ (s)+	H_3O^+ (aq) $\rightleftarrows$	Ca^{2+} (aq) +	HSO_3^- (aq) +	H_2O
Start (M)	----	0	0.125	0.125	----
Change (M)	----	$+x$	$-x$	$-x$	----
Final (M)	----	x	0.125 - x	0.125 - x	----

Now write the equilibrium expression and solve for the concentration of Ca^{2+} ions:

$$K_{eq} = 1.6 \times 10^3 = \frac{(0.125 - x)^2}{x} \qquad \text{Assume that } x \ll 0.125$$

$$1.6 \times 10^3 = \frac{0.125^2}{x} \qquad x = 9.8 \times 10^{-6}; \text{ assumption is valid.}$$

$$[Ca^{2+}] = 0.125 - 9.8 \times 10^{-6} = 0.125 \text{ M}$$

18.43 Write the chemical formulas according to the procedures in your textbook. The coordination number is the number of bonds between ligands and the metal species. Both CN^- and NH_3 are monodentate (form 1 bond with the metal center), ethylenediamine is a bidentate ligand (forms two bonds to the metal center).
(a) $[Fe(CN)_6]^{3-}$; (b) $[Zn(NH_3)_4]^{2+}$; (c) $[V(en)_3]^{3+}$

18.45 This is a complexation equilibrium problem. The problem gives information about the amounts of both starting materials, so this is a limiting reactant situation. Because K for complexation generally is large, we do a concentration table that takes the reaction to completion and then brings it back to equilibrium. We must calculate the initial concentration of each species, construct a table of amounts, and use the results to determine the final solution concentrations.

The reaction is Zn^{2+} (aq) + 4 NH_3 (aq) $\rightleftarrows$ $[Zn(NH_3)_4]^{2+}$ (aq) $K_f = 4.1 \times 10^8$

Calculate the initial concentration of Zn^{2+}:

$$n_{Zn^{2+}} = \frac{m}{MM} = \frac{0.275 \text{ g}}{136.29 \text{ g/mol}} = 2.02 \times 10^{-3} \text{ mol}$$

$$M = \frac{n}{V} = \frac{2.02 \times 10^{-3} \text{ mol}}{(375 \text{ mL})(10^{-3} \text{ L/mL})} = 5.4 \times 10^{-3} \text{ mol/L}$$

Complete a concentration table after taking the reaction to completion:

Reaction:	Zn^{2+} (aq) +	4 NH_3 (aq) $\rightleftarrows$	$[Zn(NH_3)_4]^{2+}$ (aq)
Start (M)	5.4×10^{-3}	0.250	0
Change (M)	-5.4×10^{-3}	$-4(5.4 \times 10^{-3})$	$+ 5.4 \times 10^{-3}$
Completion (M)	0	0.228	5.4×10^{-3}
Change (M)	$+x$	$+4x$	$-x$
Equilibrium (M)	x	$0.228 + 4x$	$5.4 \times 10^{-3} - x$

$$K_f = 4.1 \times 10^8 = \frac{0.0054 - x}{x(0.228 + 4x)^4} \qquad \text{Assume that } x \ll 0.0054$$

$$x(0.228)^4(4.1 \times 10^8) = 0.0054$$

$$x = \frac{0.0054}{(0.228)^4 (4.1 \times 10^8)} = 4.9 \times 10^{-9}; \text{ assumption is valid.}$$

Here are the concentrations at equilibrium:

$[NH_3]_{eq} = 0.228 \text{ M}$ $[Zn^{2+}]_{eq} = 4.9 \times 10^{-9} \text{ M}$ $[[Zn(NH_3)_4]^{2+}]_{eq} = 5.4 \times 10^{-3} \text{ M}$

18.47 This problem describes a common ion effect solubility reaction. We are asked to determine the final ion concentrations in solution. Begin by analyzing the chemistry. The starting materials are NaCl, a salt, and Pb^{2+}. In addition to water, the major species in solution are Na^+, Cl^-, and Pb^{2+}.

The presence of lead(II) and chloride ions together as major species in solution will result in the reaction:

$$Pb^{2+} (aq) + 4Cl^- (aq) \rightleftharpoons PbCl_4^{2-} (aq) \qquad K_f = \frac{[PbCl_4^{2-}]}{[Pb^{2+}][Cl^-]^4} = 2.5 \times 10^{15}$$

Because K_f for the reaction is so large, assume that the reaction goes to completion and then do the equilibrium calculations. The problem gives information about the amounts of both starting materials, so this is a limiting reactant situation. We must calculate the number of moles of each species, construct a table of amounts, and use the results to determine the final solution concentrations.

Calculations of initial amounts:

$n_{Na^+} = n_{Cl^-} = 0.25 \text{ moles}$

$n_{Pb^{2+}} = (7.5 \times 10^{-3} \text{ M})(1.50 \text{ L}) = 0.0113 \text{ mol}$

Na^+ is a spectator ion:

$$[Na^+] = \frac{0.25 \text{ mol}}{1.50 \text{ L}} = 0.167 \text{ M}$$

Complete a concentration table that takes the reaction to completion:

Reaction:	$Pb^{2+} +$	$4 Cl^- \rightleftharpoons$	$PbCl_4^{2+}$
Start (mol)	0.0113	0.25	0
Change (mol)	-0.0113	-4(0.0113)	+ 0.0113
Final (mol)	0	0.205	0.0113

Compute the concentrations by dividing by the volume, 1.50 L:

$$[Cl^-] = \left(\frac{0.205 \text{ mol}}{1.50 \text{ L}} \right) = 0.137 \text{ M}$$

$$[PbCl_4^{2+}] = \left(\frac{0.0113 \text{ mol}}{1.50 \text{ L}} \right) = 0.0075 \text{ M}$$

Use these amounts to complete the concentration table at equilibrium:

Reaction:	$Pb^{2+} +$	$4 Cl^- \rightleftharpoons$	$PbCl_4^{2+}$
Start (M)	0	0.137	0.0075
Change (M)	+x	+4x	-x
Equilibrium (M)	x	0.137 + 4x	0.0075-x

$$K_f = 2.5 \times 10^{15} = \frac{0.0075 - x}{x(0.137 + 4x)^4} \qquad \text{Assume that } x \ll 0.0075$$

$$x = \frac{0.0075}{(0.137)^4(2.5 \times 10^{15})} = 8.52 \times 10^{-15}; \text{ assumption valid}$$

Here are the final concentrations:

$[Cl^-] = 0.14$ M $\qquad [Pb^{2+}] = 8.5 \times 10^{-15}$ M

$[Na^+] = 0.17$ M $\qquad [PbCl_4^{2-}] = 7.5 \times 10^{-3}$ M

18.49 $Mn^{2+} + en \rightleftharpoons [Mn(en)]^{2+}$

$[Mn(en)]^{2+} + en \rightleftharpoons [Mn(en)_2]^{2+}$

$[Mn(en)_2]^{2+} + en \rightleftharpoons [Mn(en)_3]^{2+}$

18.51 This problem describes a multiple equilibrium situation. We are asked to determine the amount of $CaSO_3$ that will dissolve in the solution. Begin by analyzing the chemistry. The starting materials are $CaSO_3$ (s), an insoluble salt, and nitrilotriacetate. In addition to water, the major species in solution is NTA^{3-}.

The solid salt has a solubility equilibrium:

$CaSO_3$ (s) $\rightleftharpoons Ca^{2+}$ (aq) $+ SO_3^{2-}$ (aq) $\qquad K_{sp} = 1.0 \times 10^{-4}$ (See Problem 18.39)

The resulting calcium ions can coordinate with the NTA^{3-} to form a complex:

Ca^{2+} (aq) $+ 2NTA^{3-}$ (aq) $\rightleftharpoons [Ca(NTA)_2]^{4-}$ (aq) $\qquad K_f = 3.2 \times 10^{11}$

Determine the overall chemical reaction and the equilibrium expression:

$CaSO_3$ (s) $\rightleftharpoons Ca^{2+}$ (aq) $+ SO_3^{2-}$ (aq)

$\underline{Ca^{2+} \text{ (aq)} + 2NTA^{3-} \text{ (aq)} \rightleftharpoons [Ca(NTA)_2]^{4-} \text{ (aq)}}$

$CaSO_3$ (s) $+ 2NTA^{3-}$ (aq) $\rightleftharpoons [Ca(NTA)_2]^{4-}$ (aq) $+ SO_3^{2-}$ (aq)

$$K_{eq} = \frac{\left[[Ca(NTA)_2]^{4-}\right][SO_3^{2-}]}{[NTA^{3-}]^2} = [Ca^{2+}][SO_3^{2-}]\frac{\left[[Ca(NTA)_2]^{4-}\right]}{[NTA^{3-}]^2[Ca^{2+}]} = K_{sp}K_f$$

$K_{eq} = (1.0 \times 10^{-4})(3.2 \times 10^{11}) = 3.2 \times 10^7$

Because the equilibrium constant is large, take the reaction to completion and then return to equilibrium. At completion, all the ligands have reacted to form the two products in 1>2 stoichiometry.

Here is the concentration table for the return from completion to equilibrium:

Reaction:	$CaSO_3$ (s)+	$2 NTA^{3-}$ $\rightleftharpoons$	$[Ca(NTA)_2]^{4-} +$	SO_3^{2-} (aq)
Start (M)	----	0	0.125	0.125
Change (M)	----	$+2x$	$-x$	$-x$
Final (M)	----	$2x$	$0.125 - x$	$0.125 - x$

$$K_{eq} = 3.2 \times 10^7 = \frac{(0.125 - x)^2}{(2x)^2}$$

Take the square root of each side: $5.65 \times 10^3 = \dfrac{0.125 - x}{2x}$

$2x(5.65 \times 10^3) = 0.125 - x \qquad x = 1.1 \times 10^{-5}$

$[SO_3^{2-}] = 0.125 - 1.1 \times 10^{-5} = 0.125$ M

The amount of sulfite ions formed will equal the amount of solid dissolved:

$$n_{CaSO_3} = n_{(SO_3)^{2-}} = 0.125 \text{ M } (0.50 \text{ L}) = 6.25 \times 10^{-2} \text{ moles}$$

$$MM_{CaSO_3} = 40.08 \text{ g/mol} + 32.07 \text{ g/mol} + 3(16.00 \text{ g/mol}) = 120.1 \text{ g/mol}$$

$$m_{CaSO_3} = 6.25 \times 10^{-2} \text{ moles} \left(\frac{120.1 \text{ g}}{1 \text{ mol}} \right) = 7.5 \text{ g}$$

18.53   A molecular picture of a buffer solution should show molecules/ions of the conjugate acid/base pair in the correct proportions, as determined by the pH of the buffer solution. For formic acid, $pK_a = 3.75$. Use the buffer equation to calculate the base/acid ratio of a formic acid buffer with pH = 4.04:

$$\log \left\{ \frac{(\text{mol } A^-)}{(\text{mol } HA)} \right\} = pH - pK_a = 4.04 - 3.75 = 0.29 \qquad \left\{ \frac{(\text{mol } A^-)}{(\text{mol } HA)} \right\} = 2$$

The picture should show twice as many formate ions as formic acid molecules. The solution is acidic, so there should also be some hydronium ions, but the exact proportion depends on the total concentration of the buffer species.

Here is a view showing 6 formate ions, 3 formic acid molecules, and 1 hydronium ion:

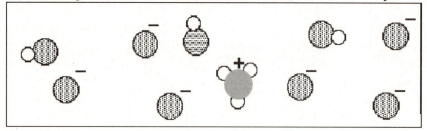

18.55   To determine if all the solid dissolves, calculate the value that Q would have if all the solid dissolves, and compare its value with the K_{sp} value. If $Q < K_{sp}$, the spontaneous direction of reaction is to the right even after all of the solid dissolves, so all of the solid dissolves, while if $Q > K_{sp}$, not all of the solid dissolves.

The reaction is $PbCl_2 (s) \rightleftharpoons Pb^{2+} (aq) + 2 \text{ Cl}^- (aq)$

$K_{sp} = 2 \times 10^{-5}$ and $Q = [Pb^{2+}][Cl^-]^2$

Use mole-mass conversions to determine concentrations:

$$n_{Pb^{2+}} = \frac{m}{MM} = 0.50 \text{ g} \left(\frac{1 \text{ mol}}{278.1 \text{ g}} \right) = 1.80 \times 10^{-3} \text{ mol}$$

$$[Pb^{2+}] = \frac{1.80 \times 10^{-3} \text{ mol}}{0.300 \text{ L}} = 6.0 \times 10^{-3} \text{ M}$$

$[Cl^-] = 2[Pb^{2+}] = 1.2 \times 10^{-2} \text{ M}$

$Q = (6.0 \times 10^{-3})(1.2 \times 10^{-2})^2 = 8.6 \times 10^{-7}$; $Q < K_{sp}$, so all of the solid dissolves.

18.57 This problem describes the $H_2PO_4^-/HPO_4^{2-}$ buffer solution.

(a) Use the buffer equation to calculate the pH of a buffer solution:

The conjugate acid/base pair is $H_2PO_4^-/HPO_4^{2-}$, $pK_a = pK_{a2} = 7.21$

$$pH = pK_a + \log\left\{\frac{[A^-]}{[HA]}\right\} = 7.21 + \log\left\{\frac{(0.20\ M)}{(0.50\ M)}\right\} = 7.21 - 0.40 = 6.81$$

(b) Calculate the change in molarity due to the added base, then use the buffer equation to calculate the new pH. The added hydroxide reacts completely with the weak acid and generates the conjugate weak base:

$$n_{NaOH} = \frac{m}{MM} = 0.120\ g\left(\frac{1\ mol}{40.0\ g}\right) = 3.00 \times 10^{-3}\ mol$$

$$\Delta M = \frac{n_{NaOH}}{V} = \frac{3.00 \times 10^{-3}\ mol}{0.15\ L} = 0.020\ M$$

$[HPO_4^{2-}] = 0.20 + 0.020 = 0.22\ M$

$[H_2PO_4^-] = 0.50 - 0.020 = 0.48\ M$

$$pH = pK_a + \log\left\{\frac{[A^-]}{[HA]}\right\} = 7.21 + \log\left\{\frac{(0.22\ M)}{(0.48\ M)}\right\} = 7.21 - 0.34 = 6.87$$

$\Delta pH = 6.87 - 6.81 = 0.06$

(c) Adding acid reduces the concentration of weak base, increases the concentration of weak acid, and reduces the pH. Use the buffer equation to calculate the conjugate base/acid ratio at the limit of the pH range. Then calculate the change in molarity that this ratio represents, and convert to moles:

$$pH_{min} = 6.81 - 0.10 = 6.71 = 7.21 + \log\left\{\frac{[A^-]}{[HA]}\right\}$$

$$\log\left\{\frac{[A^-]}{[HA]}\right\} = 6.71 - 7.21 = -0.50$$

$$\left\{\frac{[A^-]}{[HA]}\right\} = 0.32 = \left\{\frac{(0.20 - \Delta M)}{(0.50 + \Delta M)}\right\}$$

$(0.20 - \Delta M) = 0.32(0.50 + \Delta M) = 0.16 + 0.32\Delta M$, so $0.04 = 1.32\Delta M$

$\Delta M = 3.0 \times 10^{-2}\ M$

moles of acid neutralized $= \Delta MV = (3.0 \times 10^{-2}\ M)(0.25\ L) = 7.5 \times 10^{-3}\ mol$

18.59 The bromocresol purple indicator is purple when pH < 8.5 and yellow when pH > 8.5:

(a) HCl is a strong acid (pH << 5) therefore indicator color is yellow.

(b) NaOH is a strong base (pH >> 7) therefore indicator color is purple.

(c) KCl is a neutral ionic solution (pH =7) therefore indicator color is purple.

(d) NH_3 is a weak base (pH >7) therefore indicator color is purple.

18.61 In a titration of a weak acid by a strong base, pH = pK_a at the midpoint of the titration. Thus, pK_a = 2.36 for leucine and K_a = 4.4 x 10^{-3}.

18.63 To determine if a precipitate forms, evaluate Q for the solution after mixing and compare its value for the value for K_{eq}.   If $Q < K_{eq}$, the spontaneous direction of reaction is to the right after the solutions are mixed and a precipitate forms, while if $Q > K_{eq}$ no precipitate forms.

The reaction is: $Ca^{2+} (aq) + SO_4^{2-} (aq) \rightleftharpoons CaSO_4 (s)$

$$K_{eq} = \frac{1}{K_{sp}} = \frac{1}{4.9 \times 10^{-5}} = 2.0 \times 10^4 \text{ and } Q = \frac{1}{[Ca^{2+}][SO_4^{2-}]}$$

The total volume of the solution is 350 mL + 150 mL = 500 mL

$$[Ca^{2+}] = 2.00 \times 10^{-2} M \left(\frac{350 \text{ mL}}{500 \text{ mL}}\right) = 0.0140 \text{ M}$$

$$[SO_4^{2-}] = 1.50 \times 10^{-2} M \left(\frac{150 \text{ mL}}{500 \text{ mL}}\right) = 0.00450 \text{ M}$$

$$Q = \frac{1}{(0.0140)(0.00450)} = 1.59 \times 10^4; \ Q < K_{eq}, \text{ so the precipitate forms.}$$

18.65 To calculate an equilibrium constant when experimental data concerning concentrations are available, identify the reaction, then complete an amounts table and substitute into the equilibrium constant expression.

To complete the amounts table, use stoichiometric reasoning and the fact that the change is 2.0 x 10^{-3} M for Pb^{2+}:

Reaction:	$PbF_2 (s) \rightleftharpoons$	$Pb^{2+} (aq) +$	$2 F^- (aq)$
Start (M)	solid	0	0
Change (M)	solid	2.0 x 10^{-3}	4.0 x 10^{-3}
Final (M)	solid	2.0 x 10^{-3}	4.0 x 10^{-3}

Now substitute into the equilibrium constant expression and evaluate K_{sp}:

$$K_{sp} = [Pb^{2+}]_{eq}[F^-]_{eq}^2 = (2.0 \times 10^{-3})(4.0 \times 10^{-3})^2 = 3.2 \times 10^{-8}$$

18.67 The buffer equation is used for calculations involving buffer solutions. Concentrations of acid and conjugate base are needed. The weak acid in the ammonia buffer system is the ammonium ion, pK_a = 9.25;

(a) [NH$_3$] = 1.00 M

$$n_{NH_4^+} = 35.0 \text{ g}\left(\frac{1 \text{ mol}}{53.49 \text{ g}}\right) = 0.654 \text{ mol} \qquad [NH_4^+] = \frac{n}{V} = \frac{0.654 \text{ mol}}{1.00 \text{ L}} = 0.654 \text{ M}$$

$$pH = pK_a + \log\left\{\frac{[A^-]}{[HA]}\right\} = 9.25 + \log\left\{\frac{(1.00 \text{ M})}{(0.654 \text{ M})}\right\} = 9.25 + 0.18 = 9.43$$

(b) First calculate the ratio of base to acid in the new solution, then use the new ratio to determine the amount of change from the original ratio.

The new pH is $9.43 - 0.05 = 9.38$

$$\log\left\{\frac{[A^-]}{[HA]}\right\} = pH - pK_a = 9.38 - 9.25 = 0.13 \qquad \left\{\frac{[A^-]}{[HA]}\right\} = 10^{0.13} = 1.35$$

Thus, $[NH_3] = 1.35[NH_4^+]$; from before, the total concentration is

$[NH_4^+] + [NH_3] = 1.654$ M

$[NH_4^+] + 1.35[NH_4^+] = 1.654$ M, so $2.35[NH_4^+] = 1.654$ M and $[NH_4^+] = 0.704$ M

The moles of acid required for this change are

$\Delta n = \Delta M \, V = (0.704 \text{ M} - 0.654 \text{ M})(1.00 \text{ L}) = 0.050$ mol

(c) It is easiest to work with moles in this calculation. 250 mL of buffer solution contains:

$$250 \text{ mL}\left(\frac{10^{-3}\text{L}}{1 \text{ mL}}\right)\left(\frac{1.00 \text{ mol}}{1 \text{ L}}\right) = 0.250 \text{ mol NH}_3$$

$$250 \text{ mL}\left(\frac{10^{-3}\text{L}}{1 \text{ mL}}\right)\left(\frac{0.654 \text{ mol}}{1 \text{ L}}\right) = 0.163 \text{ mol NH}_4^+$$

The amount of added acid is $5.0 \text{ mL}\left(\frac{10^{-3}\text{L}}{1 \text{ mL}}\right)\left(\frac{12.0 \text{ mol}}{1 \text{ L}}\right) = 0.060$ mol

The new amounts are:

$(0.163 + 0.060) = 0.223$ mol NH_4^+

$(0.250 - 0.060) = 0.190$ mol NH_3

$$pH = pK_a + \log\left\{\frac{(\text{mol NH}_3)}{(\text{mol NH}_4^+)}\right\} = 9.25 + \log\left\{\frac{(0.223 \text{ mol})}{(0.190 \text{ mol})}\right\} = 9.25 + 0.07 = 9.32$$

18.69 To calculate an equilibrium constant when experimental data concerning concentrations are available, identify the reaction, then complete a concentration table and substitute into the equilibrium constant expression.

The reaction is $Ca_3(PO_4)_2 \, (s) \rightleftharpoons 3 \, Ca^{2+} \, (aq) + 2 \, PO_4^{3-} \, (aq)$

The solubility in mol/L is $\left(\frac{3.5 \times 10^{-5}\text{g}}{1 \text{ L}}\right)\left(\frac{1 \text{ mol}}{310.174 \text{ g}}\right) = 1.13 \times 10^{-7}$ M

Reaction:	$Ca_3(PO_4)_2 \rightleftharpoons$	$3 \, Ca^{2+}$ +	$2 \, PO_4^{3-}$
Initial (M)	solid	0	0
Change (M)	solid	$3(1.13 \times 10^{-7})$	$2(1.13 \times 10^{-7})$
Equilibrium (M)	solid	3.39×10^{-7}	2.26×10^{-7}

Now substitute into the equilibrium constant expression and evaluate K_{sp}:

$$K_{sp} = [Ca^{2+}]_{eq}^3[PO_4^{3-}]_{eq}^2 = (3.39 \times 10^{-7})^3(2.26 \times 10^{-7})^2 = 2.0 \times 10^{-33}$$

18.71 The major species present are different at various points during a titration, so there are different dominant equilibria which must be identified before doing an equilibrium calculation to determine pH.

(a) Before titration begins, the major species are a weak base, formate (HCO_2^-) and H_2O, and the dominant equilibrium is:

$$HCO_2^- (aq) + H_2O (l) \rightleftharpoons HCO_2H (aq) + OH^- (aq)$$

$$K_b = \frac{K_w}{K_a} = \frac{1.0 \times 10^{-14}}{1.8 \times 10^{-4}} = 5.6 \times 10^{-11}$$

Reaction: H_2O +	$HCO_2^- \rightleftharpoons$	HCO_2H +	OH^-
Initial concentration (M)	0.200	0	0
Change in concentration (M)	$-x$	$+x$	$+x$
Final concentration (M)	0.200 - x	x	x

Now substitute into the equilibrium constant expression and solve for x:

$$K_b = 5.6 \times 10^{-11} = \frac{[HCO_2H]_{eq}[OH^-]_{eq}}{[HCO_2^-]_{eq}} = \frac{x^2}{(0.200-x)} \quad \text{Assume that } x \ll 0.200$$

$x^2 = 1.1 \times 10^{-11}$ $x = 3.3 \times 10^{-6}$ $[OH^-] = 3.3 \times 10^{-6}$ M

pOH = -log(3.3×10^{-6}) = 5.48 and pH = 14.00 – 5.48 = 8.52

(b) At the midpoint of the titration, the buffer equation applies and the concentrations of acid and conjugate base are equal, so pH = pK_a = 3.74;

(c) At the stoichiometric point, the major acid-base species present are HCO_2H and H_2O and the dominant equilibrium is:

$$HCO_2H (aq) + H_2O (l) \rightleftharpoons HCO_2^- (aq) + H_3O^+ (aq) \quad K_a = 1.8 \times 10^{-4}$$

Correct the initial concentration of HCO_2H for the volume added during titration:

$$[HCO_2H] = \frac{0.0600 \text{ mol } HCO_2H}{0.300 \text{ L} + 0.010 \text{ L}} = 0.193 \text{ M}$$

Here is the concentration table:

Reaction: H_2O +	$HCO_2H \rightleftharpoons$	HCO_2^- +	H_3O^+
Initial concentration (M)	0.193	0	0
Change in concentration (M)	$-x$	$+x$	$+x$
Final concentration (M)	0.193 - x	x	x

Now substitute into the equilibrium constant expression and solve for x:

$$K_a = 1.8 \times 10^{-4} = \frac{[HCO_2^-]_{eq}[H_3O^+]_{eq}}{[HCO_2H]_{eq}} = \frac{x^2}{(0.193-x)}$$

$x^2 = (0.193 - x)(1.8 \times 10^{-4}) = (3.5 \times 10^{-5}) - (1.8 \times 10^{-4})x$

$x^2 + (1.8 \times 10^{-4})x - (3.5 \times 10^{-5}) = 0$

$$x = \frac{-b \pm \sqrt{b^2 - 4ac}}{2a} = \frac{-(1.8 \times 10^{-4}) \pm \sqrt{(1.8 \times 10^{-4})^2 + 4(3.5 \times 10^{-5})}}{2} = 5.8 \times 10^{-3}$$

$[H_3O^+]_{eq} = 5.8 \times 10^{-3}$ M, and pH = 2.23

(d) A suitable indicator for this titration must change color around pH = 2. Thymol blue, pK_{In} = 1.75, would be the best choice.

18.73   When a precipitate forms upon mixing aqueous solutions, the equilibrium reaction is the reverse of the solubility reaction. Follow the standard procedure for a reaction with a large equilibrium constant.

Reaction is $Mn^{2+} (aq) + CO_3^{2-} (aq) \rightleftharpoons MnCO_3 (s)$

"Initial" concentrations are after mixing but before reaction:

$$[Mn^{2+}] = \left(\frac{0.750 \text{ L}}{0.750 + 0.150 \text{ L}} \right)(0.0250 \text{ M}) = 0.0208 \text{ M}$$

$$[CO_3^{2-}] = \left(\frac{0.150 \text{ L}}{0.750 + 0.150 \text{ L}} \right)(0.500 \text{ M}) = 0.0833 \text{ M}$$

Allow the reaction to go to completion:

$[Mn^{2+}] = 0.000$ M $[CO_3^{2-}] = 0.0625$ M

These are the initial concentrations to enter in the concentration table:

Reaction:	$MnCO_3$ (s) $\rightleftharpoons$	CO_3^{2-} +	Mn^{2+}
Initial (M)	solid	0.0625	0
Change (M)	solid	+x	+x
Equilibrium (M)	solid	0.0625 + x	x

Assume that $x \ll 0.0625$:

$$K_{sp} = 2.2 \times 10^{-11} = (x)(0.0625) \qquad x = \frac{2.2 \times 10^{-11}}{0.0625} = 3.5 \times 10^{-10}$$

Thus, $[Mn^{2+}] = 3.5 \times 10^{-10}$ M

To calculate the mass of precipitate, use standard stoichiometric methods. The reaction goes virtually to completion, so:

$$n_{MnCO_3} = n_{Mn^{2+}} = MV = (0.0250 \text{ mol/L})(0.750 \text{ L}) = 0.01875 \text{ mol}$$

$$m = n \, MM = 0.01875 \text{ mol} \left(\frac{114.95 \text{ g}}{1 \text{ mol}} \right) = 2.16 \text{ g}$$

18.75   "What mass will dissolve" means "calculate the amount present in solution at equilibrium." Initial amounts are zero, so it is easy to complete a concentration table. Calculate molarity from the solubility product expression, then convert to mass using standard stoichiometric methods. Let x be the number of moles of salt dissolving in 1 L of solution.

Reaction:	$Zn(OH)_2$ (s) $\rightleftharpoons$	Zn^{2+} +	2 OH^-
Initial (M)	solid	0	0
Change (M)	solid	+x	+2x
Equilibrium (M)	solid	x	2x

$K_{sp} = 3.0 \times 10^{-17} = [Zn^{2+}]_{eq}[OH^-]_{eq}^2 = 4 \, x^3$

$x = 2.0 \times 10^{-6}$

$$m = M \, V \, MM = (2.0 \times 10^{-6} \text{ M})(1.00 \text{ L}) \left(\frac{99.40 \text{ g}}{1 \text{ mol}} \right) = 2.0 \times 10^{-4} \text{ g dissolves}$$

18.77 The best choice for a buffer solution generally is the weak acid whose pK_a is closest to the desired pH of the buffer solution. For a pH = 4.80 buffer solution, acetic acid/acetate, $pK_a = 4.75$, would be the best choice. To prepare the buffer solution, add the appropriate amounts of sodium acetate (NaAc) and acetic acid (HAc) solution to water and make up to 1.0 L with additional water. Calculate the acetate/acetic acid ratio using the buffer equation:

$$\log\left\{\frac{[Ac^-]}{[HAc]}\right\} = pH - pK_a = 4.80 - 4.75 = 0.05 \qquad\qquad \left\{\frac{[Ac^-]}{[HAc]}\right\} = 10^{0.05} = 1.12$$

Thus, $[Ac^-] = 1.12\,[HAc]$

$[Ac^-] + [HAc] = 0.35\,M$ $[HAc] = 0.35\,M - [Ac^-]$

$[Ac^-] = 1.12(0.35\,M - [Ac^-]) = 0.39 - 1.12\,[Ac^-]$

$2.12\,[Ac^-] = 0.39$ and $[Ac^-] = 0.18\,M$ $n_{Ac^-} = (0.18\,M)(1.0\,L) = 0.18\,mol$

$m_{NaAc} = n\,MM = (0.18\,mol)(82.0\,g/mol) = 15\,g$

$[HAc] = 0.35 - 0.18 = 0.17\,M$

$n_{HAc} = (0.17\,M)(1.0\,L) = 0.17\,mol$

$$V_{HAc} = \frac{n}{M} = 0.17\ mol\left(\frac{1\,L}{1.00\ mol}\right) = 0.17\,L$$

Mix together 15 g sodium acetate, 0.17 L 1.0 M acetic acid, and enough water to make 1.0 L.

18.79 This is a complexation equilibrium problem. Because K_{eq} for complexation generally is large, we do a concentration table that takes the reaction to completion and then brings it back to equilibrium. The problem gives information about the amounts of both starting materials, so this is a limiting reactant situation. We must calculate the initial concentration of each species, construct a table of amounts, and use the results to determine the final solution concentrations.

$$Zn^{2+}\,(aq) + 3\,C_2O_4{}^{2-}\,(aq) \rightleftharpoons [Zn(C_2O_4)_3]^{4-}\,(aq) \qquad K_f = 1.4 \times 10^8$$

Calculate the initial concentration of metal cations:

$$n_{Zn^{2+}} = \frac{m}{MM} = \frac{0.275\,g}{136.29\,g/mol} = 2.02 \times 10^{-3}\,mol$$

$$M = \frac{n}{V} = \frac{2.02 \times 10^{-3}\,mol}{(0.450\,L)} = 4.49 \times 10^{-3}\,mol/L$$

Complete a concentration table after taking the reaction to completion:

Reaction:	$Zn^{2+}\,(aq)\,+$	$3\,C_2O_4{}^{2-}\,(aq) \rightleftharpoons$	$[Zn(C_2O_4)_3]^{4-}\,(aq)$
Start (M)	4.49×10^{-3}	0.250	0
Change (M)	-4.49×10^{-3}	$-3(4.49 \times 10^{-3})$	$+4.49 \times 10^{-3}$
Completion (M)	0	0.237	4.49×10^{-3}
Change (M)	$+x$	$+3x$	$-x$
Equilibrium (M)	x	$0.237 + 3x$	$4.49 \times 10^{-3} - x$

$$K_f = 1.4 \times 10^8 = \frac{0.00449 - x}{x(0.237 + 3x)^3}$$ Assume that $x \ll 0.00449$

$$x(0.237)^3(1.4 \times 10^8) = 0.00449$$

$$x = \frac{0.00449}{(0.237)^3(1.4 \times 10^8)} = 2.4 \times 10^{-9}; \text{ assumption is valid.}$$

Here are the final concentrations:

$[C_2O_4{}^{2-}] = 0.237$ M $[Zn^{2+}] = 2.4 \times 10^{-9}$ M $[[Zn(C_2O_4)_3]^{4-}] = 4.5 \times 10^{-3}$ M

18.81 "Saturated aqueous solution" identifies this as a solubility equilibrium.

(a) Reaction is MgF_2 (s) $\rightleftharpoons$ Mg^{2+} (aq) + 2 F⁻ (aq), $K_{sp} = [Mg^{2+}]_{eq}[F^-]_{eq}^2$

(b) If $[Mg^{2+}]_{eq} = 1.14 \times 10^{-3}$ M, by stoichiometry $[F^-]_{eq} = 2(1.14 \times 10^{-3}$ M)

$K_{sp} = (1.14 \times 10^{-3})[2(1.14 \times 10^{-3})]^2 = 5.93 \times 10^{-9}$;

(c) To estimate the equilibrium constant at a temperature different from 298 K, calculate $\Delta H^o{}_{reaction}$ and $\Delta S^o{}_{reaction}$ at 298 K and then use Equations 13-9 and 15-3:

$\Delta G^o{}_{reaction} = \Delta H^o{}_{reaction} - T\Delta S^o{}_{reaction}$ $\Delta G^o = -RT \ln K_{eq}$

$\Delta H^o{}_{reaction} = 1 \text{ mol}(-467.0 \text{ kJ/mol}) + 2 \text{ mol}(-335.4 \text{ kJ/mol})$

$- 1 \text{ mol}(-1124.2 \text{ kJ/mol}) = -13.6 \text{ kJ}$

$\Delta S^o{}_{reaction} = 1 \text{ mol}(-137 \text{ J/molK}) + 2 \text{ mol}(-13.8 \text{ J/molK})$

$- 1 \text{ mol}(57.2 \text{ J/molK}) = -222 \text{ J/K}$

$$\Delta G^o{}_{reaction, 373K} = (-13.6 \text{ kJ/mol}) - (373 \text{ K})(-222 \text{ J/mol K})\left(\frac{10^{-3} \text{ kJ}}{1 \text{ J}}\right)$$

$$= 69.2 \text{ kJ/mol}$$

$$\ln K_{sp} = -\frac{6.92 \times 10^4 \text{J mol}^{-1}}{(8.314 \text{ J mol}^{-1} \text{ K}^{-1})(373 \text{ K})} = -22.3$$ $K_{sp} = e^{-22.3} = 2.1 \times 10^{-10}$

18.83 To calculate equilibrium concentrations, set up a concentration table:

Reaction:	HgS (s) $\rightleftharpoons$	Hg²⁺ (aq) +	S²⁻ (aq)
Initial (M)	solid	0	0
Change (M)	solid	+x	+x
Equilibrium (M)	solid	x	x

$4.0 \times 10^{-53} = [Hg^{2+}]_{eq}[S^{2-}]_{eq} = x^2$, from which $x = 6.3 \times 10^{-27}$

$[Hg^{2+}]_{eq} = 6.3 \times 10^{-27}$ M (a very small concentration)

To calculate the volume that would be expected to contain a single Hg^{2+} cation:

$$V = \frac{n}{M}, \text{ with } n = \frac{1 \text{ cation}}{N_A}$$

$$V/\text{cation} = \frac{1}{N_A M} = \left(\frac{1 \text{ mol}}{6.022 \times 10^{23} \text{ cations}}\right)\left(\frac{1 \text{ L}}{6.3 \times 10^{-27} \text{ mol}}\right) = 2.6 \times 10^2 \text{ L/cation}$$

18.85 (a) To determine the ion concentrations in a solution, follow the five step procedure for solving an equilibrium problem:

1.) The species in solution are Ca^{2+}, Cl^-, PO_4^{3-}, and H_2O.

2. & 3.) The reaction is formation of $Ca_3(PO_4)_2$ precipitate:

$$3\ Ca^{2+}\ (aq) + 2\ PO_4^{3-}\ (aq) \rightleftharpoons Ca_3(PO_4)_2\ (s)\quad K_{eq} = \frac{1}{K_{sp}} = \frac{1}{[Ca^{2+}]_{eq}^3[PO_4^{3-}]_{eq}^2}$$

4.) Initial concentrations are those present after adding solid but before reaction occurs:

$$[Ca^{2+}] = \frac{(120\ mol)}{(3.00 \times 10^3 L)} = 4.0 \times 10^{-2}\ M \qquad [PO_4^{3-}] = 2.2 \times 10^{-3}\ M$$

Take the reaction to completion and then return to equilibrium. At completion:

$[Ca^{2+}] = 0.037\ M \qquad [PO_4^{3-}] = 0.000\ M$

Reaction:	$Ca_3(PO_4)_2\ (s) \rightleftharpoons$	$3\ Ca^{2+}\ (aq) +$	$2\ PO_4^{3-}\ (aq)$
Initial (M)	solid	0.037	0
Change (M)	solid	$+3x$	$+2x$
Equilibrium (M)	solid	$0.037 + 3x$	$2x$

5.) Assume that $3x \ll 3.7 \times 10^{-2}$, substitute into the solubility product expression, and solve:

$$K_{sp} = 2.0 \times 10^{-33} = [Ca^{2+}]_{eq}^3[PO_4^{3-}]_{eq}^2 = (3.7 \times 10^{-2})^3(2x)^2$$

$$x^2 = \frac{(2.0 \times 10^{-33})}{(3.7 \times 10^{-2})^3(4)} = 9.9 \times 10^{-30} \qquad x = 3.1 \times 10^{-15}$$

$$[PO_4^{3-}]_{eq} = 2x = 6.2 \times 10^{-15}\ M$$

(b) Because this reaction goes essentially to completion, the calculation of the mass of precipitate can be done using standard stoichiometric methods:

For phosphate ions, $n = 3.00 \times 10^3\ L \left(\frac{2.2 \times 10^{-3} mol}{1\ L} \right) = 6.6\ mol$

$$n_{Ca_3(PO_4)_2} = \frac{1}{2}n_{PO_4^{3-}} = 3.3\ mol$$

$$m = n\ MM = 3.3\ mol \left(\frac{310.18\ g}{1\ mol} \right) = 1.0 \times 10^3\ g$$

18.87 To calculate an equilibrium constant when experimental data concerning concentrations are available, identify the reaction, then complete an amounts table and substitute into the equilibrium constant expression.

The reaction is $Sr(IO_3)_2\ (s) \rightleftharpoons Sr^{2+}\ (aq) + 2\ IO_3^-\ (aq)$

At 25 °C, The solubility in mol/L is $\dfrac{0.030\ g}{(0.100\ L)(437.42\ g/mol)} = 6.86 \times 10^{-4}\ M$

Use this value to complete a concentration table at 25 °C:

Reaction:	$Sr(IO_3)_2$ (s) ⇌	Sr^{2+} (aq)	$2\ IO_3^-$ (aq)
Initial (M)	solid	0	0
Change (M)	solid	$+6.86 \times 10^{-4}$	$+2(6.86 \times 10^{-4})$
Equilibrium (M)	solid	6.86×10^{-4}	1.37×10^{-3}

Now substitute into the equilibrium constant expression and evaluate K_{sp}:

$$K_{sp} = [Sr^{2+}]_{eq}[IO_3^-]_{eq}^2 = (6.86 \times 10^{-4})(1.37 \times 10^{-3})^2 = 1.29 \times 10^{-9}$$

$$\Delta G^o = -RT \ln(K_{eq}) = -(8.314 \times 10^{-3}\ kJ/mol\ K)(298\ K) \ln(1.29 \times 10^{-9}) = 50.7\ kJ/mol$$

At 100.0 °C, the solubility in mol/L is $\dfrac{0.80\ g}{(0.100\ L)(437.42\ g/mol)} = 1.83 \times 10^{-2}\ M$

Use this value to complete a concentration table at 100 °C:

Reaction:	$Sr(IO_3)_2$ (s) ⇌	Sr^{2+} (aq)	$2\ IO_3^-$ (aq)
Initial (M)	solid	0	0
Change (M)	solid	$+1.83 \times 10^{-2}$	$+2(1.83 \times 10^{-2})$
Equilibrium (M)	solid	1.83×10^{-2}	3.66×10^{-2}

Now substitute into the equilibrium constant expression and evaluate K_{sp}:

$$K_{sp} = [Sr^{2+}]_{eq}[IO_3^-]_{eq}^2 = (1.83 \times 10^{-2})(3.66 \times 10^{-2})^2 = 2.5 \times 10^{-5}$$

$$\Delta G^o = -RT \ln(K_{eq}) = -(8.314 \times 10^{-3}\ J/mol\ K)(373\ K) \ln(2.45 \times 10^{-5}) = 33\ kJ/mol$$

18.89   To determine equilibrium concentrations, set up the appropriate concentration table. In this problem, one equilibrium concentration is given and the other two are stoichiometrically related, so the table can be relatively simple (note that the volume of the atmosphere is so large relative to the volume of ground water that the change in CO_2 pressure is negligible):

Reaction: $CaCO_3$ (s) + H_2O (l)	+ CO_2 (g) ⇌	Ca^{2+} (aq) +	$2\ HCO_3^-$ (aq)
Initial concentration	3.2×10^{-4} atm	0 M	0 M
Change in concentration	~ 0	+x M	+2x M
Concentration at equilibrium	3.2×10^{-4} atm	x M	2x M

Substitute equilibrium values into the equilibrium constant expression and solve for x:

$$K_{eq} = 1.56 \times 10^{-8} = \frac{[Ca^{2+}]_{eq}[HCO_3^-]_{eq}^2}{(p_{CO_2})_{eq}} = \frac{(x)(2x)^2}{(3.2 \times 10^{-4})}$$

$(1.56 \times 10^{-8})(3.2 \times 10^{-4}) = 4\,x^3$, so $x^3 = 1.25 \times 10^{-12}$ and $x = [Ca^{2+}] = 1.1 \times 10^{-4}\ M$.

18.91   Salts that are more soluble in acidic solution are those whose anions have weak conjugate acids. These are Ag_2CO_3, Ag_2SO_4, and Ag_2S. Those that are independent of pH have anions that are conjugate bases of strong acids, AgBr and AgCl.

18.93   Use the buffer equation to carry out calculations on buffer solutions:

(a) $pH = pK_a + \log\left\{\dfrac{[TRIS]}{[TRISH^+]}\right\} = (14.00 - 5.91) + \log\left[\dfrac{(0.30\ M)}{(0.60\ M)}\right] = 7.79$

(b) Adding HCl provides H_3O^+ ions, which react quantitatively with TRIS to generate TRISH$^+$.

The amount of added acid is $n = 5.0 \text{ mL} \left(\dfrac{10^{-3} \text{ L}}{1 \text{ mL}} \right) \left(\dfrac{12 \text{ mol}}{1 \text{ L}} \right) = 0.060 \text{ mol.}$

Using moles in the buffer equation saves us from calculating the dilution effect of adding 5.0 mL of the acid. If the initial solution is 1.0 L of buffer, then the amounts in moles are the same as the molarity of each species:

$$pH = (14.00 - 5.91) + \log \left[\frac{(0.30 - 0.06) \text{ moles}}{(0.60 + 0.06) \text{ moles}} \right] = 8.09 + \log \left[\frac{0.24 \text{ mol}}{0.66 \text{ mol}} \right] = 7.65$$

18.95 At different points during a titration, different major species are present, because the titration reaction consumes formic acid and produces formate:

$$HCO_2H \, (aq) + OH^- \, (aq) \rightleftharpoons HCO_2^- \, (aq) + H_2O \, (l)$$

Point A is before titration begins, when the major species are H_2O and HCO_2H.

Point B is near the midpoint of the titration, when the major species are H_2O, HCO_2H, and HCO_2^-; The dominant equilibrium for both points A and B is the acid reaction:

$$H_2O \, (l) + HCO_2H \, (aq) \rightleftharpoons HCO_2^- \, (aq) + H_3O^+ \, (aq) \qquad K_a = \frac{[HCO_2^-][H_3O^+]}{[HCO_2H]}$$

Point C is the stoichiometric point where all formic acid has been consumed, so the major species are H_2O and HCO_2^-. The dominant equilibrium for point C is:

$$H_2O \, (l) + HCO_2^- \, (aq) \rightleftharpoons HCO_2H \, (aq) + OH^- \, (aq) \qquad K_b = \frac{[HCO_2H][OH^-]}{[HCO_2^-]}$$

Point D is beyond the stoichiometric point, so excess hydroxide is present and the major species are H_2O, OH^-, and HCO_2^-. The dominant equilibrium for point D is:

$$H_2O \, (l) + HCO_2^- \, (aq) \rightleftharpoons HCO_2H \, (aq) + OH^- \, (aq) \qquad K_b = \frac{[HCO_2H][OH^-]}{[HCO_2^-]}$$

The cation of the strong base will also be present as a major species at points B, C, and D.

18.97 Molecular pictures show the correct relative numbers of the various species in the solution. The starting condition shows 9 molecules of acetic acid and 3 of acetate anions.
(a) Use the buffer equation to calculate the pH of this solution:

$$pH = pK_a + \log \left\{ \frac{[A^-]}{[HA]} \right\} = 4.75 + \log \left[\frac{3}{9} \right] = 4.75 - 0.48 = 4.27$$

(b) A hydroxide ion reacts with an acetic acid molecule to form a water molecule and an acetate ion, so the new picture shows 8 acetic acid molecules, 4 acetate ions, and 1 H_2O:

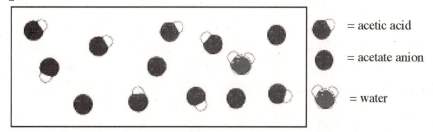

(c) HSO_4^- is a weak acid, so it transfers a proton to an acetate ion:

$$C_2H_3O_2^- (aq) + HSO_4^- (aq) \rightleftharpoons SO_4^{2-} (aq) + HC_2H_3O_2 (aq)$$

The new picture shows 10 molecules of acetic acid, 2 of acetate anions, and 1 sulfate:

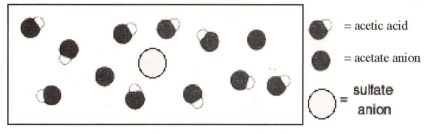

18.99   The key to working this problem is to recognize that the reaction can be written as the sum of two reactions for which K_{eq} values are available:

$$Ni^{2+} (aq) + 4\ CN^- (aq) \rightleftharpoons [Ni(CN)_4]^{2-} (aq) \qquad K_f = 1 \times 10^{22}$$
$$Ni(OH)_2 (s) \rightleftharpoons Ni^{2+} (aq) + 2\ OH^- (aq) \qquad K_{sp} = 5.5 \times 10^{-16}$$
$$Ni(OH)_2 (s) + 4\ CN^- (aq) \rightleftharpoons [Ni(CN)_4]^{2-} (aq) + 2\ OH^- (aq)$$
$$K_{eq} = K_f K_{sp} = 5.5 \times 10^6$$

We can use this equilibrium to calculate the molarity of ions at equilibrium. The equilibrium constant is large, so take the reaction to completion and then return to equilibrium:

Reaction: $Ni(OH)_2$ (s) +	4 CN^- (aq) $\rightleftharpoons$	$[Ni(CN)_4]^{2-}$ (aq) +	2 OH^- (aq)
Initial concentration (M)	0.500	0	0
Change (M)	− 0.500	+ 0.125	+ 0.250
Conc. at completion (M)	0	0.125	0.250
Change (M)	+ 4x	− x	− 2x
Equilibrium conc. (M)	4x	0.125 - x	0.250 - 2x

Substitute into the equilibrium constant expression:

$$K_{eq} = 5.5 \times 10^6 = \frac{(0.125 - x)(0.250 - 2x)}{(4x)^4}$$

Make the approximation that x is small relative to 0.125 and solve for x:

$$5.5 \times 10^6 = \frac{(0.125)(0.250)}{(4x)^4} \qquad 64\,x^4 = \frac{0.031}{5.5 \times 10^6} \qquad x^4 = 8.9 \times 10^{-11}$$

$x = 3.1 \times 10^{-3}$ This is only 2.5% of 0.125, so the approximation is valid.
Because the value for K_f is only known to one significant figure, x must be rounded to
one significant figure: $x = 0.003$
$[CN^-] = 4x = 0.01$ M $[Ni(CN)_4]^{2-} = 0.125 - 0.003 = 0.122$ M

$[OH^-] = 0.250 - 0.003 = 0.244$ M

Use the concentration of the complex to calculate the amount of $Ni(OH)_2$ that will
dissolve, and then use molar mass to convert to mass:
$n = MV = (0.122$ mol/L$)(225$ mL$)(10^{-3}$ L/mL$) = 0.0275$ mol
$m = n\,MM = (0.0275$ mol$)(92.7$ g/mol$) = 2.54$ g

19.1   Oxidation numbers are determined by applying the rules given in your textbook:
(a) ionic compound containing OH^- anions, with O -2, H $+1$, and Fe $+3$ to give overall neutrality;
(b) F (most electronegative) is -1, so N must be $+3$ to give overall neutrality;
(c) O is -2, H is $+1$, and C is -2 to give overall neutrality;
(d) ionic compound, with $K^+ = +1$ and CO_3^{2-}; in the anion, O (more electronegative) is -2 and C is $+4$ to give the -2 net charge on the ion;
(e) ionic compound; in NH_4^+, H is $+1$ and N is -3, and in NO_3^-, O is -2 and N is $+5$;
(f) Cl (more electronegative) is -1, so Ti is $+4$ to give overall neutrality;
(g) ionic, Pb must be $+2$ to balance -2 charge on sulfate; O is -2, so S is $+6$ to give the -2 charge on the anion;
(h) When it is a pure element, P is 0.

19.3   Identify redox reactions by determining whether or not oxidation numbers change:
(a) no change in oxidation numbers (Br remains -1), so not redox;
(b) redox, because Fe changes from $+2$ to $+3$ (O also changes);
(c) no change in oxidation numbers (Fe remains $+2$), so not redox;
(d) redox, because O in O_2 changes from 0 to -2 (C also changes);
(e) redox, because N changes from 0 to -3 and H changes from 0 to $+1$.

19.5   Oxidation numbers are determined by applying the rules given in your textbook:
(a) F is -1, so Cl is $+5$ to give overall neutrality;
(b) Cl is more electronegative, so it is -1;
(c) ionic, ClO_4^- has net charge of -1, and O is -2, so Cl is $+7$ to give net charge of -1;
(d) Cl is 0 because this is a pure element;
(e) ionic, ClO_2^- has net charge of -1, and O is -2, so Cl is $+3$ to give net charge of -1;
(f) ionic, ClO^- has net charge of -1, and O is -2, so Cl is $+1$ to give net charge of -1.

19.7   Determine half-reactions by inspection, making use of oxidation numbers if necessary. Balance each half-reaction following the standard steps in order (balance all but H and O by inspection, balance O by adding H_2O, balance H by adding H_3O^+ cations to the side that is deficient in hydrogen and an equal number of H_2O to the other side, balance charge by adding e^-):
(a) The reactants are Na and H_2O, and one product is H_2; Na^+ and OH^- must be the other products: $Na\,(s) \rightarrow Na^+\,(aq) + e^-$ and $2\,H_2O\,(l)^+ + 2\,e^- \rightarrow H_2\,(g) + 2\,OH^-\,(aq)$
(b) The reactants are Au and NO_3^-, and the products are $[AuCl_4]^-$ and NO; aqua regia also contains Cl^- anions:
$Au\,(s) + 4\,Cl^-\,(aq) \rightarrow [AuCl_4]^-\,(aq) + 3\,e^-$ and
$$NO_3^-\,(aq) + 4\,H_3O^+\,(aq) + 3\,e^- \rightarrow NO\,(g) + 6\,H_2O\,(l)$$
(c) The reactants are MnO_4^- and $C_2O_4^{2-}$, and the products are Mn^{2+} and CO_2:
$MnO_4^-\,(aq) + 8\,H_3O^+\,(aq) + 5\,e^- \rightarrow Mn^{2+}\,(aq) + 12\,H_2O\,(l)$ and
$$C_2O_4^{2-}\,(aq) \rightarrow 2\,CO_2\,(g) + 2\,e^-$$

(d) The reactants are C and H_2O, and the products are H_2 and CO:

$$C (s) + 3 H_2O (l) \rightarrow CO (g) + 2 H_3O^+ (aq) + 2e^- \text{ and}$$

$$2 H_3O^+ (aq) + 2e^- \rightarrow H_2 (g) + 2 H_2O (l)$$

19.9   Balance each half-reaction following the standard steps in order. 1.) balance all but H and O by inspection; 2.) balance O by adding H_2O; 3.) balance H by adding H_3O^+ cations to the side that is deficient in hydrogen and an equal number of H_2O to the other side. If the solution is basic, add an H_2O for each deficient H to the side that is deficient in hydrogen and an equal number of OH^- anions to the other side; 4.) balance charge by adding e^-:

(a) (1) Cu is balanced; (2) add H_2O on the left, (3) add 2 H_3O^+ on the right and 2 water on the left; (4) add 1 e^- on the right: $Cu^+ (aq) + 3 H_2O (l) \rightarrow CuO (s) + 2 H_3O^+ (aq) + e^-$

(b) (1) S is balanced; (2) no O is present; (3) add 2 H_3O^+ on the left and 2 water on the right; (4) add 2 e^- on the left: $S (s) + 2 H_3O^+ (aq) + 2 e^- \rightarrow H_2S (g) + 2 H_2O (l)$

(c) (1) Ag is balanced; add Cl^- on the right; (2 & 3) there are no H or O atoms; (4) add 1 e^- on the left: $AgCl + e^- \rightarrow Ag + Cl^-$
    Cancel duplicated species: $Ag_2O (s) + H_2O (l) + 2 e^- \rightarrow 2 Ag (s) + 2 OH^- (aq)$

(d) (1) I is balanced; (2) add 3 H_2O on the left; (3) add 6 H_2O on the right and 6 OH^- on the left; (4) add 6 e^- on the right: $I^- + 3 H_2O + 6 OH^- \rightarrow IO_3^- + 6 H_2O + 6 e^-$
    Cancel duplicated species: $I^- (aq) + 6 OH^- (aq) \rightarrow IO_3^- (aq) + 3 H_2O (l) + 6 e^-$

(e) (1) I is balanced; (2) add 2 H_2O on the right; (3) add 4 H_2O on the left and 4 OH^- on the right; (4) add 4 e^- on the left: $IO_3^- + 4 H_2O + 4 e^- \rightarrow IO^- + 2 H_2O + 4 OH^-$
    Cancel duplicated species: $IO_3^- (aq) + 2 H_2O (l) + 4 e^- \rightarrow IO^- (aq) + 4 OH^- (aq)$

(f) (1) C is balanced; (2) add H_2O on the left; (3) add 4 H_3O^+ on the right and 4 H_2O on the left; (4) add 4 e^- on the right:
$H_2CO (aq) + 5 H_2O (l) \rightarrow CO_2 (g) + 4 H_3O^+ (aq) + 4 e^-$

19.11   To combine half-reactions into a net redox reaction, multiply by appropriate integers so that the electrons will cancel:

(a) $2 [Cu^+ + 3 H_2O \rightarrow CuO + 2 H_3O^+ + e^-]$
           $+ [S + 2 H_3O^+ + 2 e^- \rightarrow H_2S + 2 H_2O]$
        $2 Cu^+ + 4 H_2O + S \rightarrow 2 CuO + 2 H_3O^+ + H_2S$

(b) $6 [AgCl + e^- \rightarrow Ag + Cl^-]$
           $+ [I^- + 6 OH^- \rightarrow IO_3^- + 3 H_2O + 6 e^-]$
              $6 AgCl + I^- + 6 OH^- \rightarrow 6 Ag + 6 Cl^- + 3 H_2O + IO_3^-$

(c) $2 [I^- + 6 OH^- \rightarrow IO_3^- + 3 H_2O + 6 e^-]$
           $+ 3 [IO_3^- + 2 H_2O + 4 e^- \rightarrow IO^- + 4 OH^-]$
              $2 I^- + IO_3^- \rightarrow 3 IO^-$

(d) $2 [S + 2 H_3O^+ + 2 e^- \rightarrow H_2S + 2 H_2O]$

$+ [H_2CO + 5 H_2O \rightarrow CO_2 + 4 H_3O^+ + 4 e^-]$

$2 S + H_2CO + H_2O \rightarrow 2 H_2S + CO_2$

19.13   The species that loses electrons is oxidized and acts as the reducing agent, while the species that gains electrons is reduced and acts as the oxidizing agent. In this reaction, Cl^- loses electrons (oxidized, reducing agent) to become Cl_2, and Mn in MnO_4^- gains electrons (reduced, oxidizing agent) to become Mn^{2+}.

19.15   Balance redox reactions following the standard procedure. Break into half-reactions, balance each half-reaction using the stepwise technique, then multiply by appropriate integers so the electrons cancel:

(a) $CN^- \rightarrow CNO^-$

(1) C and N are already balanced; (2) add 1 H_2O on the left; (3) add 2 H_2O on the right and 2 OH^- on the left; (4) add 2 e^- on the right:

$CN^- + H_2O + 2 OH^- \rightarrow CNO^- + 2 H_2O + 2 e^-$;

Cancel duplicated species: $CN^- + 2 OH^- \rightarrow CNO^- + H_2O + 2 e^-$

$MnO_4^- \rightarrow MnO_2$

(1) Mn is balanced; (2) add 2 H_2O on the right; (3) add 4 H_2O on the left and 4 OH^- on the right; (4) add 3 e^- on the left:

$MnO_4^- + 4 H_2O + 3 e^- \rightarrow MnO_2 + 2 H_2O + 4 OH^-$

Cancel duplicated species: $MnO_4^- + 2 H_2O + 3 e^- \rightarrow MnO_2 + 4 OH^-$

Multiply first reaction by 3, second by 2, and add:

$3 [CN^- + 2 OH^- \rightarrow CNO^- + H_2O + 2 e^-]$

$+ 2 [MnO_4^- + 2 H_2O + 3 e^- \rightarrow MnO_2 + 4 OH^-]$

$\overline{3 CN^- + 2 MnO_4^- + H_2O \rightarrow 3 CNO^- + 2 MnO_2 + 2 OH^-}$

(b) $As \rightarrow HAsO_2$

(1) As is balanced; (2) add 2 H_2O on the left; (3) add 3 H_3O^+ on the right and 3 H_2O on the left; (4) add 3 e^- on the right: $As + 5 H_2O \rightarrow HAsO_2 + 3 H_3O^+ + 3 e^-$

$O_2 \rightarrow H_2O$

(1) No elements other than O and H; (2) add 1 H_2O on the right; (3) add 4 H_3O^+ on left and 4 H_2O on the right; (4) add 4 e^- on the left: $O_2 + 4 H_3O^+ + 4 e^- \rightarrow 6 H_2O$

Multiply first reaction by 4, second by 3, and add:

$4 [As + 5 H_2O \rightarrow HAsO_2 + 3 H_3O^+ + 3 e^-]$

$+ 3 [O_2 + 4 H_3O^+ + 4 e^- \rightarrow 6 H_2O]$

$\overline{4 As + 3 O_2 + 2 H_2O \rightarrow 4 HAsO_2}$

(c) $Br^- \rightarrow BrO_3^-$

(1) Br is balanced; (2) add 3 H_2O on the left; (3) add 6 H_2O on the right and 6 OH^- on the left; (4) add 6 e^- on the right: $Br^- + 3\ H_2O + 6\ OH^- \rightarrow BrO_3^- + 6H_2O + 6\ e^-$

Cancel duplicated species: $Br^- + 6\ OH^- \rightarrow BrO_3^- + 3\ H_2O + 6\ e^-$

$MnO_4^- \rightarrow MnO_2$, same as reaction in part (a):

$$MnO_4^- + 2\ H_2O + 3\ e^- \rightarrow MnO_2 + 4\ OH^-$$

Multiply second reaction by 2 and add:

$$[Br^- + 6\ OH^- \rightarrow BrO_3^- + 3\ H_2O + 6\ e^-]$$
$$+\ 2\ [MnO_4^- + 2\ H_2O + 3\ e^- \rightarrow MnO_2 + 4\ OH^-]$$
$$\overline{Br^- + 2\ MnO_4^- + H_2O \rightarrow BrO_3^- + 2\ MnO_2 + 2\ OH^-}$$

(d) $NO_2 \rightarrow NO_3^-$

(1) N is balanced; (2) add 1 H_2O on the left; (3) add 2 H_3O^+ on the right and 2 H_2O on the left; (4) add 1 e^- on the right: $NO_2 + 3H_2O \rightarrow NO_3^- + 2\ H_3O^+ + e^-$

$NO_2 \rightarrow NO$

(1) N is balanced; (2) add 1 H_2O on the right; (3) add 2 H_3O^+ on the left and 2 H_2O on the right; (4) add 2 e^- on the left: $NO_2 + 2\ H_3O^+ + 2\ e^- \rightarrow NO + 3\ H_2O$

Multiply first reaction by 2 and add:

$$2\ [NO_2 + 3H_2O \rightarrow NO_3^- + 2\ H_3O^+ + e^-]$$
$$+\ [NO_2 + 2\ H_3O^+ + 2\ e^- \rightarrow NO + 3\ H_2O]$$
$$\overline{3\ NO_2 + 3\ H_2O \rightarrow 2\ NO_3^- + NO + 2\ H_3O^+}$$

(e) $ClO_4^- \rightarrow ClO^-$

(1) Cl is balanced; (2) add 3 H_2O on the right; (3) add 6 H_3O^+ on the left and 6 H_2O on the right; (4) add 6 e^- on the left: $ClO_4^- + 6\ H_3O^+ + 6\ e^- \rightarrow ClO^- + 9\ H_2O$

$Cl^- \rightarrow Cl_2$

(1) multiply Cl^- by 2; (2,3) no H or O present; (4) add 2 e^- on the right:

$$2\ Cl^- \rightarrow Cl_2 + 2\ e^-$$

Multiply second reaction by 3 and add:

$$[ClO_4^- + 6\ H_3O^+ + 6\ e^- \rightarrow ClO^- + 9\ H_2O]$$
$$+\ 3\ [2\ Cl^- \rightarrow Cl_2 + 2\ e^-]$$
$$\overline{ClO_4^- + 6\ Cl^- + 6\ H_3O^+ \rightarrow ClO^- + 9\ H_2O + 3\ Cl_2}$$

(f) $AlH_4^- \rightarrow Al^{3+}$

(1) Al is balanced; (2) no O present; (3) add 4 H_2O on the right and 4 OH^- on the left; (4) add 8 e^- on the right: $AlH_4^- + 4\ OH^- \rightarrow Al^{3+} + 4\ H_2O + 8\ e^-$

$H_2CO \rightarrow CH_3OH$

(1) C is balanced; (2) O is balanced; (3) add 2 H_2O on the left and 2 OH^- on the right; (4) add 2 e^- on the left: $H_2CO + 2\ H_2O + 2\ e^- \rightarrow CH_3OH + 2\ OH^-$

Multiply second reaction by 4 and add:

$$[AlH_4^- + 4\ OH^- \rightarrow Al^{3+} + 4\ H_2O + 8\ e^-]$$
$$+ 4\ [H_2CO + 2\ H_2O + 2\ e^- \rightarrow CH_3OH + 2\ OH^-]$$
$$\overline{AlH_4^- + 4\ H_2CO + 4\ H_2O \rightarrow Al^{3+} + 4\ CH_3OH + 4\ OH^-}$$

19.17   To determine spontaneity using standard thermodynamic values, calculate $\Delta G^o_{reaction}$ from tabulated values for ΔG^o_f. The reaction is spontaneous if $\Delta G^o_{reaction}$ is negative:

(a) $\Delta G^o_{reaction}$ = 2 mol(-129.7 kJ/mol) – 0 kJ/mol = –259.4 kJ; spontaneous;

(b) $\Delta G^o_{reaction}$ = 2 mol(-58.5 kJ/mol) – 0 kJ/mol = –117.0 kJ; spontaneous;

(c) $\Delta G^o_{reaction}$ = 1 mol(-300.1 kJ/mol)
          – [1 mol(-53.6 kJ/mol) + 1 mol (0 kJ/mol)] = –246.5 kJ; spontaneous;

(d) $\Delta G^o_{reaction}$ = [1 mol(-300.1 kJ/mol) + 1 mol(0 kJ/mol)]
          – [1 mol(-100.4 kJ/mol) + 1 mol(0 kJ/mol)] = –199.7 kJ; spontaneous.

19.19   A molecular view of a process occurring at an electrode should show the species involved in charge transfer and indicate the direction of movement of electrons. See Figure 19-5 for an example. At a silver-silver chloride electrode undergoing reduction, solid AgCl gains an electron from the electrode to release chloride anions into the solution and produce solid Ag metal:

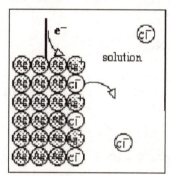
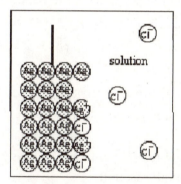

19.21   A passive electrode and a supply of gas are needed to study a redox reaction that includes gases. See Figure 19-8 for a sketch of a hydrogen electrode. A chlorine electrode is exactly analogous:

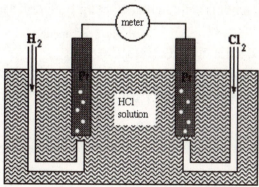

19.23 Active electrodes participate in redox reactions, while passive electrodes only provide or accept electrons. The Ag-AgCl electrode is active, while the Pt electrodes are passive.

19.25 Standard electrode potentials are calculated using Equation 19-1:
$$E^o_{cell} = E^o_{cathode} - E^o_{anode}$$

19.15 (a) MnO_4^- undergoes reduction to MnO_2 (cathode) $E^o_{cathode} = 0.595$ V

CN^- undergoes oxidation to CNO^- (anode) $E^o_{anode} = -0.970$ V

$E^o_{cell} = 0.595$ V $- (-0.970$ V$) = 1.565$ V

19.15 (b) O_2 undergoes reduction to H_2O (cathode) $E^o_{cathode} = 1.229$ V

As undergoes oxidation to $HAsO_2$ (anode) $E^o_{anode} = 0.248$ V

$E^o_{cell} = 1.229$ V $- (0.248$ V$) = 0.981$ V

19.15 (e) ClO_4^- undergoes reduction to ClO^- (cathode) $E^o_{cathode} = 1.36$ V

Cl^- undergoes oxidation to Cl_2 (anode) $E^o_{anode} = 1.35827$ V

$E^o_{cell} = 1.36$ V $- (1.35827$ V$) = 0.002$ V, which rounds to 0.00 V

19.27 Balance redox reactions following the standard procedure. Break into half-reactions, balance each half-reaction using the stepwise technique, then multiply by appropriate integers so the electrons cancel:

$NO \rightarrow N_2O$

(1) multiply NO by 2 to balance N, (2) add 1 H_2O on right, (3) add 2 H_3O^+ on left and 2 H_2O on the right, (4) add 2 e^- on left:
$$2 NO + 2 H_3O^+ + 2 e^- \rightarrow N_2O + 3 H_2O$$

$NO \rightarrow NO_3^-$

(1) N already balanced, (2) add 2 H_2O on left, (3) add 4 H_3O^+ on right and 4 H_2O on the left, (4) add 3 e^- on right:
$$NO + 6 H_2O \rightarrow NO_3^- + 4 H_3O^+ + 3 e^-$$

Multiply the first reaction by 3 and the second reaction by 2, then add to obtain the overall reaction:
$$3 [2 NO + 2 H_3O^+ + 2 e^- \rightarrow N_2O + 3 H_2O]$$
$$+ 2 [NO + 6 H_2O \rightarrow NO_3^- + 4 H_3O^+ + 3 e^-]$$
$$\overline{8 NO + 3 H_2O \rightarrow 3 N_2O + 2 NO_3^- + 2 H_3O^+}$$

Standard electrode potentials are calculated using Equation 19-1:
$$E^o_{cell} = E^o_{cathode} - E^o_{anode}$$

NO undergoes reduction to N_2O (cathode) $E^o_{cathode} = 1.591$ V

NO undergoes oxidation to NO_3^- (anode) $E^o_{anode} = 0.957$ V

$E^o_{cell} = 1.591$V $- 0.957$ V $= 0.634$ V

19.29   The direct way to measure a standard reduction potential is with a cell containing a standard hydrogen electrode as a reference. Because F⁻ is the conjugate base of a weak acid, the F_2/F^- portion of the cell must be separated from the H_2/H_3O^+ portion, requiring a porous plate. In addition, the fluoride solution could be made basic to prevent formation of HF. Set up a standard hydrogen electrode on one side and a Pt electrode immersed in a 1 M solution of NaF with F_2 bubbling over the electrode on the other side:

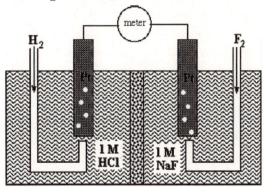

The standard reduction potential of F_2 is + 2.866 volts, so F_2 will be reduced in this cell and the H_2/H^+ electrode will be the anode.

19.31   To act as a reducing agent under standard conditions, a substance must have a more negative standard reduction potential than the substance that is to be reduced. The standard reduction potential of Be is – 1.847 V. Scanning the table, we find the following metals whose standard potentials are more negative than this:
Ba (–2.912 V), Ca (–2.868 V), Cs (–3.026 V), Li (–3.0401 V), Mg (–2.37 V), K (–2.931 V), and Na (–2.71 V). All these metals lie in the s block of the periodic table and easily lose 1 or 2 electrons.

19.33   The standard free energy change of a reaction is related to the standard reduction potential through Equation 19-3: $\Delta G^o = -nFE^o$
The standard potentials for the reactions were calculated in Problem 19.25:
19.25 (a) $E^o_{cell} = 1.565$ V, $n = 6$:

$$\Delta G^o = -(6 \text{ mol})(9.6485 \times 10^4 \text{ C/mol})(1.565 \text{ V})\left(\frac{1 \text{ J}}{1 \text{ V C}}\right)\left(\frac{10^{-3} \text{ kJ}}{1 \text{ J}}\right) = -906.0 \text{ kJ}$$

19.25 (b) $E^o_{cell} = 0.981$ V, $n = 12$:

$$\Delta G^o = -(12 \text{ mol})(9.6485 \times 10^4 \text{ C/mol})(0.981 \text{ V})\left(\frac{1 \text{ J}}{1 \text{ V C}}\right)\left(\frac{10^{-3} \text{ kJ}}{1 \text{ J}}\right)$$
$$= -1.14 \times 10^3 \text{ kJ}$$

19.25 (e) $E^o_{cell} = 0.002$ V (rounds to 0.00 V), $n = 6$:

$$\Delta G^o = -(6 \text{ mol})(9.6485 \times 10^4 \text{ C/mol})(0.002 \text{ V})(1 \text{ J/V C})(10^{-3} \text{ kJ/J}) = -1 \text{ kJ (rounds to 0 kJ)}$$

19.35   Operating potentials are related to standard cell potentials through Equation 19-6:

$$E = E^{o} - \frac{RT}{nF} \ln Q$$

The reaction for the nickel-cadmium battery is:

$2 \, NiO(OH) \, (s) + 2 \, H_2O \, (l) + Cd \, (s) \rightarrow 2 \, Ni(OH)_2 \, (s) + Cd(OH)_2 \, (s)$

Here, all the substances are solids or pure liquids, so $Q = 1$, $\ln Q = 0$, and $E = E^{o} = 1.35$ V. Hence, E is independent of $[OH^{-}]$.

19.37   The connection between electric current and chemical amounts is provided by Equation 19-7, $n = \dfrac{It}{F}$, where n is moles of electrons flowing.

$$n = 15 \, s \left(\frac{5.9 \, C}{1 \, s} \right) \left(\frac{1 \, mol}{9.6485 \times 10^4 \, C} \right) = 9.2 \times 10^{-4} \text{ mol electrons}$$

The half-reactions in a lead storage cell are:

$$H_2O + Pb \, (s) + HSO_4^{-} \, (aq) \rightarrow PbSO_4 \, (s) + H_3O^{+} \, (aq) + 2 \, e^{-}$$

$$PbO_2 \, (s) + HSO_4^{-} \, (aq) + 3 \, H_3O^{+} \, (aq) + 2 \, e^{-} \rightarrow PbSO_4 \, (s) + 5 \, H_2O \, (l)$$

$$n_{Pb} = n_{PbO_2} = 9.2 \times 10^{-4} \, mol \left(\frac{1 \, mol \, Pb}{2 \, mol \, e^{-}} \right) = 4.6 \times 10^{-4} \, mol$$

Finally convert moles to mass using molar masses:

$$m_{Pb} = 4.6 \times 10^{-4} \, mol \left(\frac{207.2 \, g}{1 \, mol} \right) = 9.5 \times 10^{-2} \, g$$

$$m_{PbO_2} = 4.6 \times 10^{-4} \, mol \left(\frac{239.2 \, g}{1 \, mol} \right) = 0.11 \, g$$

19.39   The connection between electric current and chemical amounts is provided by Equation 19-7, $n = \dfrac{It}{F}$, where n is moles of electrons flowing. Thus, $t = \dfrac{nF}{I}$:

The reaction being driven by the alternator is :

$$PbSO_4 \, (s) + H_3O^{+} \, (aq) + 2 \, e^{-} \rightarrow H_2O + Pb \, (s) + HSO_4^{-} \, (aq)$$

When 0.850 g of $PbSO_4$ is converted into Pb in each cell, the amount of electrons that flows is as follows:

$$n_e = 2 \, n_{PbSO_4} = 2(0.850 \, g) \left(\frac{1 \, mol}{303.3 \, g} \right) = 5.6 \times 10^{-3} \, mol$$

$$I = (1.750 \, A - 1.350 \, A) = 0.400 \, C/s$$

$$t = \frac{(5.6 \times 10^{-3} \, mol)(9.6485 \times 10^4 \, C \, mol^{-1})}{0.400 \, C \, s^{-1}} = 1.35 \times 10^3 \, s$$

19.41   Balance redox reactions following the standard procedure. Break into half-reactions, balance each half-reaction using the stepwise technique, then multiply by appropriate integers so the electrons cancel:

$Cr_2O_7^{2-} \rightarrow Cr^{3+}$

   (1) multiply Cr^{3+} by 2; (2) add 7 H_2O on the right; (3) add 14 H_3O^+ on the left and 14 H_2O on the right; (4) add 6 e^- on the left:

$$Cr_2O_7^{2-} + 14\ H_3O^+ + 6\ e^- \rightarrow 2\ Cr^{3+} + 21\ H_2O$$

$CH_3CHO \rightarrow CH_3CO_2H$

   (1) C is balanced; (2) add 1 H_2O on the left; (3) add 2 H_3O^+ on the right and 2 H_2O on the left; (4) add 2e^- on the right:

$$CH_3CHO + 3\ H_2O \rightarrow CH_3CO_2H + 2\ H_3O^+\ (aq) + 2\ e^-$$

Multiply the second reaction by 3 and add:

$$[Cr_2O_7^{2-} + 14\ H_3O^+\ (aq) + 6\ e^- \rightarrow 2\ Cr^{3+} + 21\ H_2O]$$
$$+ 3[CH_3CHO + 3\ H_2O \rightarrow CH_3CO_2H + 2\ H_3O^+\ (aq) + 2\ e^-]$$
$$\overline{Cr_2O_7^{2-} + 3\ CH_3CHO + 8\ H_3O^+\ (aq) \rightarrow 2\ Cr^{3+} + 3\ CH_3CO_2H + 12\ H_2O}$$

It takes 2 mol of electrons to oxidize 1 mol of CH_3CHO:

$$n = 1.00\ g \left(\frac{1\ mol}{44.1\ g}\right)\left(\frac{2\ mol\ e^-}{1\ mol\ CH_3CHO}\right) = 0.0454\ mol\ of\ electrons\ needed$$

Each mole of $Cr_2O_7^{2-}$ delivers 6 mol of electrons:

$$n_{dichromate} = 0.0454\ mol\left(\frac{1\ mol\ Cr_2O_7^{2-}}{6\ mol\ e^-}\right) = 7.57\ x\ 10^{-3}\ mol$$

$$m = 7.57\ x\ 10^{-3}\ mol\left(\frac{262\ g}{1\ mol}\right) = 1.98\ g\ of\ sodium\ dichromate\ needed$$

19.43   The Nernst equation is Equation 19-6, $E = E^o - \dfrac{RT}{nF}\ \ln Q$. To determine Q, use the balanced redox equation for the standard dry cell:

$$2MnO_2\ (s) + H_2O\ (l) + Zn\ (s) \rightarrow Mn_2O_3\ (s) + Zn(OH)_2\ (s)$$

All reactants and products are either solid or liquid, therefore $Q = 1$, $\ln Q = 0$, indicating that the potential does not decrease with use. The voltage actually does decrease with time, because of the changes in hydroxide ion concentrations associated with each half-reaction:

   Cathode: 2 MnO_2 (s) + H_2O (l) + 2$e^- \rightarrow Mn_2O_3$ (s) + 2 OH^- (aq)

   Anode: Zn (s) + 2 OH^- (aq) $\rightarrow Zn(OH)_2$ (s) + 2e^-

Your textbook states that the paste is not fluid enough to allow the hydroxide ions generated at the cathode to travel readily to the anode for consumption. Therefore the net reaction can be written as:

$2MnO_2$ (s) + H_2O (l) + Zn (s) + 2 OH^- (anode) $\rightarrow$

$$Mn_2O_3\ (s) + Zn(OH)_2\ (s) + 2\ OH^-\ (cathode)$$

$$Q = \frac{[OH^-]_{cathode}}{[OH^-]_{anode}}$$

The concentrations at the anode and cathode change as the battery is used. Thus, as the battery operates, the anode concentration decreases while the cathode concentration increases, $Q > 1$ and $\ln Q > 0$, and the potential of the battery decreases with use.

19.45   When two metals are in contact, the one with the more negative standard reduction potential corrodes first. Here are the values for Cr and Fe, from Appendix F:

$Cr^{2+} + 2\ e^- \rightarrow Cr$ $E^o = -0.913$ V $Fe^{2+} + 2\ e^- \rightarrow Fe$ $E^o = -0.447$ V

Thus, Cr oxidizes more easily and will preferentially corrode.

19.47   Zinc–air batteries are characterized by a stable potential and compact size but limited current capacity, making them well suited for use where large currents are not needed, such as in pacemakers and cameras. They cannot supply the large current needed to start an automobile engine; moreover, they are irreversible (not rechargeable) so they would have to be replaced frequently.

19.49   The connection between electric current and chemical amounts is provided by Equation 19-7, $n = \dfrac{It}{F}$, where n is moles of electrons flowing. Examine the balanced half-reaction to determine the stoichiometric relationship between moles of electrons and moles of chemical species. Here, the reaction is $2\ Cl^- \rightarrow Cl_2 + 2\ e^-$, so 1 mol of Cl_2 is formed by 2 mol of electrons:

$$t = 200.0\ \text{min}\left(\frac{60\ s}{1\ \text{min}}\right) = 1.200\ \text{x}\ 10^4\ s$$

$$n = 1.200\ \text{x}\ 10^4\ s\left(\frac{4.50\ C}{1\ s}\right)\left(\frac{1\ \text{mol}\ e^-}{9.6485\ \text{x}\ 10^4\ C}\right)\left(\frac{1\ \text{mol}\ Cl_2}{2\ \text{mol}\ e^-}\right) = 0.280\ \text{mol}\ Cl_2$$

$$m = n\ MM = 0.280\ \text{mol}\left(\frac{70.906\ g}{1\ \text{mol}}\right) = 19.9\ g$$

19.51   When an external potential is applied to a galvanic cell, the reduction that occurs is the one with the least negative reduction potential, and the oxidation that occurs is the one with the least positive reduction potential. The oxidation reaction during recharging is the reverse of the galvanic reduction:

$Cu \rightarrow Cu^{2+} + 2\ e^-$.

However, instead of $Zn^{2+} + 2\ e^- \rightarrow Zn$ ($E^o = -0.7618$ V), during recharging the reduction of H_3O^+ occurs:

$2\ H_3O^+\ (aq) + 2\ e^- \rightarrow H_2 + 2\ H_2O$ ($E^o = 0$ V)

In neutral H_2O, the H_3O^+ ion concentration is only 10^{-7} M, but the resulting E is still less negative than E^o for Zn^{2+}/Zn.

19.53 (a) To determine where to attach the negative wire from the charger, examine the half-reactions of the battery when it is producing current:

Cathode: $NiO(OH)$ (s) + H_2O (l) + e- → $Ni(OH)_2$ (s) + OH- (aq)

Anode: Cd (s) + 2 OH- (aq) → $Cd(OH)_2$ (s) + 2 e-

To recharge, electrons must be supplied to the cadmium electrode, so this is the electrode to which the negative wire (anode) from the charger should be attached. The reaction is $Cd(OH)_2$ (s) + 2 e- → Cd (s) + 2 OH- (aq)

(b) The connection between electric current and chemical amounts is provided by Equation 19-7, $n = \dfrac{It}{F}$, where n is moles of electrons flowing. Thus $t = \dfrac{nF}{I}$:

$$n_{Cd(OH)_2} = 1.55 \text{ g} \left(\frac{1 \text{ mol}}{146.4 \text{ g}} \right) = 0.0106 \text{ mol}$$

$$n_{e-} = 2\ n_{Cd(OH)_2} = 0.0212 \text{ mol}$$

$$t = 0.0212 \text{ mol} \left(\frac{9.6485 \times 10^4 \text{ C}}{1 \text{ mol}} \right) \left(\frac{1 \text{ s}}{125 \times 10^{-3} \text{ C}} \right) = 1.64 \times 10^4 \text{ s}$$

Convert to hours:

$$1.64 \times 10^4 \text{ s} \left(\frac{1 \text{ min}}{60 \text{ s}} \right) \left(\frac{1 \text{ hr}}{60 \text{ min}} \right) = 4.56 \text{ hr}$$

19.55 Quantity of charge can be calculated once the amount of electron flow has been determined. Use the half-reaction to determine the relationship between amount of chemical change and amount of electron flow: $Ag^+ + e^- → Ag$

$$n_e = n_{Ag} = (12.89 \text{ g} - 10.77 \text{ g}) \left(\frac{1 \text{ mol}}{107.9 \text{ g}} \right) = 0.0196 \text{ mol e}^-$$

charge = nF = (0.0196 mol)(9.6485 × 10^4 C/mol) = 1.89 × 10^3 C
Determine the current by dividing the charge by the time in seconds:

$$t = 15.0 \text{ min} \left(\frac{60 \text{ s}}{1 \text{ min}} \right) = 900 \text{ s} \qquad i = \frac{\text{charge}}{t} = \frac{1.89 \times 10^3 \text{ C}}{900 \text{ s}} = 2.10 \text{ A}$$

19.57 A molecular picture must show the species undergoing redox reactions and the direction of electron flow. In a silver coulometer, both electrodes are silver. Ag is oxidized to Ag^+ at the anode, and Ag^+ is reduced to Ag at the cathode.

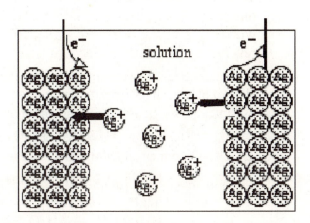

19.59   Standard electrode potentials allow calculation of equilibrium constants if the half-reactions can be combined to give the equilibrium reaction whose constant is desired. In this case, subtract the second reaction from the first and divide by 4:

$$O_2 + 2\ H_2O + 4\ e^- \rightleftharpoons 4\ OH^- \qquad\qquad E^\circ = 0.401\ V$$

$$-\ [O_2\ (g) + 4\ H_3O^+\ (aq) + 4\ e^- \rightleftharpoons 6\ H_2O \qquad E^\circ = 1.229\ V]$$

Net: $8\ H_2O\ (l) \rightleftharpoons 4\ OH^-\ (aq) + 4\ H_3O^+\ (aq)$

$$E^\circ_{reaction} = 0.401\ V - 1.229\ V = -0.828\ V$$

Dividing by 4 changes n_e but not E°:

$$2\ H_2O\ (l) \rightleftharpoons OH^-\ (aq) + H_3O^+\ (aq) \qquad n_e = 1 \qquad E^\circ = -0.828\ V$$

Use Equation 19-5, rearranged:

$$\log K_{eq} = \left(\frac{nE^\circ}{5.916 \times 10^{-2}\ V}\right) = \left(\frac{1(-0.828\ V)}{5.916 \times 10^{-2}\ V}\right) = -14.00$$

$$K_{eq} = 1.0 \times 10^{-14} = [OH^-][H_3O^+]$$

19.61   An equilibrium constant can be calculated from standard reduction potentials if half-reactions can be combined to give the equilibrium reaction whose constant is desired.

The equilibrium is $\qquad Zn^{2+} + 4\ NH_3 \rightleftharpoons [Zn(NH_3)]_4^{2+}$

Appendix F provides these standard potentials:

$$Zn^{2+} + 2\ e^- \rightleftharpoons Zn \qquad\qquad\qquad E^\circ = -0.7618\ V$$

$$[Zn(NH_3)]_4^{2+} + 2\ e^- \rightleftharpoons Zn + 4\ NH_3 \qquad\qquad E^\circ = -1.04\ V$$

Subtract the second half-reaction from the first to give the standard reduction potential corresponding to the desired equilibrium:

$$E^\circ = (-0.7618\ V) - (-1.04\ V) = +0.2782\ V$$

Now use the relationship between K_{eq} and E° to determine the equilibrium constant:

$$\log K_{eq} = \left(\frac{nE^\circ}{5.916 \times 10^{-2}\ V}\right) = \left(\frac{2(0.2782\ V)}{5.916 \times 10^{-2}\ V}\right) = 9.405$$

$$K_{eq} = 10^{9.405} = 2.5 \times 10^9$$

19.63   (a) Balance redox reactions following the standard procedure. Break into half-reactions, balance each half-reaction using the stepwise technique, then multiply by appropriate integers so the electrons cancel when the half-reactions are combined:

$H_2 \rightarrow H_2O$ (1) no elements except O and H; (2) add 1 H_2O on the left; (3) add 2 H_2O on the right and 2 OH^- on the left; (4) add 2 e^- on the right:

$$H_2 + 2\ OH^- + H_2O \rightarrow 3\ H_2O + 2\ e^-$$

Cancel duplicated species:

$$H_2 + 2\ OH^- \rightarrow 2\ H_2O + 2\ e^-$$

$Cr(OH)_3 \rightarrow Cr$

(1) Cr is balanced; (2) add 3 H_2O on the right; (3) add 3 H_2O on the left and 3 OH^- on the right; (4) add 3 e^- on the left:

$$Cr(OH)_3 + 3\ H_2O + 3\ e^- \rightarrow Cr + 3\ H_2O + 3\ OH^-$$

Cancel duplicated species:

$$Cr(OH)_3 + 3\ e^- \rightarrow Cr + 3\ OH^-$$

Multiply the first reaction by 3 and the second reaction by 2, then add:

$$3\ [H_2 + 2\ OH^- \rightarrow 2\ H_2O + 2\ e^-]$$
$$+ 2\ [Cr(OH)_3 + 3\ e^- \rightarrow Cr + 3\ OH^-]$$
$$\overline{3\ H_2 + 2\ Cr(OH)_3 \rightarrow 6\ H_2O + 2\ Cr}$$

(b) To determine the standard potential, consult values in Appendix F, and subtract E^o for the oxidation from E^o for the reduction:

$2\ H_2O + 2\ e^- \rightarrow H_2 + 2\ OH^-$ $E^o = -0.828$ V

$Cr(OH)_3 + 3\ e^- \rightarrow Cr + 3\ OH^-$ $E^o = -1.48$ V

$E^o{}_{cell} = (-1.48\ V) - (-0.828\ V) = -0.65$ V

(c) Calculate ΔG^o from $\Delta G^o = -nFE^o{}_{cell}$:

$$\Delta G^o = -(6\ mol)(9.6485 \times 10^4\ C/mol)(-0.65\ V)(10^{-3}\ kJ/J) = 3.8 \times 10^2\ kJ$$

19.65   (a) Balance the half-reaction using the stepwise technique:

$Cr_2O_7{}^{2-} \rightarrow Cr^{3+}$

(1) multiply Cr^{3+} by 2; (2) add 7 H_2O on the right; (3) add 14 H_3O^+ on the left and 14 H_2O on the right; (4) add 6 e^- on the left:

$$Cr_2O_7{}^{2-} + 14\ H_3O^+\ (aq) + 6\ e^- \rightarrow 2\ Cr^{3+} + 21\ H_2O$$

(b) It takes 2 mol K_2CrO_4 to generate 1 mol $Cr_2O_7{}^{2-}$, which in turn consumes 6 mol of e^-, so 1 mol K_2CrO_4 consumes 3 mol e^-:

$$0.250\ mol \left(\frac{1\ mol\ K_2CrO_4}{3\ mol\ e^-}\right) = 0.0833\ mol\ K_2CrO_4\ required$$

$$m = n\ MM = 0.0833\ mol \left(\frac{194\ g}{1\ mol}\right) = 16.2\ g\ of\ K_2CrO_4$$

19.67   Identify what is taking place in a galvanic cell by identifying the species present:

(a) On the left: $Ni^{2+} + 2\ e^- \rightarrow Ni$ $E^o = -0.257$ V

On the right: $Fe^{2+} + 2\ e^- \rightarrow Fe$ $E^o = -0.447$ V

(b) To generate a positive overall potential, the iron half-reaction operates as oxidation:

$Fe + Ni^{2+} \rightarrow Ni + Fe^{2+}$ $E^o = -0.257\ V - (-0.447\ V) = 0.190$ V

(c) Oxidation occurs at the anode, so the Fe electrode is the anode and Ni is the cathode.

(d) A molecular picture of an electrochemical cell must show the species undergoing redox reactions and the direction of electron flow:

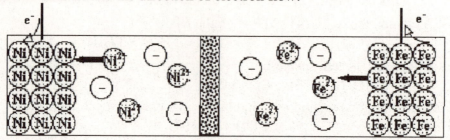

19.69 Use the Nernst equation to determine the concentration at which $E = 0.0$ V:

$$Fe + Ni^{2+} \rightleftharpoons Ni + Fe^{2+} \qquad E^o = -0.257 \text{ V} - (-0.447 \text{ V}) = 0.190 \text{ V}$$

$$E = 0.0 \text{ V} = E^o - \frac{(0.05916)}{n} \log\left(\frac{[Fe^{2+}]}{[Ni^{2+}]}\right) \qquad \log\left(\frac{[Fe^{2+}]}{[Ni^{2+}]}\right) = \left(\frac{nE^o}{0.05916 \text{ V}}\right)$$

The concentration of reactant must be reduced to reach a potential of 0.0 V:

$$\log\left\{\frac{(1.00)}{[Ni^{2+}]}\right\} = \frac{(2)(0.190 \text{ V})}{(0.05916 \text{ V})} = 6.42 \qquad [Ni^{2+}] = 10^{-6.42} = 3.8 \times 10^{-7} \text{ M}$$

19.71 Balance redox reactions following the standard procedure. Break into half-reactions, balance each half-reaction using the stepwise technique, then multiply by appropriate integers so the electrons cancel when the half-reactions are combined:
$MnO_4^- \rightarrow Mn^{2+}$

(1) Mn is already balanced; (2) add 4 H_2O on the right; (3) add 8 H_3O^+ on the left and 8 H_2O on the right; (4) add 5 e⁻ on the left:

$$MnO_4^- + 8 \text{ } H_3O^+ \text{ } (aq) + 5 \text{ e}^- \rightarrow Mn^{2+} + 12 \text{ } H_2O$$

(a) $H_2SO_3 \rightarrow HSO_4^-$

(1) S is already balanced; (2) add 1 H_2O on the left; (3) add 3 H_3O^+ on the right and 3 H_2O on the left; (4) add 2 e⁻ on the right:

$$H_2SO_3 + 4 \text{ } H_2O \rightarrow HSO_4^- + 3 \text{ } H_3O^+ \text{ } (aq) + 2 \text{ e}^-$$

Multiply the Mn reaction by 2 and the S reaction by 5, and add:

$$2 \text{ } [MnO_4^- + 8 \text{ } H_3O^+ \text{ } (aq) + 5 \text{ e}^- \rightarrow Mn^{2+} + 12 \text{ } H_2O]$$
$$+ 5 \text{ } [H_2SO_3 + 4H_2O \rightarrow HSO_4^- + 3 \text{ } H_3O^+ \text{ } (aq) + 2 \text{ e}^-]$$
$$\overline{2 \text{ } MnO_4^- + 5 \text{ } H_2SO_3 + H_3O^+ \text{ } (aq) \rightarrow 2 \text{ } Mn^{2+} + 5 \text{ } HSO_4^- + 4 \text{ } H_2O}$$

(b) $SO_2 \rightarrow HSO_4^-$

(1) S is already balanced; (2) add 2 H_2O on the left; (3) add 3 H_3O^+ on the right and 3 H_2O on the left; (4) add 2 e⁻ on the right:

$$SO_2 + 5 \text{ } H_2O \rightarrow HSO_4^- + 3 \text{ } H_3O^+ \text{ } (aq) + 2 \text{ e}^-$$

Multiply the Mn reaction by 2 and the S reaction by 5, and add:

$$2 \; [MnO_4^- + 8 \; H_3O^+ \; (aq) + 5 \; e^- \rightarrow Mn^{2+} + 12 \; H_2O]$$

$$+ \; 5 \; [SO_2 + 5 \; H_2O \rightarrow HSO_4^- + 3 \; H_3O^+ \; (aq) + 2 \; e^-]$$

$$\overline{2 \; MnO_4^- + 5 \; SO_2 + H_2O + H_3O^+ \; (aq) \rightarrow 2 \; Mn^{2+} + 5 \; HSO_4^-}$$

(c) $H_2S \rightarrow HSO_4^-$

(1) S is already balanced; (2) add 4 H_2O on the left; (3) add 9 H_3O^+ on the right and 9 H_2O on the left; (4) add 8 e^- on the right:

$$H_2S + 13 \; H_2O \rightarrow HSO_4^- + 9 \; H_3O^+ \; (aq) + 8 \; e^-$$

Multiply the Mn reaction by 8 and the S reaction by 5, and add:

$$8 \; [MnO_4^- + 8 \; H_3O^+ \; (aq) + 5 \; e^- \rightarrow Mn^{2+} + 12 \; H_2O]$$

$$+ \; 5 \; [H_2S + 13 \; H_2O \rightarrow HSO_4^- + 9 \; H_3O^+ \; (aq) + 8 \; e^-]$$

$$\overline{8 \; MnO_4^- + 5 \; H_2S + 19 \; H_3O^+ \; (aq) \rightarrow 8 \; Mn^{2+} + 5 \; HSO_4^- + 31 \; H_2O}$$

(d) $H_2S_2O_3 \rightarrow HSO_4^-$

(1) Multiply HSO_4^- by 2; (2) add 5 H_2O on left; (3) add 10 H_3O^+ on right and 10 H_2O on the left; (4) add 8 e^- on right:

$$H_2S_2O_3 + 15 \; H_2O \rightarrow 2 \; HSO_4^- + 10 \; H_3O^+ \; (aq) + 8 \; e^-$$

Multiply the Mn reaction by 8 and the S reaction by 5, and add:

$$8 \; [MnO_4^- + 8 \; H_3O^+ \; (aq) + 5 \; e^- \rightarrow Mn^{2+} + 12 \; H_2O]$$

$$+ \; 5 \; [H_2S_2O_3 + 15 \; H_2O \rightarrow 2 \; HSO_4^- + 10 \; H_3O^+ \; (aq) + 8 \; e^-]$$

$$\overline{8 \; MnO_4^- + 5 \; H_2S_2O_3 + 14 \; H_3O^+ \; (aq) \rightarrow 8 \; Mn^{2+} + 10 \; HSO_4^- + 21 \; H_2O}$$

19.73   The connection between electric current and chemical amounts is provided by Equation 19-7, $n = \dfrac{It}{F}$, where n is moles of electrons flowing. Thus, $t = \dfrac{nF}{I}$:

The reaction is $Cu^{2+} + 2 \; e^- \rightarrow Cu$

$$n_{Cu} = 0.250 \; L \left(\frac{0.245 \; mol}{1 \; L} \right) = 0.06125 \; mol \qquad\qquad n_{electron} = 2 \; n_{Cu} = 0.1225 \; mol$$

$$t = 0.1225 \; mol \left(\frac{9.6485 \times 10^4 \; C}{1 \; mol} \right) \left(\frac{1 \; s}{2.45 \; C} \right) \left(\frac{1 \; min}{60 \; s} \right) = 80.4 \; min$$

19.75   The process in Problem 19.73 is electrodeposition, presumably with Cu serving as both electrodes. Reduction of Cu^{2+} to Cu occurs at the cathode, and oxidation of Cu to Cu^{2+} occurs at the anode:

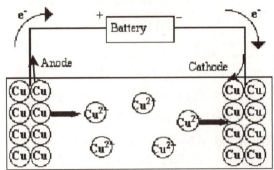

19.77   The better oxidizing agent under standard conditions is the reactant having the more positive reduction potential:
        (a) MnO_4^- (1.679 vs. 1.232 V)
        (b) O_2 (0.401 vs. 0.109 V)
        (c) Sn^{2+} (–0.137 vs. –0.447 V)

19.79   Under standard conditions, a substance can oxidize any other substance that appears on the RIGHT side of a half-reaction whose reduction potential is less positive: For O_2, $E^o = 1.229$ V; all the substances listed in Problem 19.78 (except for Co^{2+}) have less positive reduction potentials, so O_2 can oxidize Cu, Ag, Fe^{2+}, H_2, and I^- (not surprisingly, oxygen is a good oxidizing agent!).

19.81   Although the anions are different in the two solutions, the redox process involves only Zn metal and Zn^{2+} cations, so this is a concentration cell:
        (a) In the $Zn(NO_3)_2$ solution, $Zn^{2+} + 2\ e^- \rightarrow Zn$
            In the $ZnCl_2$ solution, $Zn \rightarrow Zn^{2+} + 2\ e^-$
        (b) A molecular picture shows Zn^{2+} cations depositing on the electrode from the 1.25 M solution and dissolving off the electrode into the 0.250 M solution:

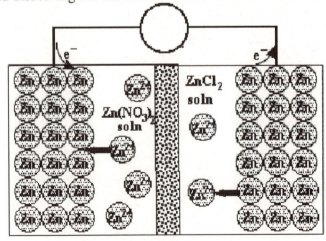

(c) To calculate a cell potential from concentration measurements, use the Nernst equation. For a concentration cell, $E^o = 0$ V, so $E = -\dfrac{(0.05916)}{n} \log Q$:

Both reactions are $Zn^{2+} + 2\ e^- \rightarrow Zn$, and the concentrations are 0.250 M and 1.25 M:

$$Zn^{2+}{}_{(1.25\ M)} \rightarrow Zn^{2+}{}_{(0.250\ M)} \qquad n = 2 \qquad Q = \dfrac{0.250\ M}{1.25\ M}$$

$$E = -\dfrac{(0.05916)}{2} \log \dfrac{0.250\ M}{1.25\ M} = 0.0207\ V$$

19.83 The connection between time and amount of electrons is provided by Equation 19-7:

$$n = \dfrac{It}{F} \qquad\qquad t = \dfrac{nF}{I}$$

From the half-reaction, $n_{electron} = n_{MnO_2}$. Only 90% of the MnO_2 is available before the battery fails:

$$n_{MnO_2} = 4.0\ g\left(\dfrac{1\ mol}{86.94\ g}\right)\left(\dfrac{90\%}{100\%}\right) = 0.0414\ mol$$

$$t = 0.0414\ mol\left(\dfrac{9.6485 \times 10^4\ C}{1\ mol}\right)\left(\dfrac{1\ s}{0.0048\ C}\right)\left(\dfrac{1\ min}{60\ s}\right)\left(\dfrac{1\ hr}{60\ min}\right) = 2.3 \times 10^2\ hr$$

19.85 Use the Nernst equation to determine a concentration from cell voltages. First determine the net reaction in order to find n, E^o, and Q:

Reactions: $Fe^{3+} + e^- \rightleftharpoons Fe^{2+}$ $E^o = 0.771$ V

$Cu^{2+} + 2\ e^- \rightleftharpoons Cu$ $E^o = 0.3419$ V

Multiply the first reaction by 2 and subtract the second:

$2\ Fe^{3+} + Cu \rightleftharpoons 2\ Fe^{2+} + Cu^{2+}$ $E^o = 0.771\ V - 0.3419\ V = 0.429\ V$

$$E = E^o - \dfrac{(0.05916)}{n} \log Q = E^o - \dfrac{(0.05916)}{n} \log\left(\dfrac{[Fe^{2+}]^2[Cu^{2+}]}{[Fe^{3+}]^2}\right)$$

Substitute values and solve for the unknown concentration:

$$0.00\ V = 0.429\ V - \dfrac{(0.05916)}{2} \log\left\{\dfrac{(1.00)^2(1.00)}{[Fe^{3+}]^2}\right\}$$

$$-\log\{[Fe^{3+}]^2\} = \dfrac{2(0.429)}{(0.05916)} = 14.50 \qquad [Fe^{3+}]^2 = 3.2 \times 10^{-15} \quad [Fe^{3+}] = 5.6 \times 10^{-8}\ M$$

19.87 (a) In molten NaCl, the only species present are Na^+ cations and Cl^- anions; the Pt electrodes are passive, so the reactions are $Na^+ + e^- \rightarrow Na$ and $2\ Cl^- \rightarrow Cl_2 + 2\ e^-$.

(b) Reduction is driven by electrons supplied at the negative terminal, so the cathode is the Pt electrode connected to the negative pole of the battery, and the anode is the Pt electrode connected to the positive pole of the battery.

(c)

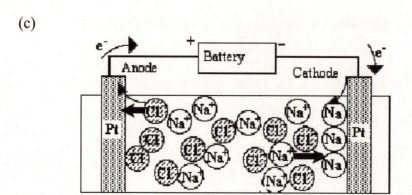

19.89 A sketch of a cell that shows molecular processes should show the molecular species undergoing redox reactions and the direction of electron flow:

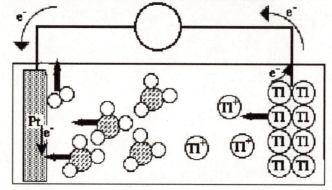

19.91 The standard potential for a redox reaction is easily calculated from the equilibrium constant for the reaction by using Equation 19-5, $E^o = \left(\dfrac{5.916 \times 10^{-2} \text{ V}}{n} \right) \log K_{eq}$:

$$E^o = \frac{1}{2}(5.916 \times 10^{-2} \text{ V}) \log(2.7 \times 10^{12}) = (0.02958)(12.43) = 0.368 \text{ V}$$

These data can be used to calculate the standard reduction potentials of the corresponding metal cations:

$$E^o = \left(\frac{5.916 \times 10^{-2} \text{ V}}{n} \right) \log K_{eq}$$

$E^o = E^o$ (hydroxide) $- E^o$ (metal cation) E^o (metal cation) $= E^o$ (hydroxide) $- E^o$
For each metal, $n = 2$

Cd: $E^o = \left(\dfrac{5.916 \times 10^{-2} \text{ V}}{2} \right) \log (7.2 \times 10^{-15}) = -0.45 \text{ V}$

E^o (Cd^{2+}/Cd) $= -0.809 + 0.45 = -0.36 \text{ V}$

Ni: $E^o = \left(\dfrac{5.916 \times 10^{-2} \text{ V}}{2} \right) \log (5.5 \times 10^{-16}) = -0.418 \text{ V}$

E^o (Ni^{2+}/Ni) $= -0.72 + 0.418 = -0.30 \text{ V}$

20.1   Oxidation states are determined by applying the rules given in Section 19.1. The
procedure can be simplified when a polyatomic ion of known charge is present:
(a) CO_3^{2-} has a charge of –2, so Mn must be +2 to give overall neutrality.
(b) Cl (more electronegative) is –1, so Mo must be +5 to give overall neutrality.
(c) Na is +1, so VO_4 has overall charge –3; each O is –2, so V is +5.
(d) O is –2, so Au must be +3 to give overall neutrality.
(e) H_2O has zero charge, SO_4^{2-} has a charge of –2, so Fe must be +3 to give overall
neutrality.

20.3   Use the periodic table to locate and identify elements from their valence configurations:
(a) There are 6 valence electrons, so the element is in column 6; $3d$ orbital is filling: Cr
(b) There are full s and d blocks, so the element is in column 12; $4d$ has just filled: Cd
(c) There are 11 valence electrons, so the element is in column 11; $3d$ orbital is full: Cu

20.5   Use the periodic table to locate a transition metal and determine the principal quantum
numbers of its valence electrons. Then remove electrons to give the appropriate cation,
remembering that when cations form, the valence s electrons are the first ones removed:
(a) $3d^5$; (b) $5d^6$; (c) $3d^8$; and (d) $4d^4$

20.7   The properties of transition metals vary regularly with their valence configurations, so
predict relative properties based on locations in the d block:
(a) Pd is in column 10, Cd is in column 12, both in the n = 4 row. Beyond the middle of
the d block, melting point decreases with Z because electrons are placed in
antibonding orbitals, so Pd melts at higher temperature.
(b) Cu and Au are both in column 11, but Au has higher MM, so Au has higher density.
(c) Cr is in column 6, Co is in column 9, both in the n = 3 row. IE_1 increases with Z
across a row because Z_{eff} increases, so Co has the higher IE_1.

20.9   Use the charges of ligands and ions to determine the oxidation states of transition metals
in coordination complexes. Because s electrons are always removed first, the count of d
electrons is given by the number of valence electrons – the oxidation state:
(a) Each Cl is –1, and NH_3 is neutral, so Ru has oxidation state +2. Ru is in column 8
(8 valence electrons), giving d^6.
(b) Each I is –1, and en is neutral, so Cr has oxidation state +3. Cr is in column 6
(6 valence electrons), giving d^3.
(c) Each Cl is –1, and trimethylphosphine is neutral, so Pd has oxidation state +2. Pd is
in column 10 (10 valence electrons), giving d^8.
(d) Each Cl is –1, and NH_3 is neutral, so Ir has oxidation state +3. Ir is in column 9
(9 valence electrons), giving d^6.
(e) CO is a neutral molecule, so Ni has oxidation state 0. Ni is in column 10 (10 valence
electrons), s^2d^8.

20.11 Compounds that contain coordination complexes are named following the six rules stated in your textbook: name cation first, name ligands in alphabetical order, name metal, add "o" for anions, use Greek prefixes, add "-ate" for anionic complexes, give oxidation number:

(a) hexaammineruthenium(II) chloride

(b) *trans*-bis(ethylenediamine)diiodochromium(III) iodide

(c) *cis*-dichlorobis(trimethylphosphine)palladium(II)

(d) *fac*-triamminetrichloroiridium(III)

(e) tetracarbonylnickel(0)

20.13 The structure of a metal complex usually is octahedral (6 ligands), tetrahedral (4 ligands), or square planar (4 ligands):

(a) Six NH_3 in an octahedron around the central Ru:

(b) Two I at opposite ends of one axis, two en in a square plane around the central Cr:

(c) Square planar arrangement about Pd, with two Cl adjacent each other:

(d) Octahedral arrangement around Ir, with three Cl in a triangular face:

(e) Tetrahedral arrangement of $C \equiv O$ about a central Ni:

20.15 The name of a complex contains the information needed to determine its chemical formula. Determine the charge of the complex from charges on the ligands and the oxidation number:
(a) *cis*-[Co(NH₃)₄ClNO₂]⁺; (b) [PtNH₃Cl₃]⁻; (c) *trans*-[Cu(en)₂(H₂O)₂]²⁺; (d) [FeCl₄]⁻

20.17 (a) *cis*-tetraamminechloronitrocobalt(III) has six ligands and octahedral geometry. The *cis* indicates that the chlorine and nitro ligands will have a 90 degree angle between them:

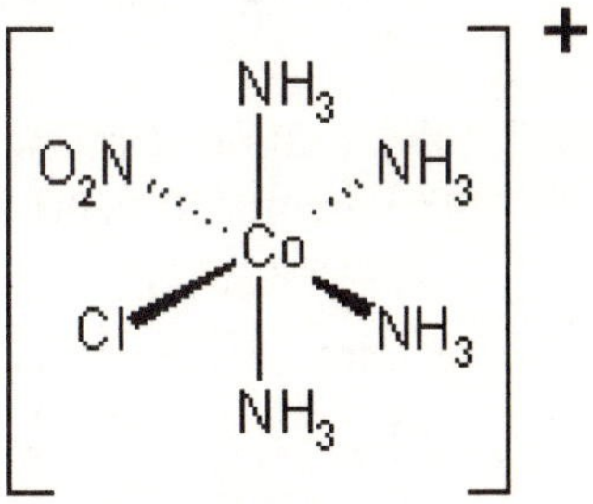

  (b) In amminetrichloroplatinate(II), platinum has eight *d* electrons in its valence shell, resulting in square planar geometry:

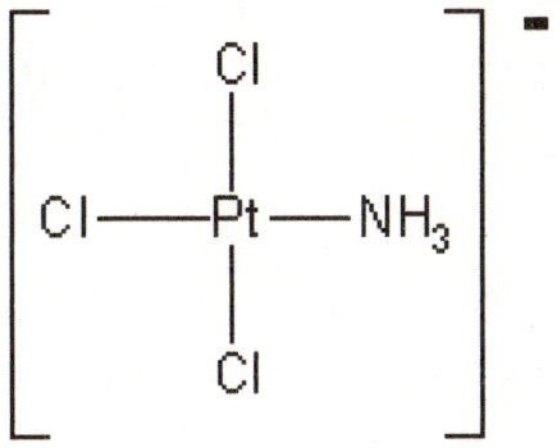

  (c) In *trans*-diaquabis(ethylenediamine) copper(II), ethylenediamine is a bidentate ligand, giving coordination number of 6 and octahedral geometry. The *trans* term means that the water ligands are opposite each other on the complex:

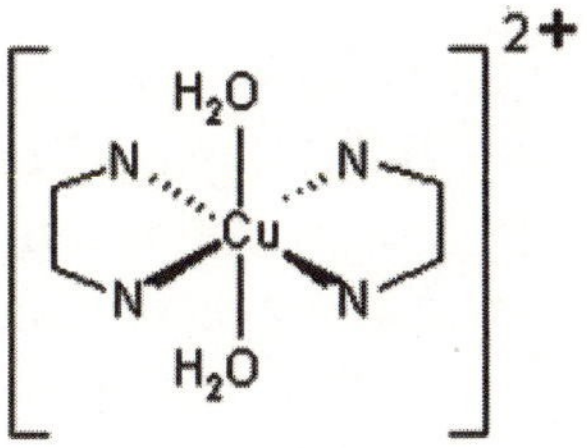

  (d) In tetrachloroferrate(III), iron(III) is a 3*d* metal with 5 valence electrons, resulting in tetrahedral geometry:

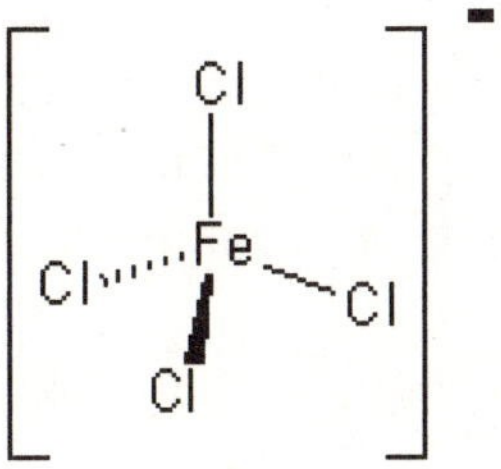

20.19 The crystal field diagram for weak and strong octahedral fields is always the same, with the populations changing depending on how many *d* electrons must be accommodated. The valence configuration provides information about the number of *d* electrons:
(a) Ti²⁺ (column 4 – 2 electrons) is *d²*; (b) Cr³⁺ (column 6 – 3 electrons) is *d³*.
    For these two ions, the low-field and high-field configurations are the same:

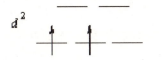

(c) Mn^{2+} (column 7 – 2 electrons) is d^5;

(d) Fe^{3+} (column 8 – 3 electrons) is d^5. These two ions have identical diagrams, with high-spin when the splitting is small and low-spin when the splitting is large:

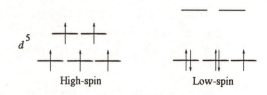

20.21   The magnetic properties of a complex are determined by its number of d electrons and the extent of its crystal field splitting energy:

(a) Ir^{3+} (column 9 – 3 electrons) is d^6, and NH_3 generates relatively large splitting. All the electrons will be paired, making the complex diamagnetic.

(b) Cr^{2+} (column 6 – 2 electrons) is d^4, and water generates relatively small splitting. The complex is paramagnetic with 4 unpaired electrons.

(c) Pt^{2+} (column 10–2 electrons) is d^8, so regardless of the splitting energy, this square planar complex is paramagnetic with 2 unpaired electrons.

(d) Pd has d^{10} configuration, so all orbitals are filled and this complex is diamagnetic.

20.23   The colors of transition metal complexes are generally determined by d-d transitions, but Zr^{4+} (column 4 – 4 electrons) has all its valence electrons removed, so there is no valence electron that can undergo a transition involving the absorption of visible light.

20.25   (a) The coordination number is the number of bonds formed between metal and ligands. The en ligand is bidentate, so three of them form six bonds, CN = 6.

(b) The oxidation number is determined by examining the net charge and correcting for charges on all species other than the transition metal. Here, en is neutral. Each Cl^- anion contributes –1 charge, so the oxidation number of Fe is +3. Fe is in column 8 of the periodic table, so its valence configuration is $3d^5$.

(c) Coordination number 6 means octahedral geometry.

(d) With an odd number of electrons, it is not possible for all of them to be paired, so this complex is paramagnetic.

(e) The complex is low-spin, so all but one of the electrons is paired.

20.27   Complexes are high spin and paramagnetic when the crystal field splitting is small, and they become low spin if the crystal field splitting is larger than the pairing energy. According to the spectrochemical series, CN^- produces greater splitting than H_2O.

Replacing the H_2O ligands with CN^- ligands increases the splitting, to the point where it takes less energy to pair the electrons than to promote them to the less stable d orbitals. Cr is in column 6 of the periodic table, so Cr^{2+} has d^4 valence configuration. There are four unpaired electrons in the high-spin configuration but only two in the low-spin configuration:

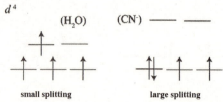

20.29 The wavelength of light that is absorbed provides a measure of the crystal field splitting energy:

$$\Delta = E = \frac{hcN_A}{\lambda} = \left(\frac{(6.626 \times 10^{-34} \text{ J s})(2.998 \times 10^8 \text{ m/s})(6.022 \times 10^{23}/\text{mol})}{465 \times 10^{-9} \text{ m}} \right)\left(\frac{10^{-3} \text{ kJ}}{1 \text{ J}} \right)$$

$\Delta = 257$ kJ/mol

The complex absorbs visible light around 465 nm, in the blue. The color of the complex will be the complementary color to blue. Consult Table 20-5 to determine that this color is orange.

20.31 Consult your textbook for the chemical reactions of various metallurgical processes:
(a) $CuFeS_2 + 3 O_2 \rightarrow CuO + FeO + 2 SO_2$
(b) $Si + O_2 + CaO \rightarrow CaSiO_3$
(c) $TiCl_4 + 4 Na \rightarrow 4 NaCl + Ti$

20.33 This is a standard stoichiometry problem. Begin by analyzing the chemistry. The reactants are Cu_2S and air and the given products are Cu metal and SO_2 gas.
Balanced reaction: $Cu_2S + O_2 \rightarrow 2 Cu + SO_2$
Start by computing the number of moles of Cu_2S, then use the appropriate mass-mole-mass and P-V-T calculations to determine the amounts of the products:

$$m_{Cu_2S} = 5.60 \times 10^4 \text{ kg}\left(\frac{10^3 \text{ g}}{1 \text{ kg}} \right)\left(\frac{2.37\%}{100\%} \right) = 1.327 \times 10^6 \text{ g}$$

$$n_{Cu_2S} = 1.327 \times 10^6 \text{g}\left(\frac{1 \text{ mol}}{159.2 \text{ g}} \right) = 8.34 \times 10^3 \text{ mol} = n_{SO_2}$$

$$m_{Cu} = 8.34 \times 10^3 \text{ mol}\left(\frac{2 \text{ mol Cu}}{1 \text{ mol Cu}_2\text{S}} \right)\left(\frac{63.55 \text{ g}}{1 \text{ mol}} \right) = 1.06 \times 10^6 \text{ g}$$

$$P_{total} = 755 \text{ torr}\left(\frac{1 \text{ atm}}{760 \text{ torr}} \right) = 0.993 \text{ atm}$$

$$V_{SO_2} = \frac{nRT}{P} = \frac{(8.34 \times 10^3 \text{mol})(0.08206 \frac{\text{L atm}}{\text{mol K}})(273.15 + 23.5 \text{ K})}{0.993 \text{ atm}} = 2.05 \times 10^5 \text{ L}$$

20.35 Standard free energy changes are calculated using standard free energies of formation found in Appendix D. $\Delta G^\circ_{reaction} = \Delta G^\circ_{products} - \Delta G^\circ_{reactants}$
ZnO (s) + C (s) $\rightarrow$ Zn (s) + CO (g)
    $\Delta G^\circ_{reaction} = [1 \text{ mol}(-137.2 \text{ kJ/mol}) + 0] - [1\text{mol}(-320.5 \text{ kJ/mol}) + 0] = 183.3$ kJ
ZnO (s) + CO (g) $\rightarrow$ Zn (s) + CO_2 (g)
    $\Delta G^\circ_{reaction} = [1(-394.4 \text{ kJ/mol}) + 1(0)] - [1(-320.5 \text{ kJ/mol}) + 1(-137.2 \text{ kJ/mol})]$
    $\Delta G^\circ_{reaction} = 63.3$ kJ

20.37 The coinage metals are those that have been used since antiquity for coins: copper, silver, and gold. All are in column 11 of the periodic table. They are characterized by high electrical conductivity, good ductility, and low chemical reactivity, in particular resistance to oxidation. Hence they are used for money (a vanishing use in technologically advanced countries), for electrical wire, for jewelry, and for other decorative objects. See your text for special examples of uses for compounds of these elements.

20.39 Titanium is used as an engineering metal because of its relatively low density, high bond strength, resistance to corrosion, and ability to withstand high temperatures, all of which make it a favored structural material.

20.41 The structural difference between hemoglobin and myoglobin is that the former has four subunits, while the latter has just one. As a consequence, hemoglobin has much more complex cooperative chemical behavior than myoglobin does.

20.43 An iron ion bonded to four sulfur atoms from cysteines is in a tetrahedral environment, so the splitting pattern is the 2-3 pattern characteristic of tetrahedral complexes. The iron cation loses one electron and is converted from Fe^{2+} to Fe^{3+}:

20.45 The number of possible isomers of a complex is determined by its geometry and the number of ligands of each type:

20.47 Bidentate ligands form complexes with two links. Each Fe ion forms an octahedral complex with three oxalate anions. When oxalate is applied to rust, it complexes and dissolves the Fe^{3+} cations. Here is a sketch showing the ligand-metal orientations:

20.49 To determine electron configurations, start from the position of the element in the periodic table and remove s electrons preferentially.
(a) Cr is in column 6, configuration $[Ar]\ 4s^1\ 3d^5$; Cr^{2+} is $[Ar]\ 3d^4$; Cr^{3+} is $[Ar]\ 3d^3$
(b) V is in column 5, configuration $[Ar]\ 4s^2\ 3d^3$; V^{2+} is $[Ar]\ 3d^3$; V^{3+} is $[Ar]\ 3d^2$; V^{4+} is $[Ar]\ 3d^1$; V^{5+} is $[Ar]$
(c) Ti is in column 4, configuration $[Ar]\ 4s^2\ 3d^2$; Ti^{2+} is $[Ar]\ 3d^2$; Ti^{4+} is $[Ar]$

20.51 Compounds that contain coordination complexes are named following the six rules stated in your textbook: name cation first, name ligands in alphabetical order, name metal, add "o" for anions, use Greek prefixes, add "-ate" for anionic complexes, list the oxidation number: (a) *cis*-tetraaquadichlorochromium(III) chloride;
(b) bromopentacarbonylmanganese(I); (c) *cis*-diamminedichloroplatinum(II)

20.53 Superoxide dismutase catalyzes the conversion of superoxide into molecular oxygen and hydrogen peroxide. This reaction occurs at a metal site that contains one Zn^{2+} ion and one Cu^{2+} ion, linked by a histidine ligand that bonds to both metal ions. An O atom binds to the Zn^{2+} ion:

$$2\ O_2^- + 2\ H_3O^+ \xrightarrow{\ SOD\ } O_2 + 2H_2O_2 + 2H_2O$$

20.55 Cu(II) in water forms an aqua complex. Addition of fluoride produces the insoluble green salt, CuF_2, while addition of chloride produces the bright green tetrachlorocopper(II) complex, $[CuCl_4]^{2-}$:
$$Cu^{2+}\ (aq) + 2\ F^-\ (aq) \rightarrow CuF_2\ (s)$$
$$Cu^{2+}\ (aq) + 4\ Cl^-\ (aq) \rightarrow [CuCl_4]^{2-}\ (aq)$$

20.57 Silver (column 11 of the periodic table) has a filled set of d orbitals, making the neutral metal difficult to oxidize. This gives silver good resistance to corrosion and makes it suitable for jewelry. Vanadium (column 5), with d^3 configuration, is readily oxidized, so it corrodes rapidly and is unsuited to jewelry.

20.59 The *mer* isomer of an octahedral complex has three like ligands arranged in a meridian plane:

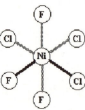

20.61 Silver tarnish is silver sulfide, Ag_2S, formed by reaction with trace amounts of H_2S in the atmosphere:

$$4\ Ag\ (s)\ +\ 2\ H_2S\ (g)\ +\ O_2\ (g) \rightarrow\ 2\ Ag_2S\ (s) + 2\ H_2O\ (g)$$

20.63 In these complexes, chromium is in the +3 oxidation state. From its location in Column 6 of the periodic table, we deduce that is has d^3 valence electron configuration. The three electrons occupy the three different t_{2g} orbitals, regardless of the magnitude of the splitting energy.

20.65 Orbital sketches show that the two orbitals experience quite different electron-electron repulsion in a square planar environment:

d_{z^2} $d_{x^2-y^2}$

20.67 (a) The charge on a complex is the charge on the transition metal cation, modified by any charges on the ligands. Ammonia and water are neutral, so when L is either of these, $n+ = 3+$; but chloride is -1, so for $L = Cl^-$, $n+ = 2+$.

(b) The color of each complex is the color that is complementary to the one corresponding to the wavelength of light that it absorbs. Consult Table 20-5 of your textbook to determine these: $L = Cl^-$, red to purple; $L = H_2O$, orange; and $L = NH_3$, yellow-orange.

(c) The wavelength of light that is absorbed provides a measure of the crystal field splitting energy:

$$\Delta = E = \frac{hcN_A}{\lambda}$$

$L = Cl^-$:

$$\Delta = \left(\frac{(6.626 \times 10^{-34} \text{ J s})(2.998 \times 10^8 \text{ m/s})(6.022 \times 10^{23}/\text{mol})}{515 \times 10^{-9} \text{ m}} \right)\left(\frac{10^{-3} \text{ kJ}}{1 \text{ J}} \right) = 232 \text{ kJ/mol}$$

$L = H_2O$:

$$\Delta = \left(\frac{(6.626 \times 10^{-34} \text{ J s})(2.998 \times 10^8 \text{ m/s})(6.022 \times 10^{23}/\text{mol})}{480 \times 10^{-9} \text{ m}} \right)\left(\frac{10^{-3} \text{ kJ}}{1 \text{ J}} \right) = 249 \text{ kJ/mol}$$

$L = NH_3$:

$$\Delta = \left(\frac{(6.626 \times 10^{-34} \text{ J s})(2.998 \times 10^8 \text{ m/s})(6.022 \times 10^{23}/\text{mol})}{465 \times 10^{-9} \text{ m}} \right)\left(\frac{10^{-3} \text{ kJ}}{1 \text{ J}} \right) = 257 \text{ kJ/mol}$$

The trend matches exactly the positions of these three ligands in the spectrochemical series: $Cl^- < H_2O < NH_3$

20.69   Tetracarbonylnickel(0), $[Ni(CO)_4]$ has tetrahedral geometry since Ni is a first-row transition element. Because of the strong field ligand, CO, Δ is large, which will result in the compound absorbing UV light and appearing colorless.
Tetracyanozinc(II) $[Zn(CN)_4]^{2-}$ has tetrahedral geometry (Zn is a first row transition metal) where Zn has a +2 charge (d^{10}). This complex is colorless because there are no possible d-d transitions.

20.71   Brass is an alloy of zinc and copper, and superoxide dismutase contains zinc and copper ions in its reaction center.

20.73   Four-coordinate complexes may be either tetrahedral or square planar. The splitting patterns for these two geometries show that the d^8 configuration can have zero spin in the square planar case but not in the tetrahedral case. Thus, the magnetic behavior indicates that $[Ni(CN)_4]^{2-}$ is square planar and $[NiCl_4]^{2-}$ is tetrahedral:

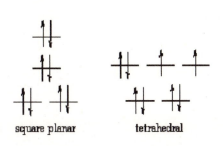

square planar   tetrahedral

20.75 When ferritin is neither empty nor filled to capacity, the protein has the capacity to provide iron as needed for hemoglobin synthesis, or to store iron if an excess is absorbed by the body.

20.77 Visible spectroscopy is useful when a compound has an energy gap between the highest occupied and lowest unoccupied orbital that matches the energy of visible light. Because Zn^{2+} has d^{10} configuration, its d orbitals are completely filled, and the lowest unoccupied orbital is quite high in energy. In contrast, Co^{2+} has d^7 configuration, giving this cation unfilled d orbitals. Consequently, metalloproteins that contain Co^{2+} absorb visible light, making it possible to study them with visible spectroscopy.

20.79 Use standard reduction potentials from Appendix F to determine which metals can be displaced by Zn. Any metal with a less negative standard potential can be displaced:

$$Zn^{2+} + 2e^- \rightleftharpoons Zn \qquad E° = -0.7618 \text{ V}$$

The following are a few examples of metals that can be displaced by Zn:

$$Co^{2+} + 2e^- \rightleftharpoons Co \qquad E° = -0.28 \text{ V}$$

$$Cu^{2+} + 2e^- \rightleftharpoons Cu \qquad E° = +0.3419 \text{ V}$$

$$Fe^{2+} + 2e^- \rightleftharpoons Fe \qquad E° = -0.447 \text{ V}$$

$$Ni^{2+} + 2e^- \rightleftharpoons Ni \qquad E° = -0.257 \text{ V}$$

When combined with the Zn half-reaction, the overall cell voltage is positive and the process is spontaneous. For example:

$$Zn + Co^{2+} \rightleftharpoons Co + Zn^{2+} \qquad E° = -0.28 \text{ V} - (-0.7618 \text{ V}) = 0.48 \text{ V}$$

21.1   Lewis acids are electron pair acceptors, and Lewis bases are electron pair donors. Thus, Lewis acids are electron-deficient, while Lewis bases have lone pairs to donate:

(a) Lewis acid is Ni, Lewis base is CO.

(b) Lewis acid is $SbCl_3$, Lewis base is Cl^-.

(c) Lewis acid is $AlBr_3$, Lewis base is $P(CH_3)_3$.

(d) Lewis acid is BF_3, Lewis base is ClF_3.

21.3   Construct Lewis structures and determine steric numbers to identify the three-dimensional structure of a molecule.

(a) AlF_3

1. There are 3+3(7) = 24 valence electrons.
2. Three electron pairs are needed for the bonding framework.
3. The remaining electrons are used to place 3 pairs on each outer F atom.

The Lewis structure shows SN = 3 for Al, so the molecule has trigonal planar geometry and there is a vacant $3p$ orbital perpendicular to the molecular plane:

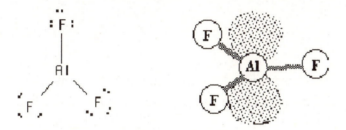

(a) SbF_5

1. There are 5 + 5(7) = 40 valence electrons.
2. Five pairs are used in the bonding framework leaving 40 – 2(5) = 30 electrons.
3. The remaining electrons are used to place 3 electron pairs on each F atom.

The Lewis structure shows SN = 5 for Sb, so the molecule has trigonal bipyramidal geometry, indicating dsp^3 hybridization. There are vacant $3d$ orbitals that do not participate in the dsp^3 hybridization:

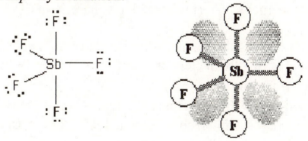

(b) SO_2

1. There are 6 + 2(6) = 18 valence electrons.
2. Two electron pairs are needed for the bonding framework, leaving 18 – 2(2) = 14.
3. Place 3 pairs of electrons on each outer O atom.
4. Place the remaining two electrons on the inner S atom.
5. The resulting structure has $FC_S = 6 – 2 – 2 = 2$, move two lone pairs from the outer O atoms to form two double bonds and reduce the formal charge to 0.

The Lewis structure shows SN = 3 for S, (so the molecule has a trigonal planar geometry with a lone pair bent) and there is a delocalized π orbital to which an electron pair can be added:

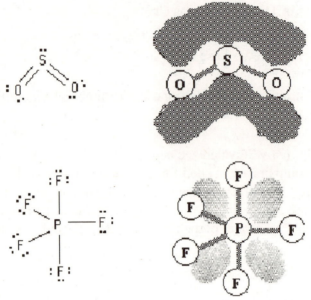

21.5 In the first step, an electron pair from the O atom of H_2O displaces a π bond in Lewis acid-base adduct formation. Then a proton from H_2O migrates to a C–O oxygen atom:

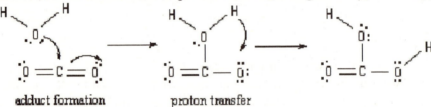

21.7 Polarizability of cations increases substantially with the value of **n** and decreases, within the same row of the periodic table, with Z and the charge on the ion:

$$Fe^{3+} \text{ (high charge)} < V^{3+} \text{ (lowest } Z, \text{ high charge)} < Fe^{2+} < Pb^{2+} \text{ (large } n)$$

21.9 Hard acids have low polarizability, and soft acids have high polarizability:
(a) The hardest acid is BF_3 (both elements from row 2), then BCl_3, and $AlCl_3$ is softest (both elements from row 3).
(b) The hardest acid is Al^{3+} (row 3), then Tl^{3+} (row 6), and Tl^+ (low charge) is softest.
(c) Polarizability increases with **n**, so hardest is $AlCl_3$, then $AlBr_3$, and AlI_3 is softest.

21.11 When an electronegative O atom bonds to a less electronegative S atom, it withdraws electron density from S, decreasing the polarizability about S and increasing the hardness of the base.

21.13 A metathesis reaction will occur if exchange of partners couples the harder Lewis acid with the harder Lewis base and the softer Lewis acid with the softer Lewis base:

(a) Al^{3+} is harder than Na^+, and Cl^- is harder than I^-, so metathesis occurs, giving $AlCl_3$ (hard-hard) and NaI (soft-soft).

(b) the Lewis acid is Ti^{4+} in each substance, so no reaction occurs.

(c) Ca^{2+} is softer than H^+, and S^{2-} is softer than O^{2-}, so metathesis occurs, giving H_2O (hard-hard) and CaS (soft-soft).

(d) C (in CH_3^-) is a soft base, and Li^+ is a hard acid, so metathesis occurs, giving $(CH_3)_3P$ (soft-soft) and $LiCl$ (hard-hard).

(e) Ag^+ and I^- are both soft, and Si^{4+} and Cl^- are both hard, so no reaction occurs.

21.15 Descriptions of bonding always begin with a Lewis structure. As your textbook describes, Al_2Cl_6 contains two "bridging" chlorine atoms. Standard procedures would predict tetrahedral geometry about all inner atoms, but the Al–Cl–Al bond angles of 91° indicate that Cl uses p orbitals. Each Al atom can be described as using sp^3 hybrids to form four σ bonds to four different Cl atoms.

In Lewis acid-base terms, the bridged molecule forms from two $AlCl_3$ units linking together in double adduct formation between the Al Lewis acid atoms and two Cl Lewis base atoms.

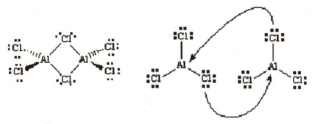

21.17 Thallium lies below indium and gallium, so its properties should be similar to those metals: valence of 3, soft Lewis acid. Like its neighbor, Pb, it is toxic.

21.19 Determine the Lewis structure using standard procedures:
$SnCl_4$

1. There are $4 + 4(7) = 32$ valence electrons.
2. Four electron pairs are needed for the bonding framework, leaving $32 – 4(2) = 24$.
3. Place 3 pairs of electrons on each outer Cl atom, leaving $24 – 4(6) = 0$.
4. No remaining electrons.
5. The resulting structure has $FC_{Sn} = 4 – 4 = 0$, so the structure is complete.

$SnCl_4$ can function as a Lewis acid because the Sn atom has empty d orbitals that can accept electrons to form more bonds.

21.21   A BN pair has the same number of valence electrons as a pair of C atoms, so the structure and bonding of $B_3N_3H_6$ is just like that of benzene, a six-membered planar ring with a delocalized set of π bonds. There are $3(3) + 3(5) + 6(1) = 30$ valence electrons, all of which are involved in the bonding network:

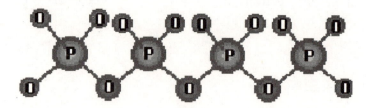

The B and N atoms have bonding and geometry that can be described using sp^2 hybrid orbitals, resulting in 3 σ bonds around each ring atom. In addition, there is a delocalized π bonding network encompassing all six ring atoms and containing 6 electrons.

21.23   The band gap decreases from top to bottom of each column of the periodic table, so Ge has a smaller band gap than Si. In orbital terms, this is because the principal quantum number of the valence orbitals increases. This makes the valence orbitals larger, leading to less effective overlap and smaller energy difference between bonding and antibonding orbitals.

21.25   Follow the example in section 21.4 of your text on polymer formation.
$$2\ C_2H_5Cl + Si(Cu) \rightarrow (C_2H_5)_2SiCl_2 + Cu$$
$$(C_2H_5)_2SiCl_2 + 2\ H_2O \rightarrow (C_2H_5)_2Si(OH)_2 + 2\ HCl$$
Condensation will eliminate water to give the polymer:
$$2\ (C_2H_5)_2Si(OH)_2 \rightarrow \text{polymer-linkage} + H_2O$$

21.27   Polyphosphates form by sequential condensation of PO_3^- units, so the tetraphosphate has chemical formula $P_4O_{13}^{6-}$:

21.29   Phosphoric acid is produced directly from apatite by reaction with sulfuric acid:

$$Ca_5(PO_4)_3X\ (s) + 5\ H_2SO_4\ (aq) \rightarrow 3\ H_3PO_4\ (aq) + 5\ CaSO_4\ (s) + HX\ (aq)$$

$(X = F^-,\ OH^-,\ or\ Cl^-)$

This is a Brønsted acid-base reaction in which protons are transferred from sulfuric acid to the phosphate anions. There are no redox reactions in this process.

21.31   White phosphorus consists of P_4 tetrahedra. To convert this form to the red form, break one bond in each tetrahedron and use the electrons to form bonds between P_4 units:

21.33   There are two industrial reactions in Section 21.6 in which sulfuric acid acts as a Brønsted acid:

$$CaF_2(s) + H_2SO_4\ (l) \rightarrow 2\ HF\ (g) + CaSO_4\ (s)$$

$$FeTiO_3\ (s) + 2\ H_2SO_4\ (aq) + 5\ H_2O\ (l) \xrightarrow{150\text{-}180\ ^{\circ}C} TiOSO_4\ (aq) + FeSO_4 \cdot 7H_2O\ (s)$$

21.35   The repeat structure of polyvinylchloride, in common with all polyethylene-type polymers, has an all-carbon backbone. There is a chlorine atom on every other carbon atom:

21.37   The reaction forming $TiCl_4$ from TiO_2 is as follows:

$$TiO_2\ (s) + C\ (s) + 2\ Cl_2\ (g) \rightarrow TiCl_4\ (l) + CO_2\ (g)$$

C goes from zero to +4 oxidation state, so it is oxidized and serves as the reducing agent. Cl goes from zero to -1 oxidation state, so it is reduced and serves as the oxidizing agent.

21.39 This is a stoichiometry problem that can be worked by determining the percentage composition of bauxite:

Bauxite is AlOOH, MM = 60.00 g/mol $\% \text{ Al} = 100\% \left(\dfrac{26.982 \text{ g/mol}}{60.00 \text{ g/mol}} \right) = 44.97\,\%$

1 kg of bauxite rock contains $1.00 \text{ kg} \left(\dfrac{85\%}{100\%} \right) \left(\dfrac{44.97\%}{100\%} \right) = 0.382$ kg Al

If the processing is 75 % efficient, each 1.00 kg of bauxite rock yields the following:
(0.382 kg)(0.75) = 0.287 kg Al

To produce 2500 kg of Al requires the following:

$2500 \text{ kg Al} \left(\dfrac{1 \text{ kg bauxite rock}}{0.287 \text{ kg Al}} \right) = 8.7 \times 10^3$ kg bauxite rock

21.41 Sulfur has d orbitals available for bond formation, allowing the formation of SF_6, in which the bonding can be described using d^2sp^3 hybrid orbitals on the S atom. Oxygen has no valence d orbitals available. In principle, SBr_6 could also form, but the Br atom is too large for six Br atoms to be accommodated around a central S atom.

21.43 Lewis acids are electron pair acceptors, and Lewis bases are electron pair donors. The As atom in $AsCl_3$ has a lone pair that it donates, giving this compound Lewis base character. The As atoms also can accommodate additional electron pairs by using valence d orbitals, giving this compound Lewis acid character:

$AsCl_3 + BF_3 \rightarrow Cl_3As–BF_3$ ($AsCl_3$ acts as Lewis base)

$AsCl_3 + Cl^- \rightarrow AsCl_4^-$ ($AsCl_3$ acts as Lewis acid)

21.45 Balance a redox half-reaction following the standard steps outlined in Chapter 19 (balance all but H and O by inspection, balance O by adding H_2O, balance H by adding H_3O^+/H_2O or OH^-/H_2O, balance charge by adding electrons):

Begin by balancing the Al reaction:

$Al + 2 H_2O \rightarrow AlO_2^-$; add 4 OH^- on the reactant side and 4 H_2O to the product side to balance H: $Al + 2 H_2O + 4 OH^- \rightarrow AlO_2^- + 4 H_2O$

Cancel 2 H_2O on each side and add 3 electrons to balance charge:

$Al + 4 OH^- \rightarrow AlO_2^- + 2 H_2O + 3 e^-$

(a) $NO_3^- \rightarrow NH_3 + 3 H_2O$; add 9 OH^- on the product side and 9 H_2O on the reactant to balance H: $NO_3^- + 9 H_2O \rightarrow NH_3 + 3 H_2O + 9 OH^-$

Cancel 3 H_2O on each side and add 8 electrons to balance the charge:

$NO_3^- + 6 H_2O + 8 e^- \rightarrow NH_3 + 9 OH^-$;

Multiply this half-reaction by 3 and the Al half-reaction by 8 to balance the electrons, and add:

$8 Al + 32 OH^- + 3 NO_3^- + 18 H_2O + 24 e^- \rightarrow 3 NH_3 + 27 OH^- + 8 AlO_2^- + 16 H_2O + 24 e^-$

Canceling duplicated species yields:

$8 Al + 5 OH^- + 3 NO_3^- + 2 H_2O \rightarrow 3 NH_3 + 8 AlO_2^-$

(b) $2 H_2O + 2 e^- \rightarrow H_2 + 2 OH^-$

Multiply this half-reaction by 3 and the Al half-reaction by 2, and add:

$2 Al + 8 OH^- + 6 H_2O + 6 e^- \rightarrow 3 H_2 + 6 OH^- + 2 AlO_2^- + 4 H_2O + 6 e^-$

Cancel duplicated species:

$2 Al + 2 OH^- + 2 H_2O \rightarrow 3 H_2 + 2 AlO_2^-$

(c) $SnO_3^{2-} \rightarrow Sn + 3 H_2O$

Add 6 OH$^-$ on the product side and 6 H$_2$O on the reactant side to balance H:

$SnO_3^{2-} + 6 H_2O \rightarrow Sn + 3 H_2O + 6 OH^-$

Cancel 3 H$_2$O on each side and add 4 electrons to balance charge:

$SnO_3^{2-} + 3 H_2O + 4 e^- \rightarrow Sn + 6 OH^-$

Multiply this half-reaction by 3 and the Al half-reaction by 4, and add:

$4 Al + 16 OH^- + 3 SnO_3^{2-} + 9 H_2O + 12 e^- \rightarrow 3 Sn + 18 OH^- + 4 AlO_2^- + 8 H_2O + 12 e^-$

Canceling duplicated species yields:

$4 Al + 3 SnO_3^{2-} + H_2O \rightarrow 3 Sn + 2 OH^- + 4 AlO_2^-$

21.47   Nitrogen, at the top of Group 15, is a non-metal showing a valence of three that forms polar bonds most readily with other non-metals (B, C, O, the halogens). Phosphorus has a similar pattern of reactivity but can also involve d orbitals in its bonding. Arsenic and antimony are metalloids with useful semiconductor properties, and Bismuth is metallic.

21.49   The production of aluminum from its ore is described in Section 21.3. First, bauxite ore is treated with strong base to produce soluble $[Al(OH)_4]^-$:

$Al(O)OH (s) + OH^- (aq) + H_2O(l) \rightarrow [Al(OH)_4]^- (aq)$

When the solution is diluted with water, aluminum hydroxide precipitates:

$[Al(OH)_4]^- (aq) \rightarrow Al(OH)_3 (s) + OH^- (aq)$

Strong heating drives off water, leaving aluminum oxide:

$2 Al(OH)_3 \xrightarrow{1250\ ^\circ C} Al_2O_3 (s) + 3 H_2O (g)$

Finally, electrolysis of aluminum oxide dissolved in cryolite reduces Al to pure metal:

$2 Al_2O_3 (melt) + 3 C (s) \rightarrow 3 CO_2 (g) + 4 Al (l)$

21.51   Silicon dioxide is first converted into silicon tetrachloride by reaction with molecular chlorine:

$SiO_2 (s) + 2 C (s) + 2 Cl_2 (g) \xrightarrow{high\ T} SiCl_4 (g) + 2 CO (g)$

The SiCl$_4$ is purified by distillation and then reduced by reaction with Mg metal:

$SiCl_4 (l) + 2 Mg (s) \rightarrow Si (s) + 2 MgCl_2 (s)$

21.53   Metal cations are Lewis acids, and the higher the principal quantum number of the valence electrons, the softer the acid. Hence, Hg is a very soft Lewis acid, whereas Zn is relatively hard. Among anions, S^{2-} is considerably softer than O-containing anions. By the HSAB principle, Hg forms compounds preferentially with S^{2-}, whereas Zn combines with harder bases that contain O.

21.55 Gaseous Al_2Cl_6 is in equilibrium with gaseous $AlCl_3$: Al_2Cl_6 (g) $\rightleftharpoons$ 2 $AlCl_3$ (g).
Le Châtelier's principle predicts that because the forward reaction is endothermic (bonds must be broken), an increase in temperature shifts the position of the equilibrium to the right. Thus as temperature increases, the number of moles of gaseous substance increases, so the pressure increases faster than would be predicted by the ideal gas equation.

21.57 The reaction can be balanced by recognizing the chemical formulas of the substances involved. The reactants are phosphoric acid (H_3PO_4), and fluoroapatite ($Ca_5(PO_4)_3F$); the products are calcium dihydrogen phosphate ($Ca(H_2PO_4)_2$) and HF:
The unbalanced reaction is:
$$Ca_5(PO_4)_3F \ (s) + H_3PO_4 \ (aq) \rightarrow Ca(H_2PO_4)_2 \ (s) + HF \ (aq)$$
$$5\ Ca + 4\ P + 16\ O + F + 3\ H \rightarrow Ca + 2\ P + 8\ O + F + 5\ H$$
Balance Ca by giving $Ca(H_2PO_4)_2$ a coefficient of 5:
$$Ca_5(PO_4)_3F \ (s) + H_3PO_4 \ (aq) \rightarrow 5\ Ca(H_2PO_4)_2 \ (s) + HF \ (aq)$$
$$5\ Ca + 4\ P + 16\ O + F + 3\ H \rightarrow 5\ Ca + 10\ P + 40\ O + F + 21\ H$$
Next balance phosphorus by giving H_3PO_4 a coefficient of 7:
$$Ca_5(PO_4)_3F \ (s) + 7\ H_3PO_4 \ (aq) \rightarrow 5\ Ca(H_2PO_4)_2 \ (s) + HF \ (aq)$$
$$5\ Ca + 10\ P + 40\ O + F + 21\ H \rightarrow 5\ Ca + 10\ P + 40\ O + F + 21\ H$$
Note that now all atoms are balanced:
$$Ca_5(PO_4)_3F \ (s) + 7\ H_3PO_4 \ (aq) \rightarrow 5\ Ca(H_2PO_4)_2 \ (s) + HF \ (aq)$$
Calculate the mass percent phosphorus:
$$MM_{compound} = \{40.078 + 2[(2)(1.008) + 30.974 + (4)(15.999)]\} = 234.05 \text{ g/mol}$$
$$\% \ P = (100\%) \ \frac{2 MM_P}{MM_{compound}} = 100\% \left(\frac{(2)(30.974 \text{ g/mol})}{234.05 \text{ g/mol}} \right) = 26.47 \ \%$$

21.59 (a) Determine the Lewis structure of $SOCl_2$ following the usual procedure. There are 26 valence electrons, which give a provisional structure in which an inner S atom has +1 formal charge. Make one π bond to minimize formal charges:

(b) Balance the chemical reaction by inspection. Each H_2O requires one $SOCl_2$:
$$6\ SOCl_2 \ (l) + FeCl_3 \cdot 6H_2O \ (s) \rightarrow 6\ SO_2 \ (g) + 12\ HCl \ (g) + FeCl_3 \ (s).$$

21.61 Polarizability increases with atomic size, and greater polarizability leads to larger dispersion forces. Thus, the larger the atoms in a molecule, the larger the intermolecular forces and the easier it is to condense the substance. In the sequence BCl_3, BBr_3, BI_3, the halogens increase in size, accounting for the different stable phases at room temperature.

21.63 The phosphate anion is the conjugate base of a weak acid, HPO_4^{2-}:

$PO_4^{3-} + H_2O \rightarrow HPO_4^{2-} + OH^-$ $pK_b = 14.00 - pK_a = 14.00 - 12.32 = 1.68$

21.65 The reactions can be balanced by inspection:

$2\ Al_2O_3 + 6\ Cl_2 \rightarrow 4\ AlCl_3 + 3\ O_2$

$AlCl_3 + 3\ Na \rightarrow Al + 3\ NaCl$

21.67 Compounds with Lewis acid-base properties will undergo metathesis reactions if a transfer of bonding partners associates harder acids with harder bases.

(a) Al^{3+} is quite hard due to its +3 charge, and Cl^- is softer than CH_3:

$AlCl_3 + 3\ LiCH_3 \rightarrow Al(CH_3)_3 + 3\ LiCl$

(b) Sulfur is a Lewis acid and a Lewis base and forms an adduct with H_2O, a Lewis base:

$SO_3 + H_2O \rightarrow H_2SO_4\ (aq)$

(c) The antimony atom can bond to one additional F^- anion: $SbF_5 + LiF \rightarrow Li(SbF_6)$;

(d) S is harder than As, and F is harder than Cl, so there is no reaction.

21.69 The reaction consumes 12 H atoms for 4 P atoms, or 3 H atoms for every P atom. Thus, the product is H_3PO_4: $P_4O_{10} + 6\ H_2O \rightarrow 4\ H_3PO_4$

22.1 The elemental symbol identifies the value of Z, and the left superscript is A. $Z + N = A$:

Part:	(a)	(b)	(c)
Z	10	82	40
A	20	205	90
N	10	123	50

22.3 Nuclides in the "belt of stability" are stable. Instability occurs if a nuclide has too few neutrons, too many neutrons, or $Z > 83$. In addition, most odd-odd nuclides are unstable. Mn has $Z = 25$. In this region of the periodic table, the belt of stability has $N:Z \sim 1.2$

A	$N:Z$	stability	reason
53	1.12	unstable	too few neutrons
54	1.16	unstable	odd-odd
55	1.20	stable	in belt of stability
56	1.24	unstable	too many neutrons, odd-odd

22.5 Energy releases are calculated using Equation 22-3: $\Delta E = (\Delta m)(8.988 \times 10^{10} \text{ kJ/g})$:

$$\Delta m = 1.00 \text{ met ton} \left(\frac{10^3 \text{ kg}}{1 \text{ met ton}} \right)\left(\frac{10^3 \text{ g}}{1 \text{ kg}} \right) = 1.00 \times 10^6 \text{ g}$$

$$\Delta E = 1.00 \times 10^6 \text{ g} \left(\frac{8.988 \times 10^{10} \text{ kJ}}{1 \text{ g}} \right) = 8.99 \times 10^{16} \text{ kJ}$$

22.7 Binding energy is calculated by determining the mass defect (difference between the mass of the atom and the sum of the masses of its individual components) and converting to energy. When an element has only one stable isotope, its molar mass is the molar mass of that isotope:

$MM_{Cs} = 132.91$ g/mol

The stable isotope has 55 protons and electrons, and $133 - 55 = 78$ neutrons:

$m_{components} = 55(1.007276 + 0.0005486) + 78(1.008665) = 134.106223$ g/mol

$\Delta m = 132.91$ g/mol $- 134.106223$ g/mol $= - 1.20$ g/mol

$$\Delta E = \left(\frac{-1.20 \text{ g}}{1 \text{ mol}} \right)\left(\frac{8.988 \times 10^{10} \text{ kJ}}{1 \text{ g}} \right) = -1.079 \times 10^{11} \text{ kJ/mol}$$

$$\Delta E \text{ (per nucleon)} = \frac{-1.079 \times 10^{11} \text{kJ/mol}}{133 \text{ nucleons}} = -8.11 \times 10^8 \text{ kJ mol}^{-1} \text{ nucleon}^{-1}$$

22.9 The repulsive barrier for fusion depends on the product of the nuclear charges and on the radii of the two nuclides. The stability of the product nuclide depends on its proton-neutron ratio: $^1\text{H} + {}^6\text{Li} \rightarrow {}^7\text{Be}$, charge product $= (1)(3) = 3$ and product has n/p $= 0.75$; $^4\text{He} + {}^4\text{He} \rightarrow {}^8\text{Be}$, charge product $= (2)(2) = 4$ and product has n/p $= 1.00$; the sum of the two nuclear radii is about the same for both reactions. Thus, the repulsive barrier is greater for the second reaction, but the product nuclide is more stable.

22.11 Symbols and names for nuclear particles appear in Table 22-3 and should be memorized:

	Description	Symbol	Name
(a)	high energy photon	γ	gamma ray
(b)	positive particle with mass number 4	α	alpha particle
(c)	positive particle with m_e	β^+	positron

22.13 To identify the products of nuclear decay processes, make use of the principles of conservation of mass number and charge:

(a) no change in mass number or charge, product is $^{125}_{52}\text{Te}$;

(b) EC changes nuclear charge by –1, no change in mass number, product is $^{123}_{51}\text{Sb}$;

(c) β decay changes nuclear charge by +1, no change in mass number, gamma decay does not change mass number or charge, product is $^{127}_{53}\text{I}$.

22.15 Nuclides with too many neutrons decay by emitting beta particles, thereby increasing Z while decreasing N. Nuclides with too few neutrons decay by emitting positrons or capturing electrons. Odd-odd nuclides may decay by any of these three processes.
^{53}Mn has too few neutrons, may emit a positron or undergo electron capture (observed mode is EC).
^{54}Mn is odd-odd, could decay by any of the three modes (observed mode is EC).
^{55}Mn is stable. ^{56}Mn has too many neutrons, decays by beta emission.

22.17 Equation 22-4 is used to determine half-lives from radioactive decay data:

$$\text{Rate} = \frac{\Delta N}{\Delta t} = \frac{-N \ln 2}{t_{1/2}}$$

Summarize the known data: rate $= -242$ decays/s; $m = 1.33 \times 10^{-12}$ g
Convert mass to number of nuclei, using $MM \cong A = 26 + 33 = 59$ g/mol

$$N = 1.33 \times 10^{-12}\text{g}\left(\frac{1 \text{ mol}}{59 \text{ g}}\right)\left(\frac{6.022 \times 10^{23} \text{ nuclei}}{1 \text{ mol}}\right) = 1.36 \times 10^{10} \text{ nuclei}$$

$$t_{1/2} = -\frac{(1.36 \times 10^{10} \text{ nuclei})(\ln 2)}{-242 \text{ decays s}^{-1}} = 3.9 \times 10^7 \text{ s}$$

Convert to a more convenient time unit:

$$3.9 \times 10^7 \text{s}\left(\frac{1 \text{ min}}{60 \text{ s}}\right)\left(\frac{1 \text{ hr}}{60 \text{ min}}\right)\left(\frac{1 \text{ day}}{24 \text{ hr}}\right)\left(\frac{1 \text{ yr}}{365 \text{ day}}\right) = 1.2 \text{ yr}$$

22.19 Identify products of nuclear decay by noting that α-decay changes A by –4 and Z by –2, while β-decay changes Z by +1 while leaving A unchanged:
$^{232}_{90}\text{Th} \rightarrow \alpha + \ ^{228}_{88}\text{Ra} \rightarrow \beta + \ ^{228}_{89}\text{Ac} \rightarrow \beta + \ ^{228}_{90}\text{Th} \rightarrow \alpha + \ ^{224}_{88}\text{Ra} \rightarrow \alpha + \ ^{220}_{86}\text{Rn}$
$^{220}_{86}\text{Rn} \rightarrow \alpha + \ ^{216}_{84}\text{Po} \rightarrow \beta + \ ^{216}_{85}\text{At} \rightarrow \alpha + \ ^{212}_{83}\text{Bi} \rightarrow \beta + \ ^{212}_{84}\text{Po} \rightarrow \alpha + \ ^{208}_{82}\text{Pb}$

22.21 Identify the compound nucleus and the final products of a nuclear reaction by applying the principles of conservation of mass number and charge:

(a) $^{12}_{6}C + ^{1}_{0}n \rightarrow \left(^{13m}_{6}C\right) \rightarrow ^{12}_{5}B + ^{1}_{1}p$

(b) $^{16}_{8}O + ^{4}_{2}\alpha \rightarrow \left(^{20m}_{10}Ne\right) \rightarrow ^{20}_{10}Ne + \gamma$

(c) $^{247}_{96}Cm + ^{11}_{5}B \rightarrow \left(^{258m}_{101}Md\right) \rightarrow ^{255}_{101}Md + 3 \, ^{1}_{0}n$

22.23 Pictures of nuclear reactions should show each nuclide with the appropriate numbers of protons and neutrons. The reaction in this problem is:

$^{14}_{7}N + ^{4}_{2}\alpha \rightarrow \left(^{18m}_{9}F\right) \rightarrow ^{18}_{8}O + ^{0}_{1}\beta^{+}$

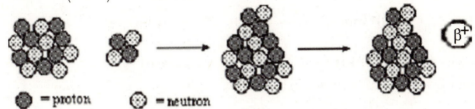

● = proton   ◉ = neutron

22.25 Stable elements with mass numbers around 135 have N/Z ratios in the range of 1.4. Use $A = N + Z$ to calculate what Z value is likely to be stable:

$N = A - Z$ $\qquad \dfrac{135 - Z}{Z} = 1.4$ $\qquad 2.4 \, Z = 135$ $\qquad Z = 56$

Thus, the most likely high-mass elements resulting from U-235 fission are Cs, Ba, La, Ce.

22.27 Calculate the mass defect from the masses of the individual reactants and products, then convert to energy released:

$\Delta m = [80.9199 + 151.9233 + 3(1.0087)] - [235.0439 + 1.0087] = -0.1833$ g/mol

$\Delta E = \left(\dfrac{-0.1833 \text{ g}}{1 \text{ mol}}\right)\left(\dfrac{8.988 \times 10^{10} \text{ kJ}}{1 \text{ g}}\right) = -1.648 \times 10^{10}$ kJ/mol

This result is somewhat less than the result of the general calculation of Section Exercise 22.4.1 for a net change of 2 neutrons.

22.29 Your description should feature the fact that the core of a nuclear reactor generates radiation that converts materials in the vicinity of the core into radioactive substances. Thus, the heat exchanger in immediate contact with the core becomes radioactive and must be separated from the turbine that generates electricity. The primary heat exchanger transfers energy to a secondary heat exchanger, which does not become radioactive and can safely drive the turbine.

22.31 The amount of energy released in a fusion reaction can be calculated from the energy per event and the total amount of matter undergoing fusion. The reactions in this example are

$$^2H + {}^3H \rightarrow {}^4He + n \qquad \Delta E = -1.7 \times 10^9 \text{ kJ/mol}$$
$$^6Li + n \rightarrow {}^4He + {}^3H \qquad \Delta E = -4.6 \times 10^8 \text{ kJ/mol}$$

$$n_H = 2.50 \text{ g}\left(\frac{1 \text{ mol}}{2.014 \text{ g}}\right) = 1.241 \text{ mol}$$

$$\Delta E = 1.241 \text{ mol}\left(\frac{-1.7 \times 10^9 \text{ kJ}}{1 \text{ mol}}\right) = -2.11 \times 10^9 \text{ kJ}$$

In addition, 0.6207 mol Li react:

$$\Delta E = 0.6207 \text{ mol}\left(\frac{-4.6 \times 10^8 \text{ kJ}}{1 \text{ mol}}\right) = -2.86 \times 10^8 \text{ kJ}$$

$\Delta E_{total} = (-2.11 \times 10^9 \text{ kJ}) + (-2.86 \times 10^8 \text{ kJ}) = -2.4 \times 10^9 \text{ kJ}$
(2 significant figures because ΔE has 2)

22.33 To determine the speed of a nucleus that can fuse with another, first determine the energy needed to surmount the repulsive barrier, as described by Equation 6-1:

$$E_{electrical} = \frac{(1.389 \times 10^5 \text{ kJ pm/mol})(Z_1)(Z_2)}{d}$$

Then calculate the speed from the equation for kinetic energy, $E_{kinetic} = 1/2\, mu^2$:

$$u = \sqrt{\frac{2E_{kinetic}}{m}}$$

Determine the nuclide radii using the equation given in the problem:
$r_{tritium\ nucleus} = 1.2(3)^{1/3} \text{ fm } (10^{-3} \text{ pm/fm}) = 1.7 \times 10^{-3} \text{ pm}$
$r_{deuterium\ nucleus} = 1.2(2)^{1/3} \text{ fm } (10^{-3} \text{ pm/fm}) = 1.5 \times 10^{-3} \text{ pm}$
Here are the data:
  $Z = 1$ for both nuclides; and $m = 3.0$ g/mol for tritium

$$E = \frac{(1.389 \times 10^5 \text{ kJ pm/mol})(1)(1)}{(1.7 \times 10^{-3} \text{ pm} + 1.5 \times 10^{-3} \text{ pm})} = 4.3 \times 10^7 \text{ kJ/mol}$$

$$u = \left[\frac{(2)(4.3 \times 10^7 \text{ kJ/mol})(10^3 \text{ J/kJ})}{(3.0 \text{ g/mol})(10^{-3} \text{ kg/g})}\right]^{1/2} = 5.3 \times 10^6 \text{ m/s}$$

(Remember that $1 \text{ J} = 1 \text{ kg m}^2/\text{s}^2$, so the units in this calculation cancel to give m/s.)

22.35 Your description should include the extremely high energies required to initiate fusion and the difficulties in containing the fusion components at the temperature required for nuclei to have these high energies.

22.37 The characteristics of first-generation stars are described in your text:

Stage	**Temperature**	**Composition**
H-burning	4×10^7 K	H, He, e⁻
He-burning	10^8 K	He, Be, C, O (H depleted)
C-burning	10^9 K	all nuclides from $Z = 6$ (C) up to $Z = 26$ (Fe)

22.39 Your description should include the fact that elements beyond $Z = 26$ are less stable than Fe, so they cannot be generated by fusion of lighter elements.

22.41 The problem asks for the energy released by a radioactive isotope. To determine this, it is necessary to calculate how much material decays during the time period. Determine the amount of the radioactive nuclide that decays in one day using Equation 22-5:

$$\ln\left(\frac{N_0}{N}\right) = \frac{t \ln 2}{t_{1/2}}$$

$$\ln\left(\frac{N_0}{N}\right) = \frac{(1 \text{ day})(0.693)}{(8.07 \text{ day})} = 0.0859 \qquad \left(\frac{N_0}{N}\right) = e^{0.0859} = 1.0897$$

$$N = \frac{7.45 \text{ pg}}{1.0897} = 6.84 \text{ pg} \qquad\qquad \Delta N = 7.45 \text{ pg} - 6.84 \text{ pg} = 0.61 \text{ pg}$$

Now convert to moles and multiply by the energy released per mole to obtain the amount of energy captured by the gland:

$$n = 0.61 \text{ pg}\left(\frac{10^{-12} \text{ g}}{1 \text{ pg}}\right)\left(\frac{1 \text{ mol}}{131 \text{ g}}\right) = 4.7 \times 10^{-15} \text{ mol}$$

$$E = 4.7 \times 10^{-15} \text{ mol}\left(\frac{9.36 \times 10^7 \text{ kJ}}{1 \text{ mol}}\right)\left(\frac{10^3 \text{ J}}{1 \text{ kJ}}\right) = 4.4 \times 10^{-4} \text{ J}$$

22.43 Use Equation 22-4 to determine the rate of emission of a radioactive nuclide:

$$N = 7.45 \text{ pg}\left(\frac{10^{-12} \text{ g}}{1 \text{ pg}}\right)\left(\frac{1 \text{ mol}}{131 \text{ g}}\right)\left(\frac{6.022 \times 10^{23} \text{ nuclei}}{1 \text{ mol}}\right) = 3.425 \times 10^{10} \text{ nuclei}$$

$$t_{1/2} = 8.07 \text{ day}\left(\frac{24 \text{ hr}}{1 \text{ day}}\right)\left(\frac{60 \text{ min}}{1 \text{ hr}}\right)\left(\frac{60 \text{ s}}{1 \text{ min}}\right) = 6.97 \times 10^5 \text{ s}$$

$$\frac{\Delta N}{\Delta t} = \frac{-N \ln 2}{t_{1/2}} = -\frac{(3.425 \times 10^{10} \text{ decays})(0.693)}{6.97 \times 10^5 \text{ s}} = -3.41 \times 10^4 \text{ decays/s}$$

22.45 Exposure to radiation results first in damage to those cells that reproduce most quickly, including the white blood cells that are responsible for fighting infection and the mucous membrane lining of the intestinal tract. Thus, the early symptoms of radiation exposure include reduced resistance to infection and nausea due to disruption of the digestive tract.

22.47 Dating techniques using radioisotopes are based on Equation 22-5: $\ln\left(\dfrac{N_0}{N}\right) = \dfrac{t\ln 2}{t_{1/2}}$.

To obtain an age estimate, the ratio N_0/N must be determined. Convert the mass ratio into the desired numerical ratio. Because both isotopes have the same value for A, their mole ratio and number ratio are the same as their mass ratio:

$$N_0 = N_{Sr} + N_{Rb} \qquad \frac{N_0}{N} = \frac{N_{Sr} + N_{Rb}}{N_{Rb}} = 1 + \frac{N_{Sr}}{N_{Rb}} = 1 + 0.0050 = 1.0050$$

$$t = \left(\frac{t_{1/2}}{\ln 2}\right)\ln\left(\frac{N_0}{N}\right) = \left(\frac{4.9 \times 10^{11} \text{ yr}}{0.693}\right)\ln(1.0050) = 3.5 \times 10^9 \text{ yr}$$

22.49 Add a small amount of a radioactive iron isotope to the manufacturer's iron waste stream before treatment, and monitor the iron content of the river downstream. If radioactivity appears in the river water, the source is the manufacturer; if not, the iron in the river comes from some other source.

22.51 Run the hydrolysis in water that is enriched with a radioactive isotope of oxygen. Isolate the two products and analyze them for radioactivity. Whichever product shows radioactivity is the one whose added oxygen atom comes from water.

22.53 Binding energy is calculated by determining the mass defect (difference between the mass of the atom and the sum of the masses of its individual components) and converting to energy. When an element has only one stable isotope, its molar mass is the mass of that isotope:

$MM_{Bi} = 208.980$ g/mol

The isotope has 83 protons and electrons, $209 - 83 = 126$ neutrons

$m_{components} = 83(1.007276 + 0.0005486) + 126(1.008665) = 210.741232$ g/mol

$\Delta m = 208.980 - 210.741232 = -1.761$ g/mol

$$\Delta E = \left(\frac{-1.761 \text{ g}}{1 \text{ mol}}\right)\left(\frac{8.988 \times 10^{10} \text{ kJ}}{1 \text{ g}}\right) = -1.583 \times 10^{11} \text{ kJ/mol}$$

$$\Delta E \text{ (per nucleon)} = \frac{-1.583 \times 10^{11} \text{ kJ/mol}}{209 \text{ nucleons}} = -7.574 \times 10^8 \text{ kJ mol}^{-1} \text{ nucleon}^{-1}$$

22.55 Information about nuclear decays can be obtained using Equations 22-4 and 22-5.

$$\ln\left(\frac{N_0}{N}\right) = \frac{t\ln 2}{t_{1/2}} \qquad \ln N = \ln N_0 - \frac{t\ln 2}{t_{1/2}}$$

$N_0 = 5.0$ mg $\qquad t_{1/2} = 138.4$ day $\qquad t = 365$ days

$$\ln N = \ln(5.0) - \frac{(365 \text{ days})(\ln 2)}{(138.4 \text{ days})} = -0.2182 \qquad N = e^{-0.2182} = 0.80 \text{ mg remain}$$

$$\frac{\Delta N}{\Delta t} = \frac{-N\ln 2}{t_{1/2}} \qquad t_{1/2} = 138.4 \text{ days}\left(\frac{24 \text{ hr}}{1 \text{ day}}\right)\left(\frac{60 \text{ min}}{1 \text{ hr}}\right)\left(\frac{60 \text{ s}}{1 \text{ min}}\right) = 1.196 \times 10^7 \text{ s}$$

$$N = 0.80 \text{ mg} \left(\frac{10^{-3} \text{ g}}{1 \text{ mg}} \right) \left(\frac{1 \text{ mol}}{210 \text{ g}} \right) \left(\frac{6.022 \text{ x } 10^{23} \text{ nuclei}}{1 \text{ mol}} \right) = 2.3 \text{ x } 10^{18} \text{ nuclei}$$

$$\frac{\Delta N}{\Delta t} = \frac{(2.3 \text{ x } 10^{18} \text{ decays})(\ln 2)}{1.196 \text{ x } 10^7 \text{s}} = 1.3 \text{ x } 10^{11} \text{ decays/s}$$

22.57 Determine the products of nuclear decay by applying conservation of charge and mass number: The other product of neutron decay must be a particle with –1 charge and 0 mass, which is an electron: $n \rightarrow p + e$.
Calculate the decay energy from the mass defect between reactant and products, using masses found in Table 22-1:
$\Delta m = (1.007276 \text{ g/mol}) + (5.486 \text{ x } 10^{-4} \text{ g/mol}) - (1.008665 \text{ g/mol}) = -0.0008404 \text{ g/mol}$

$$\Delta E = \left(\frac{-0.0008404 \text{ g}}{1 \text{ mol}} \right) \left(\frac{8.988 \text{ x } 10^{10} \text{ kJ}}{1 \text{ g}} \right) \left(\frac{10^3 \text{ J}}{1 \text{ kJ}} \right) = -7.554 \text{ x } 10^{10} \text{ J/mol}$$

$$E_{\text{kinetic, electron}} = \left(\frac{7.554 \text{ x } 10^{10} \text{ J}}{1 \text{ mol}} \right) \left(\frac{1 \text{ mol}}{6.022 \text{ x } 10^{23} \text{ electrons}} \right) = 1.25 \text{ x } 10^{-13} \text{ J}$$

22.59 Calculations of decay times make use of Equation 22-5:
$$\ln \left(\frac{N_0}{N} \right) = \frac{t \ln 2}{t_{1/2}} \qquad\qquad t = \left(\frac{t_{1/2}}{\ln 2} \right) \ln \left(\frac{N_0}{N} \right)$$

When 1% has decayed, $t = \left(\dfrac{1622 \text{ yr}}{\ln 2} \right) \ln \left(\dfrac{N_0}{0.99 \, N_0} \right) = 24 \text{ yr}$

When 1% remains, $t = \left(\dfrac{1622 \text{ yr}}{\ln 2} \right) \ln \left(\dfrac{N_0}{0.01 \, N_0} \right) = 1.1 \text{ x } 10^4 \text{ yr}$

22.61 The n/p ratio of an isotope determines where it lies with respect to the belt of stability. ^{11}C has 6 protons and $(11 - 6) = 5$ neutrons, n/p = 0.833; ^{15}O has 8 protons and $(15 - 8) = 7$ neutrons, n/p = 0.875. Both isotopes are neutron-deficient and lie below (to the right of) the belt of stability.   Isotopes that are neutron-deficient decay by positron emission, which increases their n/p ratios: $^{11}_{6}C \rightarrow {}^{0}_{1}\beta^+ + {}^{11}_{5}B$ and $^{15}_{8}O \rightarrow {}^{0}_{1}\beta^+ + {}^{15}_{7}N$

22.63 Calculations of decay times make use of Equation 22-5:
$$\ln \left(\frac{N_0}{N} \right) = \frac{t \ln 2}{t_{1/2}}, \text{ from which } t = \left(\frac{t_{1/2}}{\ln 2} \right) \ln \left(\frac{N_0}{N} \right)$$

When 1 μg is left, $t = \left(\dfrac{15 \text{ hr}}{\ln 2} \right) \ln \left(\dfrac{25 \, \mu g}{1 \, \mu g} \right) = 70 \text{ hr}$

22.65   To identify the products of nuclear decay processes, make use of the principles of conservation of mass number and charge:

(a) alpha emission changes nuclear charge by –2, mass number by –4, product is $^{234}_{90}\text{Th}$:

$$^{238}_{92}\text{U} \rightarrow {}^{4}_{2}\alpha + {}^{234}_{90}\text{Th}$$

(b) (n,p) changes nuclear charge by –1, no change in mass number, product is $^{60}_{27}\text{Co}$:

$$^{60}_{28}\text{Ni} + {}^{1}_{0}\text{n} \rightarrow {}^{1}_{1}\text{p} + {}^{60}_{27}\text{Co}$$

(c) process changes nuclear charge by +6, mass number by (+12 – 3), product is $^{248}_{99}\text{Es}$:

$$^{239}_{93}\text{Np} + {}^{12}_{6}\text{C} \rightarrow 3{}^{1}_{0}\text{n} + {}^{248}_{99}\text{Es}$$

(d) (p,α) changes nuclear charge by –1, mass number by –3, product is $^{32}_{16}\text{S}$:

$$^{1}_{1}\text{p} + {}^{35}_{17}\text{Cl} \rightarrow {}^{4}_{2}\alpha + {}^{32}_{16}\text{S}$$

(e) β decay changes nuclear charge by +1, no change in mass number, product is $^{60}_{28}\text{Ni}$:

$$^{60}_{27}\text{Co} \rightarrow {}^{0}_{-1}\beta + {}^{60}_{28}\text{Ni}$$

22.67   Determine the expected product using conservation of mass number and charge number: $^{208}_{82}\text{Pb} + {}^{48}_{22}\text{Ti} \rightarrow {}^{256}_{104}\text{Rf}$. (This nucleus probably would decay by emitting neutrons and/or β particles).

22.69   Energy releases are calculated using Equation 22-3: $\Delta E = (\Delta m)(8.988 \times 10^{10} \text{ kJ/g})$:

$\Delta m = (147.9146) + (4.00260) - 151.9205 = -0.0033 \text{ g/mol}$

$$\Delta E = \left(\frac{-0.0033 \text{ g}}{1 \text{ mol}}\right)\left(\frac{8.988 \times 10^{10} \text{ kJ}}{1 \text{ g}}\right) = -2.97 \times 10^{8} \text{ kJ/mol}$$

$$\Delta E \text{ (per nucleus)} = \left(\frac{-2.97 \times 10^{8} \text{ J}}{1 \text{ mol}}\right)\left(\frac{1 \text{ mol}}{6.022 \times 10^{23} \text{ nuclei}}\right)\left(\frac{10^{3} \text{ J}}{1 \text{ kJ}}\right) = -4.9 \times 10^{-13} \text{ J}$$

$$\text{Fraction carried off by the } \alpha \text{ particle} = \frac{3.59 \times 10^{-13} \text{ J}}{4.93 \times 10^{-13} \text{ J}} = 0.73$$

22.71   Nuclides in the "belt of stability" are stable. Instability occurs if a nuclide has too few neutrons, too many neutrons, or $Z > 83$. In addition, most odd-odd nuclides are unstable: (a) too many neutrons; (b) $Z > 83$; (c) odd-odd; (d) too many neutrons.

22.73   In a nuclear reactor, the moderator serves to slow down fast neutrons, so they are more efficiently captured by the nuclear fuel. This reduces the amount of fuel required to sustain the reaction. The control rods serve to absorb some of the neutrons, allowing the reactor to operate just below its critical point and generate large quantities of heat without heating up beyond control.

22.75 Information about nuclear decays can be obtained using Equation 22-5.

$$\ln\left(\frac{N_0}{N}\right) = \frac{t\ln 2}{t_{1/2}} \qquad\qquad \ln N = \ln N_0 - \frac{t\ln 2}{t_{1/2}}$$

$N_0 = (10^5/s)(30\ s) = 3.0 \times 10^6$

$$t = 5\ hr\left(\frac{60\ min}{1\ hr}\right)\left(\frac{60\ s}{1\ min}\right) = 1.8 \times 10^4\ s$$

If $t_{1/2} = 1100\ s$, $\ln N = \ln(3.0 \times 10^6) - \dfrac{(1.8 \times 10^4\ s)(\ln 2)}{(1100\ s)} = 3.57$

$N = e^{3.57} = 36$ nuclei remaining

If $t_{1/2} = 876\ s$, $\ln N = \ln(3.0 \times 10^6) - \dfrac{(1.8 \times 10^4\ s)(\ln 2)}{(876\ s)} = 0.67$

$N = e^{0.67} = 2$ nuclei remaining

22.77 Determine the nuclear reaction by applying conservation of mass number and charge number:

$^{10}_{5}B + n \rightarrow \left(^{11}_{5}B\right) \rightarrow ^{7}_{3}Li + \alpha$. This reaction does not pose a significant health hazard, because Li-7 is a stable isotope and the α-particles are easily stopped using an appropriate shield.

22.79 Dating techniques using radioisotopes are based on Equation 22-5: $\ln\left(\dfrac{N_0}{N}\right) = \dfrac{t\ln 2}{t_{1/2}}$

Calculate the N_0/N ratio by taking the ratio of counts $g^{-1}\ hr^{-1}$ for the new and old samples:

Fresh sample: count rate $= \dfrac{18400\ counts}{(1.00\ g)(20\ hr)} = 920$ counts $g^{-1}\ hr^{-1}$

Old sample: count rate $= \dfrac{1020\ counts}{(0.250\ g)(24\ hr)} = 170$ counts $g^{-1}\ hr^{-1}$

$N_0/N = \dfrac{920}{170} = 5.4$

$t = \left(\dfrac{t_{1/2}}{\ln 2}\right)\ln\left(\dfrac{N_0}{N}\right) = \left(\dfrac{5730\ yr}{0.693}\right)\ln(5.4) = 1.4 \times 10^4$ yr

22.81 Use Equation 22-5 to calculate the useful lifetime of the dating technique:

$$\ln\left(\frac{N_0}{N}\right) = \frac{t\ln 2}{t_{1/2}} \qquad\qquad t = \frac{t_{1/2}}{\ln 2}\ln\left(\frac{N_0}{N}\right)$$

$$t = \left(\frac{5730\ yr}{0.693}\right)\ln\left(\frac{15.3\ counts/g\ min}{0.03\ counts/g\ min}\right) = 5.2 \times 10^4\ yr$$

22.83　Determine the products of nuclear decay using conservation of charge number and mass number. Use mass-energy equivalence to determine the mass of a product, given the mass of a reactant and the energy given off in the process:

(a) Sr has $Z = 38$ and Zr has $Z = 40$, so the decay requires 2 β-particles (0 mass number, –1 charge number): $^{90}_{38}Sr \rightarrow {}^{90}_{39}Y + {}^{0}_{-1}\beta$; $^{90}_{39}Y \rightarrow {}^{90}_{40}Zr + {}^{0}_{-1}\beta$

(b) These nuclear decay processes involve emission of electrons and are exothermic, so there is a decrease in mass of the isotope in the process. Thus, Sr has the larger mass.

(c) $\Delta E = \Delta m(8.988 \times 10^{10} \text{ kJ/g})$; net reaction is $^{90}_{38}Sr \rightarrow {}^{90}_{40}Zr + 2\,{}^{0}_{-1}\beta$

$\Delta m = [89.9043 + 2(0.0005486)] - 89.9073 = -0.0019 \text{ g/mol}$

$$\Delta E = \left(\frac{-0.0019 \text{ g}}{1 \text{ mol}}\right)\left(\frac{8.988 \times 10^{10} \text{ kJ}}{1 \text{ g}}\right) = -1.7 \times 10^8 \text{ kJ/mol}$$

22.85　Calculations of decay times make use of Equation 22-5:

$$\ln\left(\frac{N_0}{N}\right) = \frac{t \ln 2}{t_{1/2}} \qquad\qquad t = \left(\frac{t_{1/2}}{\ln 2}\right)\ln\left(\frac{N_0}{N}\right)$$

First calculate the amount initially bound to the thyroid gland:

$$N_0 = (0.5 \text{ mg})\left(\frac{10^3 \mu g}{1 \text{ mg}}\right)\left(\frac{45\%}{100\%}\right) = 225 \ \mu g$$

When 0.1 μg is left, $t = \left(\dfrac{13.2 \text{ hr}}{\ln 2}\right)\ln\left(\dfrac{225 \ \mu g}{0.1 \ \mu g}\right) = 1.5 \times 10^2 \text{ hr}$

22.87　Around $Z = 43$, the n/p ratio for stable isotopes is about 1.27, so the isotope that is most likely to be stable has $(43)(1.27) = 55$ neutrons. This, however, is an odd-odd isotope, so we might expect to find 56 neutrons, $^{99}_{43}Tc$. This, in fact, is the nuclide used widely in nuclear medicine.

22.89　To determine where the oxygen atom in the water molecule comes from, prepare a sample of the alcohol that is enriched in radioactive ^{18}O and run the reaction using this sample. Separate the products and measure the radioactivity of the ester and the water. If the C–OH bond in the alcohol breaks during the condensation, the ^{18}O will appear in the water, while if the C–OH bond in the carboxylic acid breaks during the condensation, the ^{18}O will appear in the ester.

22.91　The precipitate will contain radioactive Na, because once the NaBr is dissolved in solution, its Na^+ cations mix freely with the Na^+ cations of the existing solution. When the solution cools and $NaNO_3$ precipitates, some of the Na^+ cation in the precipitate will be radioactive Na-24.

NOTES

NOTES

NOTES

NOTES

NOTES

NOTES

NOTES